AF896880

ROUTLEDGE LIBRARY EDITIONS: EVOLUTION

Volume 10

HUMAN EVOLUTION

HUMAN EVOLUTION

An Introduction for the Behavioural Sciences

GRAHAM RICHARDS

Routledge
Taylor & Francis Group
LONDON AND NEW YORK

First published in 1987 by Routledge & Kegan Paul Ltd

This edition first published in 2020
by Routledge
2 Park Square, Milton Park, Abingdon, Oxon OX14 4RN

and by Routledge
52 Vanderbilt Avenue, New York, NY 10017

Routledge is an imprint of the Taylor & Francis Group, an informa business

© 1987 Graham Richards

All rights reserved. No part of this book may be reprinted or reproduced or utilised in any form or by any electronic, mechanical, or other means, now known or hereafter invented, including photocopying and recording, or in any information storage or retrieval system, without permission in writing from the publishers.

Trademark notice: Product or corporate names may be trademarks or registered trademarks, and are used only for identification and explanation without intent to infringe.

British Library Cataloguing in Publication Data
A catalogue record for this book is available from the British Library

ISBN: 978-0-367-27938-7 (Set)
ISBN: 978-0-429-31628-9 (Set) (ebk)
ISBN: 978-0-367-25991-4 (Volume 10) (hbk)
ISBN: 978-0-429-29110-4 (Volume 10) (ebk)

Publisher's Note
The publisher has gone to great lengths to ensure the quality of this reprint but points out that some imperfections in the original copies may be apparent.

Disclaimer
The publisher has made every effort to trace copyright holders and would welcome correspondence from those they have been unable to trace.

Preface to 2019 re-issue

This book's unexpected re-issue provides an opportunity for some brief hindsight reflections. Over 30 years is a long time in any scientific field and human evolution studies are no exception. Most readers will be aware of the discoveries of the tiny *homo floresiensis* on the Javanese island of Flores (2003) and the Denisovans (2010) in the Altai Mountains of Siberia on which the web is awash with information. Another front on which huge advances have been made is in DNA (particularly mtDNA) analysis. In the 1980s the question of possible interbreeding between Neanderthals and early modern humans was hotly debated, but is now firmly established, as is the interbreeding of both with Denisovans in east Asia. The proportion of Neanderthal DNA in contemporary west Europeans is around 4% (estimates vary from 1% –7% as far as I can gather). A similar proportion, or higher, is typically found in east Asians along with up to 6% from Denisovans. Taken together new fossil finds and genetic data are creating a plethora of hypotheses about the actual number of hominin lineages over the last million years, their inter-connections, and the timings of their dispersals out of and indeed within Africa. The bearings of these post-1990 developments on the evolution of human behaviour are unclear. An improved grasp of Neanderthal life-style suggests both a closer resemblance to that of contemporary *Homo sapiens* than once assumed, and that their posture was even more upright. The new data from archaeology and palaeontology has rendered a simple plot-line harder than it once seemed regarding the origins of language, art, cooking, caring for others and the other hallmarks of 'culture', even if it clearly implies propensities for wanderlust and sexual xenophilia in all our ancestral branches!

During the late 1980s the picture was more promising. The original and insightful work of social anthropologist Tim Ingold for example, which came to my attention too late for inclusion here, but really must be checked out by new students (see especially his contributions in Kathleen R. Gibson & Tim Ingold (eds), 1993), Peter Wheeler's model of the linkage between hairlessness and bipedalism facilitating a noon-day scavenging niche also broke new ground (see P. Wheeler, 1991 for an accessible summary), along with Robert J. Blumenschine's research on scavenging and hunting (these also appeared too late for inclusion in the present work, but see Graham Richards (1989a)).

Also, in 1988, Richard W. Byrne & Andrew Whiten (eds) *Machiavellian Intelligence: Social Expertise and the Evolution of Intellect in Monkeys, Apes, and Humans* was a major research-based advance of the kind I had been hoping for, while Thomas Wynn published a book-length account of his Piagetian analyses of stone tools in 1989. A little later Duane Quiatt and Vernon Reynolds (1995) also provided an important overview from a primatological perspective. My own position was more fully stated in the 1989a paper and *On Psychological Language and the Physiomorphic Basis of Human Behaviour* (1989b). After 1990 academic circumstances returned me to a previous focus on the history of Psychology (which, in a fashion too complicated to explain here, actually led to the human evolution detour in the first place). Despite the flurry of immediate progress (only a few examples of which I just have mentioned) the multi-disciplinary 'Palaeopsychology' project which I aspired to promote never materialised. Instead we got *Evolutionary Psychology*. If not without some value this sub-discipline has failed, in my view, on two fronts. Firstly, by relying almost entirely on the reductionist conceptual repertoire of Sociobiology (as a spin off from which it began) and secondly because its central interest has been in conjuring up evolutionary explanations of currently human (especially American) behaviour of cultural concern, in terms of these concepts (such as 'inclusive fitness'). What it has not, to my awareness, done is actually collaborate with the various 'palaeodisciplines' in trying to understand human behavioural evolution itself. For a couple of later jabs at this I refer the reader to Graham Richards, 2001, 2003, but of course, see also H. & S. Rose (2000) (eds) *Alas Poor Darwin*.

Having got that off my chest, I will sum up by saying that while the field of human evolution has moved on considerably since first publication, I believe my discussions of behavioural topics continue to be relevant, serving at least as useful starting points for students, and that my critiques of aspects of Sociobiology in Chapter 5 still stand. My overall orientation, as outlined in the original Preface, also, I like to imagine, has some elements of originality. No comparable single work providing an updated overview of the range of topics I covered has, as far as I know, appeared since. Is it possible the whole subject stalled sometime in the mid-1990s? Or was it just my involvement in it? Certainly, my core observation that human uniqueness lies not in some specific thing we do, but in the fact that we can do everything, still has not, it seems, properly sunk in. (Oh, and Figure 5.2 I is useless without further explanation – which I cannot now provide, sorry about that!)

References

Blumenschine, Robert J. (1986a) *Early hominid scavenging opportunities: Implications of carcass availability in the Serengeti and Ngorongoro ecosystems*, British Archaeological Reports International Series, 283.

Blumenschine, Robert J. (1986b) ‘Carcass consumption sequences and the archaeological distinction of scavenging and hunting’, *Journal of Human Evolution*, Vol.15, pp.639–659.
Byrne, Richards W. and Whiten, Andrew (1988) *Machiavellian Intelligence: Social Expertise and the Evolution of Intellect in Monkeys, Apes, and Humans*, Oxford: Oxford University Press, Clarendon Press.
Gibson, Kathleen R. and Ingold, Tim (eds) (1993), *Tools, Language and Cognition in Human Evolution*, Cambridge: Cambridge University Press.
Ingold, Tim (1986a) *The Appropriation of Nature: Essays on Huma Ecology and Social Relations,* Manchester: Manchester University Press.
Ingold, Tim (1986b) *Evolution and Social Life*, Cambridge: Cambridge University Press.
Quiatt, Duane and Reynolds, Vernon (1995) *Primate Behaviour: Information, Social Knowledge and the Evolution of Culture*, Cambridge: Cambridge University Press.
Richards, Graham (1989a) ‘Human Behavioural Evolution: A Physiomorphic Model’, *Current Anthropology*, Vol.30 (2), pp.244–255.
Richards, Graham (1989b) *On Psychological Language and the Physiomorphic Basis of Human Behaviour*, London: Routledge.
Richards, Graham (2001) ‘Genetic possession’, *Psychology, Evolution and Gender*, 3 (2), pp.197–203.
Richards, Graham (2003) ‘Palæopsychology: the road not taken’, *Sexualities, Evolution and Gender*, 5(1), pp.23–30.
Rose, H. and Rose, S. (eds.) (2000) *Alas, Poor Darwin. Arguments against Evolutionary Psychology*, London: Jonathan Cape.
Wheeler, P. (1991) ‘Body hair reduction and tract orientation in man: Hydrodynamics or thermoregulatory aerodynamics?’ in Roede, Machteld, Wind, Jan, Patrick, John and Reynolds, Vernon (ed.). *The Aquatic Ape: Fact or Fiction? The First Scientific Evaluation of a Controversial Theory of Human Evolution*, London: Souvenir Press.
Wynn, Thomas (1989) *The Evolution of Spatial Competence*, Urbana & Chicago: University of Illinois Press, Illinois Studies in Anthropology, No.17.

Graham Richards. February 2019, Tunbridge Wells.

HUMAN EVOLUTION

An introduction for the behavioural sciences

Graham Richards

Routledge & Kegan Paul
London and New York

First published in 1987 by
Routledge & Kegan Paul Ltd
11 New Fetter Lane, London EC4P 4EE

Published in the USA by
Routledge & Kegan Paul Inc.
in association with Methuen Inc.
29 West 35th Street, New York, NY 10001

Set in Linotron Plantin, 10 on 12 pt
by Input Typesetting Ltd, London SW19 8DR
and printed in Great Britain
by Billings & Sons Ltd,
Worcester

© *Graham Richards 1987*

No part of this book may be reproduced in
any form without permission from the publisher
except for the quotation of brief passages
in criticism

Library of Congress Cataloging in Publication Data

Richards, Graham.
Human evolution.
Bibliography: p.
Includes indexes.
1. Human evolution. I. Title.
GN281.R53 1987 573.2 86–22581

British Library CIP Data also available

ISBN 0–7102–0326–8

ISBN 0-7102-1381-6(p)

This book is for my daughters
Carol and Rebecca
and my mother Dorothy

There was a child went forth every day,
And the first object he looked upon, that object he became,
And that object became part of him for the day, or a certain part of the day,
Or for many years or stretching cycles of years.
. . . . . .
The strata of coloured clouds, the long bar of maroon tint, away solitary by itself – the spread of purity it lies motionless in,
The horizon's edge, the flying sea-crow, the fragrance of salt marsh and shore mud;
These became part of that child who went forth every day, and who now goes, and will always go forth every day.

Walt Whitman

And the poor old lousy old earth, my earth and my father's and my mother's and my father's father's and my mother's mother's and my father's mother's and my mother's father's and my father's mother's father's and my mother's father's mother's and my father's mother's mother's and my mother's father's father's and my father's father's mother's and my mother's mother's father's and my father's father's father's and my mother's mother's mother's and other people's fathers' and mothers' . . . . .

Samuel Beckett, *Watt*

Contents

Illustrations

Plates

between pages 220 and 221

Figures

Tables

Preface

My interest in human evolution, though longstanding, began to gather momentum around 1980 for a combination of reasons. Firstly as an inveterate addict of *Current Contents* I found myself increasingly intrigued by the titles of papers appearing in such journals as *Current Anthropology*, *American Journal of Anthropology* and the *Journal of Human Evolution* which were unavailable at the main college library to which I had access. I thus began soliciting reprints of these from their authors, and would at this early point like to extend a general, heartfelt, acknowledgment to all those farflung palaeontologists, palaeoanthropologists and archaeologists who have generously supplied me with their papers over the last few years, and who must often have wondered why a psychologist at North East London Polytechnic wanted to know about occipital bunning, Miocene molars, microwear and the like. But idle curiosity was clearly not the only factor at work. My teaching has primarily been concerned with the historical development of Psychology and psychological theories, broadening more recently into history of science and Philosophical Psychology. The historical quest for the roots both of psychological ideas and scientific behaviour itself pushed my focus of attention ever further back. What now became the central, unifying, theme of my inquiries was the nature of human uniqueness itself, and the origins of our peculiarly human combination of genius and lunacy. I had long been nurturing certain ideas regarding this (briefly summarised below in Chapter 5), but a proper evaluation and formulation of these necessarily meant acquiring a clearer grasp of what was actually known about human evolution, especially its behavioural aspects.

It is perhaps worth stressing that this burgeoning concern with human evolution owed nothing to any sociobiological or 'nativist' predelictions at the theoretical level, but rather stemmed from an intensifying conviction that an evolutionary perspective was necessary for any balanced appraisal of the character of our species – with which, as a psychologist, I was supposed to be ultimately concerned. I soon became aware that an inchoate body of psychological work was developing in the human evolution field, notably that of Parker and Gibson, Wynn, some of Holloway's papers, Tanner's *On Becoming Human* and a variety of work by palaeoarchaeologists such as Glynn Isaac on the behavioural implications of early hominid sites.

In 1983 I was sufficiently enthused to read a paper at the British Psychological Society Summer Conference entitled 'On the Possibility of a Palaeopsychology'. The present work represents what an eighteenth-century philosopher might have called 'a prolegomena to any future Palaeopsychology'. That is to say, it attempts a preliminary survey of the current state of our understanding of human evolution in general, but with emphasis on its behavioural and psychological aspects. This book does *not* therefore attempt to present the reader with a coherent story of human evolution, but introduces her, or him, to the multiplicity of views on the issue currently in the arena (though omitting for the most part flying saucers, Fundamentalism and the Lost Continent of Mu, etc.). Such a survey seems to me to be the best service I can perform at this point to draw human evolution into the orbit of Psychology, and bring the possibility of a 'Palaeopsychology' nearer its realisation.

Having said all that, my own view of the situation necessarily emerges, pessimistic as it is, though without I hope unduly colouring the exposition. Books, even textbooks (*especially* textbooks?) are inescapably of their time, and the present time is not optimistic. Having finished it, I am aware that the label 'textbook' sits on the book rather uneasily in any case. This is genuinely due, in part at least, to the nature of the material and the task. There is a sort of ritual textbook format for those books on human evolution aimed at physical anthropology students or palaeontologists, but no such format for ones aimed at behavioural sciences. Sociobiology, incidently, has surprisingly little coverage of the actual data regarding human evolution, about which it has such firm opinions. A range of material has had to be gathered which has not been treated together at an introductory level before (much of it very recent), and this has often entailed taking a more evaluative approach than is usual in 'textbooks'.

Except, as in much of Chapter 2, where it would be anachronistic to do so, I have tried to de-genderise my style as far as possible. Human evolution has proved to be one of those areas where the feminist perspective has had considerable positive impact, and the avoidance of the inclusive masculine has actually brought to light numerous implicit assumptions about ancestral life-styles of which earlier writers were hardly aware.

The number of people who have helped in furthering this project in various ways is considerable, in addition to the kind band of off-print suppliers referred to earlier. I would like to first thank Mary Midgley, not only for her support in this particular endeavour, but for her perennial encouragement and the example of intellectual rigour and integrity which she has set for all her former students. Peter Bowler, Dean Falk, Ralph Holloway, Elaine Morgan, Elizabeth Parker, Philip Tobias, Alan Tuohy, Stephen Walker and Thomas Wynn have all helped with comments, ideas and material. My colleagues Keith Philips and Claire Fullerton helped me sharpen up the section on Altruism. I am particularly indebted to Dr Chris Stringer at the British Museum (Natural History) for his counsel at various times, his happy knack of directing my attention to just the right texts at the right moment and reducing the howler-rate in parts of Chapters 3 and 6. Dr Peter Andrews of the same institution provided a valuable clarification of the nature of some of the early fossil evidence at a crucial point.

The staff at North East London Polytechnic's library have been invaluable over the years, in increasingly difficult circumstances. The libraries of the British Museum (Natural History), the Institute of Archaeology and the British Library have also proved their worth. Stuart Collins and Sabrina Izzard have played a major role in supplying my second-hand book needs. Carol Richards deserves a mention for helping with the bibliography.

Any lack of the customary seraglio of typists, secretaries, and research assistants has been, throughout, more than offset by the fact that I *do* have Maura, who, whilst rightly performing none of these functions, has ruthlessly averted atrophy of the ego, wilting of the will and divergent dissipation of drive in the way only Irish novelists can for academic Brits.

For want of a scapegoat all errors are probably my fault.

Acknowledgments

Acknowledgments are due to the following for supplying photographs and for permission to use material:

Academic Press (Figures 3.10 & 4.5); British Museum (Natural History) (Figure 3.8, Appendix B and Plates 2–8); Cambridge University Press (Table 4.4.); Robert S. Corruccini of Southern Illinois University (Figures 3.4–3.7); Harvard University Press (Figures 5.1 & 5.2); Malcolm Kirk (Plate 11); Alexander Marshack of the Peabody Museum, Harvard University (Plate 13); National Geographic Magazine (Plate 11); Mark Newcomer of the Institute of Archaeology, London (Plate 10); Sue Parker of Sonoma State University, California (Table 4.4); H. C. Plotkin of University College, London (Figure 3.10); The Prehistoric Society (Figure 3.9 & Plate 9); Pat Shipman of The Johns Hopkins University, Maryland (Plate 12); and Thomas Wynn of the University of Colorado (Figures 4.4 & 4.5).

CHAPTER 1

The nature of the problem

We want to be free. We need an identity. Current controversies over the nature (if any) of our species derive much of their energy from the tension between these two psychological imperatives. The former directs us to emphasise the present as a continuing choice-point, to which we bring nothing but our consciousness of what we have learned from our personal experience of the world. Liberation involves raising to consciousness as much as possible of our past, and learning as much as possible of the factors affecting us in the present, in a process somewhat cumbersomely but accurately labelled by the German philosopher K. O. Apel (1977) 'critically emancipatory self-reflection'. The need for identity compels us to label ourselves, to classify and constrain ourselves as essentially belonging to groups which transcend our personal selves, groups of which we are representatives; our sex, nationality, class, ethnicity, occupation, religion, clan, or even our generation. Being free, however, forces us to strive to transcend these labels and accept only a logically prior identity as 'human'. While the quest for identity obviously involves a quest for origins, in a sense we can also only be free if we have established for ourselves our 'true identity'. (It is to the enduring credit of the 1960s anti-psychiatrists such as Laing and Cooper that they drew attention to the crucial importance of honesty in this respect for the future mental well-being of the child.) Both imperatives thus send us in search of our origin, and it is only in finding it that the apparent contradiction between them can be resolved and explained.

Jungian writers such as Neumann (1954) and von Franz (1972) were surely correct in seeing that the question of origins fused both psychological and cosmological levels. We cannot be free in the cosmos until we know where we belong in it, and it seems clear that the answer to this must lie in finding how the cosmos itself created us, what kind of parentage it gave us. (The perennial resilience of astrology testifies to precisely this.) As Neumann argues, the origins of consciousness and the universe are of necessity indistinguishable in creation myths;

> Mythological accounts of the beginning must invariably begin with the outside world, for world and psyche are still one. There is as yet no reflecting, self-conscious ego that could refer anything to itself, that is, reflect. Not only is the psyche open to the world, it is still identical with and undifferentiated from the world; it knows itself as world and in the world and experiences its own becoming as a world-becoming, its own images as the starry heavens, and its own contents as the world-creating gods. (1954, p. 6)

While this sort of language and approach is uncongenial to most contemporary psychologists, it provides the only perspective yet offered from which constructively to grapple with such material as the following;

> (Marduk has just slain the primordial hag Tiamat.) He turned back to where Tiamat lay bound, he straddled the legs and smashed her skull (for the mace was merciless), he severed the arteries and the blood streamed down the north wind to the unknown ends of the world.
>
> When the gods saw this they laughed out loud, and they sent him presents. They sent him their thankful tributes.
>
> The lord rested; he gazed at the huge body, pondering how to use it, what to create from the dead carcass. He split it apart like a cockle-shell; with the upper half he constructed the arc of the sky, he pulled down the bar and set a watch on the waters, so they should never escape.
>
> He crossed the sky to survey the infinite distance; he stationed himself . . . over the old abyss which he now surveyed, measuring out and marking in.
>
> He stretched the immensity of the firmament, he made Esharra, the Great Palace, to be its earthly image, and Anu and Enlil and Ea had each their right stations. (from 'The Babylonian Creation' in Sandars (1971), pp. 91–2)

This marks the beginning of a section in which Marduk constructs the present universe from the hag's corpse (if we bear in mind that it was the Babylonians who ultimately set the scientific programme in motion, there is something peculiarly disturbing in this, a distant adumbration of *Death in Nature* (Merchant, 1983)). Freudian matricidal frenzy, birth trauma and cosmological theorising are inextricably fused.

Thus the question of origins has exerted an almost universal and perennial fascination, and been an ever-present cultural pre-occupation. But the functions of origin myths are not restricted to the psychological. They serve, sociologically, to legitimate the *status quo*, to justify – or at any rate explain – the current ordering of society, and the right to rulership of its rulers. This is so whether the rulers are seen as genealogically descended from the founding gods or whether their eminence is itself proof of their evolutionary fitness; whether social institutions are those ordained by the Creator (like the church) or the ancestral founders of the current order (like the writers of the US constitution). Even where accounts of origins are not deliberately tailored to bolstering the present establishment, they nevertheless account for the present in some way, locating it in the cosmological scheme of things, e.g. Hinduism's account of us as living in the Kali Yuga, the last and worst in the current four-yuga cycle (Zimmer, 1972) or contemporary popular science images of us as possibly being at some phase in planetary evolution where matter becomes self-conscious as a prelude to joining the cosmic community.

John R. Durant (1981), in addressing this issue of the mythological character of theories of human evolution, points out how the 'beast in man' myth, in particular, has been carried over from Christianity into, first, Darwin's own model of human descent, and then into the twentieth century in Freud, Dart, Ardrey and Lorenz. In each case the moral is drawn that we have inherited a bloodthirsty savage dark side to our character, legacy of our species' distant ancestral past. More recently alternative interpretations more congenial to the liberal temperament have provided myths to oppose this, in which emphasis is shifted to co-operation or the role of women. The reader will encounter images of both kinds, and more, recurring throughout the following pages. By and large I endorse Durant's perception of the problem, his own remedy for which is given in the following passage:

> It seems to me that when people of many different political persuasions are all engaged in mythologising the theory of human evolution, the only way out of the resulting confusion is a determination to *stop playing this particular game*. Scientific studies of human origins are best undertaken without the subjective pressures and distortions which are introduced by the desire to see one's personal view of life confirmed in the testimony of the rocks; and political issues are best discussed without the pressing need to prove one's point of view by reference to the social life of baboons, or whatever. As things stand at the present time, we are in urgent need of the de-mythologisation of science. (1981, p. 437. Italics in original)

But is this not to abandon the field prematurely? We are up to the hilt in myth-making whether we wish it or not. As I will be arguing in Chapter 5, we cannot leave ideology out of the picture as one might be able to do in investigating electronics. Even at the motivational level, the drive behind those researching human evolution has not infrequently been broadly ideological in character. To anticipate some of the later discussion, one's entire perspective on human evolution is conditioned by whether or not one sees the present condition of the species as pathological. The researcher's ideological or political appraisal of the present defines what it is that the study of human evolution is, in the final analysis, explaining. Try to do without this and you will find it hard to draw any interesting conclusions at all. One cannot escape myth-making by fiat. And just for once I feel that the old excuse 'if we don't do it someone else will' has a certain validity.

Psychologists approaching human evolution are thus placed in a dilemma, for their interest in it is of two different kinds. Firstly we are interested, in the straight scientific sense, in the evolution of the species to which the majority of our studies are directed. The use we make of this knowledge will vary according to theoretical taste, but few would feel it to be *entirely* irrelevant to their task of understanding modern humans. We might also feel that Psychology could assist those disciplines such as Palaeoanthropology, directly concerned with studying human evolution. In short there is a conventional professional interest in a subject closely related to, or even overlapping with, our own. But secondly we cannot relinquish the reflexive role; an awareness of the psychological aspects, both individual and social, of origin myths must necessarily alert us in a special way to the effects of these on accounts of human evolution. We become aware of how contemporary issues of great psychological profundity, such as the nature of the relationship between the sexes, can be played out in the arena of theorising about the evolution of human social life and sexual behaviour. We become aware too of how the scientific study of human evolution is counterposed in contemporary culture to other accounts of origins of more archaic kinds, and how these derive their power perhaps from addressing more directly the psychological needs mentioned earlier, be they fundamentalist Creationism or God-was-an-astronaut in character.

The fact of the matter is that our species is, or believes itself to be, in crisis. But where do the origins of this crisis lie? What is its real nature? In such a climate the pressure is on to construct myths of origin which can structure the present, a task for which received myths are inadequate, myths which can endow the seeming chaos, actual or impending, with meaning. But there is a parallel pressure genuinely to diagnose the nature of the crisis in a rational fashion, and this too involves the exploration of origins, at least as part of the story. It is the difference between Danniken and the Leakeys. As a psychologist, I am concerned with the latter, not the former – or, more correctly, I am only concerned with the former insofar as it bears upon the latter. But I am also mindful that the difference between the two is in practice marginal, more so than Durant appreciates, that both are hoping to achieve similar ends, of structuring and giving meaning. The inevitable embroilment of the study of human evolution with myth is recognised by some current authorities (e.g. Isaac, 1983), and one long eminent palaeontologist has concluded

> I now believe that what we say about human evolution, what we pick as essential human attributes, and how we trace their development often tell us as much about paleoanthropologists and the times in which they live than about the course of evolutionary events. We are emphatically not the new theologians, but I will close with an appropriately theological quote from the Talmud:
>
> We do not see things as they are
> We see things as we are.
>
> (Pilbeam, 1980, p. 283)

The book that follows is addressed primarily to those studying Psychology and other behavioural sciences. It is not a textbook on physical anthropology or on evolutionary theory as such, though these matters often concern us. It is an attempt at surveying, with a bias towards what would interest such a readership, the current state of knowledge regarding human evolution, and particularly its behavioural aspects. But I have also tried to sustain an awareness of the second, reflexive, angle of interest; the study of human evolution as a psychological phenomenon in its own right, an activity carrying much of the weight hitherto loaded on creation myths and accounts of ancestral origins. Perhaps ancestor-worship is after all our 'natural' religion, and the study of human evolution but its current persona.

Some sections of the book are more opinionated than is customary or seemly for textbooks, which are supposed to exhibit 'benevolent eclecticism' (Maddi, 1976), on the grounds that one need not go anywhere since all directions are equally interesting. Some of this opinionatedness stems from the psychological implausibility and conceptual incoherence of the views prevalent in a field (e.g. the discussion of 'Altruism' in Chapter 5). On other occasions it has seemed to me necessary to make my own views on a topic explicit, though I have not done so extensively or indulged in special pleading on their behalf.

The 'reflexive' perspective lies behind the fairly lengthy historical chapter, for the vicissitudes of the study of human evolution illustrate many facets of its *psychological* character as a branch of scientific inquiry.

Writers on human evolution are of course haunted by the ever-present possibility of a dramatic find turning up the week before publication and rendering their efforts obsolete at birth. While this text was being written there have been at least two major finds in East Africa, one an extremely early *Sivapithecus* fossil and the other a nearly complete skeleton of a *Homo erectus* youth whose mature height would have been around 6 foot. They do not seem immediately to affect the behavioural evolution picture, although the *H.erectus* stature is unexpectedly tall. A third find from north eastern Russia is a different matter. Artefacts here appear to date from two million years ago yet are far in advance of anything of that date from Africa. Furthermore, as near-Arctic conditions prevailed in the region then, as now, it implies a level of cultural adaptation (clothes, fire, etc.) not usually believed to have been achieved until the last 100,000 years or so. A full evaluation of this data by international scholars is a long way off (as is the site!). It waits therefore in the wings of the present account, threatening to make a dramatic entry which would require complete revision of the behavioural evolution timetable. But its fate might, by contrast, be that of the Sunnyvale skeleton and *Hesperopithecus* instead (see below, pp. 57, 72).

In writing on human evolution for psychologists I am not to be understood as espousing a sociobiological or other biologically 'reductionist' view of human behaviour (my own tentative theoretical position is outlined in the course of Chapter 5). The intention is rather to provide broad pictures of both current scientific understanding of human evolution and the actual study of human evolution itself. Psychology and kindred disciplines cannot afford to ignore the evolutionary picture, while conversely the study of human evolution is in great need of input from Psychology. One aspect of this work then is to prepare students of Psychology for engagement in the required interdisciplinary two-way traffic. If it is considered to have made a contribution to the topic itself, that is a bonus.

CHAPTER 2

Linnaeus to the Leakeys

Introduction

A full history of the study of human evolution has yet to be published.[1] An outline of it is nevertheless essential for several reasons. Firstly, it will help the reader appreciate the difficulties which have had to be overcome in establishing the evolutionary perspective. Secondly, it enables us to see how sociohistorical factors and psychological needs of the sort just discussed can affect the scientific exploration of the topic. Thirdly, it reveals something of the central disciplines concerned raising matters of potential interest to anyone concerned with the psychological aspects of conceptual and theoretical change in the history of science. In any case, it would surely be paradoxical for the student of human evolution to ignore the evolution of its study!

This chapter falls into three principal sections: A The rise of the evolutionary perspective; B The great confusion; C Finding the 'missing links'. These represent the period up to 1872, the period 1873 to 1913, and the period 1914 to 1960 respectively. The period since 1960 is considered as the 'rearward portion of the present'. Like all such segmentations of the historical continuum this is largely artificial, but its rationale is as follows: the first period culminates in Darwin's two works on human evolution, *The Descent of Man* (1871) and *The Expression of the Emotions in Man and Animals* (1872); the second ends in a confusion due to the Piltdown hoax, an erroneous reconstruction of Neanderthal Man[2] and developments in genetics; the third ends with the discoveries at Olduvai Gorge of two fossil hominids, *Zinjanthropus* and *Homo habilis*, effectively recentering research in East Africa, where it as yet remains. These reference points are as convenient as any.

[1] Peter J. Bowler has such a work in progress, however, which we await with interest.

[2] See Preface for my policy regarding use of 'Man' and 'man', as opposed to non-genderised terms, in this book.

A The rise of the evolutionary perspective

This period may be interpreted as involving three separate conceptual problems: (1) the establishment of the evolutionary perspective itself, along with a scientifically viable theory of how evolution operated; (2) the relationship of 'Man' to this evolutionary process – whether and where humans were to be included in it; (3) the 'Antiquity of Man' (to use the title of Lyell's famous 1863 book), the time perspective in which human evolutionary events were envisaged as occurring. It is around these that this part of the story is structured, ending with a look at Darwin's two major works on the subject.

1 *The evolutionary perspective*

The evolutionary perspective was gaining ground slowly from as early as the mid-eighteenth century. Linnaeus (1707–1778), inventor of the modern system of zoological and botanical classification (Linnaeus, 1735) moved during his lifetime from believing in the eternal fixity of species to a belief in at least some degree of transformation (Tobias, 1980b). The notion of a 'Great Chain of Being' had been part of western cosmology since Aristotle (Lovejoy, 1936) and the German exponents of 'Naturphilosophie' such as Goethe (1749–1832) and Oken (1779–1851) placed Man at the top of this chain as its culminating achievement. But they were not evolutionists, adhering instead to a doctrine which saw different species as expressions of underlying archetypal forms. As Oldroyd (1980) points out, similarity between animals could 'be explained without difficulty by supposing . . . that both are separate manifestations or actualisations of the archetypal *Urtier* (animal-form) . . . . Thus can two utterly different approaches (i.e. Naturphilosophie and evolution) produce explanations of the same phenomenon' (p. 51). Identifying extraordinary similarities between species did not then lead automatically to an evolutionary theory. Versions of Naturphilosophie remained powerful until Darwin's time, in for example the work of the leading palaeontologist Richard Owen (1804–1892) (Desmond, 1982).

The most famous pre-Darwinian evolutionist is Lamarck (1744–1829). There are actually two 'Lamarckian' models of evolution; what he really said and what he is generally assumed to have said, the latter being more influential (Hull, 1982). The view commonly ascribed to him is usually summed up in the phrase 'inheritance of acquired characteristics'. Physical features acquired by parents in the course of striving to adapt to their environment are held to be passed on to their offspring. This implies that evolution is an active, creative, process and that Man has reached his present pinnacle by his own efforts across the generations.

His real view was subtly different from this and might be called an 'escalator' model; there is a fixed hierarchy of forms of increasing complexity up which organic matter ascends, certainly by its own efforts but in a more prestructured way than simple 'inheritance of acquired characteristics' implies. Each species represents a rung on the ladder, a form through which we passed in our evolution. Lamarck's younger contemporary Cuvier (1769–1832) upset this hierarchy (and the Great Chain of Being) by introducing a fourfold classification of animal life into separate 'kingdoms' each with its own hierarchy of complexity: vertebrata, mollusca, arthropoda and 'radiata'. After Lamarck's death his views fell rapidly into disregard under attack from Cuvier, who opposed evolutionary theories and espoused a form of creationism. Darwin's grandfather, Erasmus Darwin, had also proposed an evolutionary account, in verse, at the turn of the century which Darlington (1959) compares favourably with Lamarck's.

Throughout the early nineteenth century a number of works appeared discussing the nature of inheritance and countenancing the possibility of evolution, often with Man included. In some instances (Lawrence, 1819; Chambers, 1844) wrath descended upon the author's head in consequence. In other cases (Combe, 1828; Latham, 1850) more circumspect presentation kept them safe. Powerful scientific evidence was emerging from embryology, led by the work of the Estonian physiologist Baer (1792–1876) (Baer, 1828). This seemed to show the foetus passing through a number of stages in which its form resembled that of creatures 'lower' in the hierarchy of animal life; fish, reptile, amphibian and mammalian phases followed one another in the ontogenesis of the human foetus. Such evidence powerfully influenced Chambers and, more especially, a new generation of German physiologists. The leading figure among these was Ernst Haeckel (1834–1919), who saw in this embryological data proof of a 'biogenetic law' that 'ontogeny reflects phylogeny'. This 'recapitulationist' model proved to be seriously misleading, but for a while in the latter nineteenth century acquired the status of orthodoxy. Later in the book we will see that recapitulation remains a live issue. Following the publication in 1859 of Darwin's *Origin of Species* Haeckel published his *Generelle Physiologie* (Haeckel, 1866) and *History of Creation* (Haeckel, 1868) wholeheartedly supporting him. In Haeckel we see the Naturphilosophie tradition becoming evolutionary.

But serious psychological blocks to the acceptance of evolution persisted, both in general and as applied to humans. Firstly there was a pervasive reluctance to challenge orthodox religion except as a last resort. Privately extreme fundamentalist accounts of Creation may have been rejected by the educated, but there was a broader Christian view of science as 'Natural Theology' (expounded by Paley, 1802) which had sustained a truce between science and religion from the late eighteenth century to the mid-nineteenth. Respect for this inhibited writers in challenging the basic assumption that God designed and created all living forms in a single brilliant exercise in divine engineering. Secondly, as we will see, there is the time-perspective problem. Though being steadily eroded by geologists, myopic bias persisted and estimates of the time since life's origin were by and large inadequate down to the end of the nineteenth century and beyond. Darwin's friend and, later, rather lukewarm supporter Sir Charles Lyell (1797–1875) was the geologist most influential in extending the time-scale sufficiently for evolution by natural selection to be viable (Lyell, 1863).

Thirdly, what was a 'species'? A typical passage:

> Providence has distributed the animated world into a number of distinct species, and has ordained that each shall multiply according to its kind, and propagate the stock to perpetuity, none of them ever transgressing their own limits, or approximating in any great degree to others, or ever in any case passing into each other. Such a confusion is contrary to the established order of Nature. (Prichard, 1813, p. 7)

Such a picture permitted no possibility of species transformation or division. Furthermore, even as late as 1850, Latham puts forward a grossly oversimplified model of each species descending from 'a single protoplast pair'. Although 'multiple protoplast' origins might be possible this is very confused. It was this idea that animals could be classified into absolutely different, non-interacting species slots that Darwin's notion of continual variation was to so profoundly challenge. To see species as clusterings in a continuum of variation required a major psychological change, and one which even since Darwin has not been fully understood.

Fourthly, there is the notion of hierarchy itself, the assumption that life is organised like a pyramid with Mankind at the top, that it has a moral dimension of excellence opposed to inferiority. Although this was not entirely relinquished until the present century, if at all, one of the merits of Darwin's model was that in principle it was devoid of value judgments regarding the merits of different species. No niche was any better or worse than any other, form was determined by the survival requirements of the niche occupied. This even applied to humans, though most writers on human evolution could not swallow equality until the present century, depicting the races as hierarchically arranged with white Europeans highest and Australian aborigines, 'Hottentots' and Fuegians at the bottom. Even so, the age-old projection of moral values on to the entire animal realm had to be forsaken if a scientific account of evolution was to emerge, and as far as non-humans were concerned it attenuated rapidly after 1859.

The blocks to accepting evolution were not all psychological. The fossil record was far from unambiguous and it is worth noting that T. H.

Huxley considered the major evolutionary steps to have occurred *prior* to the start of the fossil record, since when persistance, rather than transformation, had been dominant (see Desmond (1982) for the complexities of this issue, and Irvine (1955) for more background).

Technically, the topic awaited identification of a scientifically acceptable mechanism by which evolution could be explained without invoking metaphysical principles of striving 'life-forces' and the like. It was here that the Darwin–Wallace model triumphed. Its assumptions are three-fold: (a) that population has a tendency to outstrip food-supply (as shown by Malthus 1803); (b) that there is a continuous range of spontaneous variations among offspring in their physical attributes; (c) that offspring nevertheless tend to resemble their parents. The details of how inheritance operated were as yet obscure, and when further examined the picture changed seriously (see section B, below). Taking these three assumptions in conjunction the theory of evolution as occurring by 'Natural Selection' from spontaneous variants could be persuasively advocated. A. R. Wallace (1823–1913) remained far more single-mindedly devoted to the principle of Natural Selection subsequently than did Darwin. The principle ensured that those variants best fitted to survive in their environment would reproduce more successfully, leading over the aeons of geological time to the appearance of new species and the extinction of old ones. The phrase 'survival of the fittest', coined by Darwin's contemporary evolutionist Herbert Spencer (1820–1903), is actually rather deceptive (although Darwin eagerly adopted it). As it stands it is tautologous, the fittest being by definition those who survive. Unfortunately it smuggled the moral dimension back into the debate, for 'fittest' was also understood to be 'healthiest', health meant hygiene, and cleanliness of course was next to godliness. Spencer's usage prepared the way for what is known as 'Social Darwinism' (see section B below). Even the word 'evolution' was Spencer's rather than Darwin's – who preferred the more neutral 'transformation'. Tennyson's line 'Nature red in tooth and claw' was also felt to enshrine Darwin's viewpoint. This too was a misrepresentation, concentrating on only one of an almost infinite variety of ways in which life conducted its 'struggle'. In fact face-to-face physical conflict is relatively rare as later ethologists discovered, and its intention is not often to slaughter the opponent. Within species it serves to establish dominance hierarchies and accessibility to mates while between species it is clearly not the predator's aim to exterminate its prey, predator and prey are *not* in competition, indeed elimination of prey means elimination of predator. Later in the century the Russian, Prince Kropotkin, found such a dearth of evidence for struggle, and so much for co-operation, on his Siberian expedition that he wrote a book entitled *Mutual Aid* (Kropotkin, 1939 [1902]) promoting an idealistic anarchism quite opposed to the pro-competition ideology of social Darwinists.

What should be noted is that such phrases as 'survival of the fittest' represent the immediate response of Darwin's contemporaries to his theory. It was being socially reconstrued right from the start in the light of contemporary economic doctrines and psychological assumptions. Such expressions do not accurately reflect the scientific core of the theory, but neither, it must be admitted, did Darwin devote much energy to combating them, nor was he immune from similar interpretations. The fact that the theory could be used to provide a rationale for the economic and political *status quo* was undoubtedly a factor in its final acceptance, just as hitherto the belief that evolutionary theories would undermine the moral basis of civilisation had been a factor in their suppression.

2 *Evolution and Man*

In the 1st edition of *The Origin of Species* Darwin barely mentions human evolution, and then only on the penultimate page in one of science's great throwaway lines:

> In the future I see open fields for far more important researches. Psychology will be securely based on the foundation already well laid by Mr. Herbert Spencer, that of the necessary acquirement of each mental power and capacity by gradation. Light will be thrown on the origin of man and his history. (Darwin, 1859)

This was disingenuous of him. 'Man's Place in Nature' (as T. H. Huxley called his 1863 Essays) had been a perennial riddle. Even within Christianity doctrines varied; some saw us as the privileged focus of Creation for whose benefit the whole natural world was made, others saw Man as the locus of sin and wretchedness in an otherwise perfect and immutable cosmos (Kidd, 1833; Chapman, 1837 respectively). German Naturphilosophie put us at the pinnacle of Creation. Oldroyd (1980) quotes Oken; 'Man is the summit, the crown of nature's development, and must comprehend everything that has preceded him, even as the fruit includes within itself all the earlier developed parts of the plant. In a word, Man must represent the whole world in miniature' (Oken, 1847, p. 2). This is a restatement of a position going back to the Hermetic philosophers of the Renaissance (Yates, 1964). Such accounts are resolutely anthropocentric and in 1986 I rather sourly see in them a sort of collective macho narcissism masquerading as a moral imperative, but that's by the by.

The possibility of including Man in the natural scheme of things, of treating us for scientific purposes anyway as an animal species, aroused extraordinary squeamishness. Linnaeus had courageously classified *Homo sapiens* (his term) in the *Anthropoidea* along with some other primates, although his knowledge of these was poor. Cuvier rejected this, drawing a spurious distinction between 'quadramanous' (four-handed) and 'bimanous' (two-handed) anthropoidea, the latter including only Man. Although by the mid-nineteenth century this had been reversed by comparative anatomists, other candidates for 'anatomically unique feature' were proposed, Richard Owen stuck out for the part of the brain known as the hippocampus minor, to be roundly trounced by T. H. Huxley. The acceptability of treating humans as animals was a problem throughout this early period. Both Lawrence (1819) and Chambers (1844) had been vilified for proposing that the zoological study of man as animal was the only proper foundation for research. The famed Swiss physiognomist Lavater (1741–1801), whose position was a pious, popular, version of Naturphilosophie had written in his Essays (1781–1787):

> It is well known, that of all animals, the monkey approaches nearest to the human form; yet what distance between monkey and man! – But the more enormous this distance is, the more is man bound to rejoice at it. Let him carefully guard against that false humility which would degrade his being, by an exaggeration of the relation which it bears to a creature so much his inferior!
>
> Can anyone find in the monkey, the majesty which sits enthroned on the human forehead, when the hair is turned backward? Is it not a profanation of the word hair, to apply it to the mane of the monkey? (Lavater, 1797, pp. 122, 123)

Wendt (1972) quotes a similar, but madly rabid, passage from Oken. Latham (Latham, 1851) is still acutely aware of the difficulties in making the comparison palatable:

> Unless the subject be handled with excessive delicacy, there is something revolting to fastidious minds in the cool contemplations of the *differentiae* of the Zoologist 'Who shows a Newton as he shows an ape'. Yet, provided there be no morbid gloating over the more dishonourable points of similarity, no pleasurable excitement derived from the lowering of our nature, the study is *not* ignoble. (p. 6)

Nevertheless the combined evidence of comparative anatomy and embryology was forcing the ape-human resemblance into the open. It had been a point of interest ever since Tyson (1699) had compared what he thought was a 'Pygmie' (in fact a chimpanzee) with a Monkey, an Ape and a Man. Unfortunately western thought had always seen apes and monkeys as symbols of moral degeneracy, epitomes of lust, mischievousness and even devilishness. They were essentially wicked species. Perhaps this was no coincidence, it was the sheer force of the *resemblance* which now rendered so intense the anxiety raised by this new prospect of some form of actual *identity* between us and apes. Clearcut physical differences proving increasingly elusive the orthodox case began to rest ever more precariously on moral and spiritual distinctions.

In the wake of *The Origin of Species* T. H. Huxley began extrapolating the implications of the theory of evolution for humans, reviewing the issue in a series of three essays published in January 1863, collectively entitled *Man's Place in Nature*. He derides the contemporary reluctance to face facts for fear of moral consequences. The difference from Latham's tone is dramatic; 'it is not I who seek to base Man's dignity upon his great toe, or insinuate that we are lost if an Ape has a hippocampus minor' (Huxley, 1901, p. 152).

Another issue facing those concerned with the physical status of Man was the long-running controversy over *monogenism* versus *polygenism*.[3] Was the human race a single species? Christians were implicitly committed to monogenism since we are all descended from Noah's sons, but anthropological studies from Blumenbach at the turn of the century onwards had shown humanity's amazing physical diversity (Blumenbach, 1865). Combine this with a belief in species immutability, or a short time-span since Creation, or both, and such diversity becomes nearly inexplicable in a monogenist framework. Darwin's theory resolved this but not in an entirely unambiguous fashion. By adopting Lyell's time-scales the old time-perspective is rendered obsolete, while the entire theory is aimed at proving immutability mistaken. But the resulting monogenism was now placed on such an extended time-scale that racial divergences could be placed far enough back for the polygenist spirit to win out in practice, if not in theory. For the remainder of the century, and into this, books on human evolution regularly presented readers with speculative family trees of the human race which left no doubt as to the gulf between European and 'Negro'.

By the 1860s then, it was finally possible for scientists to treat humans as members of the animal kingdom, subject to the same kinds of physio-

[3] The Monogenism versus Polygenism issue continued to preoccupy anthropologists for the remainder of the century, though the connotations of the two positions can become somewhat convoluted: Topinard for example (1877, rep. 1890), proposes a version of polygenism as being the most consistent with 'transformationism' (i.e. evolutionary theory) and considers monogenism in its classic form as outdated Christian dogma. The notion that present human 'races' evolved directly from the primate species native to their present homelands has not yet entirely died out; its most widely read twentieth-century exponent was probably F. G. Crookshank, whose extraordinary *The Mongol in Our Midst* went through three editions between 1924 and 1931, while an even more bizarre version appeared in 1973; Oscar Kiss Maeth's *The Beginning was the End*. (Both of these use the technique of publishing alongside one another photographs of a supposedly 'typical' human inhabitant of the region and an indigenous primate, to display their 'similarity'. Judiciously used such a technique could probably prove us to have descended from parrots! The admittedly curious resemblances which can occur between animal and human physiognomies have inspired a discernible genre of publications from Della Porta (1623) onwards, to which Lavater in part belongs.) The genuine element in the controversy at present seems to be the nature of the shift from *H.erectus* to *H.sapiens*, discussed in Chapter 6, but this is far removed from nineteenth-century polygenism, however interpreted.

logical laws as all other species, and displaying nothing at the biological level to elevate them to the different plane whereon Christians in particular had previously insisted on placing them. But was there yet any real evidence that we *had* evolved? Even if the new time-perspective was correct, were there any empirical data showing that it did in fact apply to humans? The issue of *antiquity*, though logically distinct from that of Darwinian *evolution*, could not be kept separate from it in this context since evolution implies antiquity even if the converse does not apply.

3 *The antiquity of Man*

As a curious prelude to this I must mention a book published in 1655 in London and Amsterdam bearing, in English, the title *A Theological Systeme upon that presupposition that Men were before Adam*. Its author, Isaac de la Peyrère, claimed that Adam, though the first Jew, was not the first man, and that there were men existing before Adam. He based his case in part on archaeology. Peyrère was promptly imprisoned for heresy, forced to recant, and died in a convent in 1679 (Casson, 1940, pp. 114–18). He deserves rescuing from obscurity, his sufferings being no doubt as great as Galileo's and ultimately in the same cause.

Evidence of human antiquity is of two kinds; fossilised bones and artefacts. The former are far rarer than the latter. What evidence was there of either kind at the time Darwin was writing *The Descent of Man*? Two best-selling summaries of the data appeared in the 1860s; Lyell's *The Antiquity of Man* (1863) and John Lubbock's *Prehistoric Times* (1865). Both remained in print well into the present century and although Lyell's is the better known, it drew heavily in places on articles published by Lubbock in the *Natural History Review* and later incorporated in his own book. The difference between them is one of emphasis, Lyell stressing the geological evidence, Lubbock the archaeological finds. Of the two Lubbock was the more orthodox Darwinian (being in fact a protégé of Darwin's), while Lyell's hedging on the 'transformation' issue somewhat rankled with the great man. Sixty years later Sir Arthur Keith could still describe his own two-volume *The Antiquity of Man* as 'in many respects' 'supplementary' to Lubbock's 'classic work' (Keith, 1925, p. xxi).

The fossil evidence for human antiquity was minimal but tantalising. Table 2.1 lists the major fossil finds down to 1940. It will be seen that only 8 precede 1871, and only 7 the works by Lyell and Lubbock. Descriptions of one of these, the Gibraltar skull, were not available until the late 1860s (Busk, 1865; Broca, 1869). The two earliest finds had been lost. This left only three clearly human fossils; the Engis skull, the 'Fossil Man of Denise' and, most famous of all, the Neanderthal remains from

TABLE 2.1 *Major hominid fossil finds prior to World War Two*

Date	*Location*	*Details*	*Finders/describers*
1822	Paviland Cave, Swansea	'Red Lady of Paviland'. Male *H.s.sapiens*	Buckland
1828	Bize Cavern, France	'human bones and teeth'. Now lost	Tournal
1828	'the cavern of Pondres' Nîmes, France	'some human bones' with extinct fauna. Now lost	Christol
1831	Liège, Belgium	The 'Engis skull'. Now classified *H.s.neanderthal*	Schmerling
1844	Denise, Le Puy en Velay, France	'The Fossil Man of Denise'. Skull fragments and other bones of two individuals. *H.s.sapiens*	Aymard
1848	Gibraltar	'The Gibraltar Skull'. *H.s.neand.*	Lt. Flint, Busk, Broca
1856	Sarsan, French Pyrenees	*Dryopithecus* mandible. (Type spec.)	E. Lartet
1857	Neanderthal Valley, nr. Dusseldorf, Germany	'Neanderthal Man'. Skull and much of skeleton. *H.s.neand.* (Type spec.)	Fühlrott
1868	Les Eyzies, France	'Cro-Magnon Man'. Five skeletons. *H.s.sap.*	L. Lartet
1874–1875	Grottes de Grimaldi, Monaco	Two children's skeletons. *H.s.sap.*	E. Rivière
1880	Sipka, Czechoslovakia	Mandible. *H.s.neand.*	Maška
1886	Spy, Belgium	Crania and numerous bones from two skeletons. *H.s.neand.*	de Puydt, Lohest, Fraipont
1887	Banolas, Spain	Mandible. *H.s.neand.*	Pacheco, Obermaier
1887 and 1892	Taubach, Germany	Teeth. *H.s.neand.*	Weiss, Nehring
1888	Chancelade	Skull and most of skeleton	Raymonden, Testut

Date	*Location*	*Details*	*Finders/describers*
1889	Wadjak, Java	Two crania. *H.s.sap.* (*H. wadjakensis*)	van Rietschoten, Dubois
1891–1898	Trinil, Java	Cranium, femur and fragments. '*Pithecanthropus*', 'Java Man'. *H.erectus*	Dubois
1899–1905	Krapina, Yugoslavia	c. 800 fragments, mostly sub-adult. *H.s.neand.*	Gorjanovic-Kramberger
1901	Grottes de Grimaldi, Monaco	4 skeletons of 'Cro-Magnon' and 'negroid' 'Grimaldi Race'. *H.s.sap.*	Villeneuve
1907	Mauer, Germany	'Heidelberg Man', 'Mauer Jaw'. Mandible. *H.erectus*, *Archaic H.sap.*?	Schoetensack
1908–1925	Weimar-Ehringsdorf, Germany	Various fragments inc. 2 mandibles; 1 adult, 1 child. *H.s.neand.*	Nehring, H. Virchow
1908	Chapelle-aux-Saints, France	Complete skeleton. 'Old Man of Chapelle-aux-Saints'. *H.s.neand.*	A. and J. Bouysonnie, Boule
1909–1921	La Ferrassie, France	Numerous fragments. *H.s.neand.*	Peyrony, Capitan
1910	Combe-Capelle, France	Skull. *H.s.sap.*	Hauser
1910–1911	La Cotte de St Brelade, Jersey	Teeth. *H.s.neand.*	Marett
1911	La Quina, S. France	Skeleton. *H.s.neand.*	Martin
1911	Siwalik Hills, India	Fragments of mandible and teeth. *Sivapithecus*. (Type spec.)	Pilgrim
1912	Piltdown, Surrey	'Piltdown Man', '*Eoanthropus dawsoni*'. *H.s.sap.* cranium + Orang utan mandible	Dawson, Smith Woodward

Date	*Location*	*Details*	*Finders/describers*
1913	Olduvai Gorge, Tanzania	O.H.1 skeleton. *H.s.sap.*	Reck
1921	Broken Hill, Rhodesia (Zimbabwe)	'Rhodesia Man'. Skull and much material from 2 skeletons. *H.rhodesiensis*	Barren, Smith Woodward
1922	Ordos Desert, China	1 *H.s.neand.* tooth	
1923	Dire-Dawa, Ethiopia	Mandible fragment. *H.s.neand.*	Vallois
1921–1937	Choukoutien, China	Remains of c. 40 individuals. *H.erectus.* (Initially class. *s. pekinensis*, 'Peking Man'). Now lost except three teeth	Andersson, Zdansky, Black, Pei, Bohlin, Weidenreich
1924	Taungs, Bechuanaland (South Africa)	Infant skull and brain endocast. *Australopithecus africanus.* (Type spec.)	Dart
1925	London, England	'Lloyd's skull'. *H.s.sap.*	Keith
1926	Devil's Tower Cave, Gibraltar	Skull. *H.s.neand.*	Garrod
1929–1934	Mt. Carmel, Palestine (Israel)	(a) Tabūn Cave: *H.s.neand.* 4 individuals' skeletons or parts (b) Skhūl Cave: *H.s.palestinus.* 10 individuals' skeletons or parts. (Type specs)	Garrod, Bates, McCown, Keith
1929, 1935	Saccopastore, Italy	Remains of two individuals. *H.s.neand.*	
1931–1932	Kanjera, Kenya	'Kanjera skull'. *H.s.sap.*	L. S. B. Leakey
1931–1932	Kanam, Kenya	'The Kanam Jaw'. Mandible of *H.habilis*	L. S. B. Leakey

Date	*Location*	*Details*	*Finders/describers*
1931–1933	Ngandong, Java	'Solo Man'. 11 cranial and numerous other remains. *H.s.soloensis*. (Neanderthaloid)	von Koenigswald, ter Haar
1933	Steinheim, Germany	'Steinheim skull'. Late *H.erectus* or Archaic *H.sap.*	Berkhemer, Sigrist
1935	Olduvai Gorge, Tanzania	O.H.2 *H.sap.*	L. S. B. Leakey
1935–1936	Swanscombe, Kent, England	'The Swanscombe Skull', 'Swanscombe Man'. Partial cranium, another fragment found in 1955. Late *H.erectus* or Archaic *H.sap.*	Marston
1936	Sterkfontein, South Africa	Skull. *A. africanus*	Broom
1936	Modjokerto, Java	Child's skull. *H. erectus*	Cosijn
1937	Eyasi, Tanzania	'Eyasi skull'. *A. africanus*	Kohl-Larsen
1937–1939	Sangiran, Java	3 skulls. *H. erectus*	von Koenigswald
1938	Kromdraai, South Africa	Large part of skull and jaw. *A. robustus*. (Orig. class *Paranthropus robustus*)	Broom, Terblanche
1938	Siwalik Hills, India	Teeth. *Ramapithecus* (Type spec.)	Lewis

Germany. The Engis skull, surviving remnant of a much larger original cache, whilst belonging, it was agreed, 'to an individual of small intellectual development' could not, Huxley claimed, be clearly differentiated from that of an 'Ethiopian' or Australian aborigine. 'Skull' is in any case an exaggeration as but a small part of the cranium remained. Its importance lay in the fact that Schmerling had found it in association with stone tools and fossils of extinct mammals. The Denise find was also indistinguishable from modern human remains, but had also been found with remains of vanished fauna. Neanderthal Man was a different kettle of fish, 'the most brutal of all known human skulls' Huxley called it. With prominent brow-ridges, a long low cranium, and unusual bone-thickness, it seemed more ape-like than any living person. Yet its brain capacity, as Huxley calculated it, was 1230 cc – twice that of a gorilla, typically human (1234 cc is now accepted). It aroused tremendous controversy, being variously ascribed to an idiot, or a Cossack, or a freak by those unable to accept its antiquity. Among its severest critics was Rudolf Virchow (1821–1902), an eminent pathologist whose persistent scorning of all subsequent hominid fossils hampered progress for the rest of the century. The resemblance to the Gibraltar skull became apparent after the work by Busk and Broca (known to psychologists as discoverer of the eponymous speech area of the brain). By 1900 at least five other European sites, from Spain to Czechoslovakia, had yielded Neanderthal-type remains, but Virchow's scepticism prevented them from receiving serious attention until Schwalbe (Schwalbe, 1904). All the same, the brutal Neanderthal had entered the European imagination, an ancestral Mr Hyde to the modern urbane Dr Jekyll. Whatever else its authenticity implied, it confirmed that a now extinct human-like creature had once lived in Europe, whilst other finds implied a co-existence of tool-making human-like creatures with mammoths, woolly rhinoceri and cave bears.

Artefact evidence was, by contrast, plentiful. Flint tools had been found in London as long ago as 1715 (associated with elephant remains) and (at Hoxne, Suffolk) in 1797 by Frere. Schmerling reported numerous such tools in caves near Liège, Belgium, while in France, Boucher de Perthes was finding large quantities of them in gravel at Abbéville, Picardy, from as deep as 30 feet. In Britain attention focused on Kent's Hole cavern in Devon, where artefacts and fossil mammals had been found since the 1820s. In each case the reports met dogmatic objections, worked flints were explained as later intrusions, fakes, or natural objects. This stubborness, epitomised in the work of Dean Buckland (Buckland 1823, 1836), eventually had to yield. An expedition by Pengelly in 1858 established beyond doubt the contemporaneity of tools and fossils in caves at Windmill Hill, near Brixham, Kent's Hole itself now being too disturbed. Boucher de Perthes' claims (Perthes, 1847) were vindicated, and new sites found, by Dr Rigollot (a converted opponent) at St Acheul (whence the present term Acheulean, or Acheulian, derives).

More recent prehistory had meanwhile opened up in Denmark and Switzerland. Danish peat-bogs and shell-middens had yielded such a quantity of artefacts, so clearly stratified, that their archaeologists, such as Thomsen, founder of Copenhagen Museum, and Worsaae, had long identified the familiar sequence of Stone, Bronze and Iron Ages. Indeed this scheme was first suggested by P. F. Suhm in 1776 (Wenke, 1980). In 1864 Copenhagen Museum alone had 4,480 stone artefacts (Lubbock, 1865). The 'three ages' were taken up by Nilsson and then extended by Lubbock (1865; S. Nilsson, 1869) who divided the Stone Age into Palaeolithic and Neolithic (Old and New) to incorporate the far cruder French and British finds. This usage persists with the addition of Mesolithic (middle Stone Age) (Brown, 1893), and Chalcolithic, a short-lived 'Copper Age' between Neolithic and Bronze. The French preferred to classify prehistory according to predominant fossil mammals; Age of Mammoths, Age of Cave-Bears, Age of Reindeer and Age of the Auroch (Lartet, E., 1858). Unsatisfactory for various reasons, this was replaced in 1869 by Mortillet's scheme (Tables 2.2 and 2.3). In Switzerland remains of lake-dwellings were discovered in Lake Zurich during the winter of 1853–54, followed by similar finds at other lakes. These exposed an extensive prehistoric lake-dwelling culture. Though not so neatly stratified as the Danish material, a Stone to Iron Age succession was still discernible in the great number of artefacts brought to light. Additionally, some sites lent themselves to ingenious calculations of the time elapsed since their occupation. Based on sedimentation and marsh-formation rates these gave figures as great as 7,000 years, not huge by modern standards but enough to upset the fundamentalist Creation date of 4,004 BC. Latham and Baron Bunsen tried to calculate Man's age on the basis of linguistic evidence, calculating the time required for divergence of modern languages from a common root. Bunsen arrived at 20,000 years. By the 3rd edition of *Antiquity of Man* in 1869, however, Lyell is proposing 100,000 years for some of the implementiferous Somme alluvium.

The end of this first period coincides with several archaeological classics; Dupont on the Belgian material (Dupont, 1872), Evans' *Ancient Stone Implements . . . of Great Britain* (Evans, 1872), *Cave Hunting* (Dawkins, 1874) and Mortillet's *Essai de Classification* (Mortillet, 1869). The first, unrecognised, find of Palaeolithic art was a bone engraving of hinds from a cave at Chaffaud, France in 1834. In 1860 E. Lartet (1801–1871), one of the greatest French pioneers, found pierced shells and worked bone at Aurignac. These were associated with implements more advanced than those of St Acheul, yet clearly older than the Danish and Swiss Neolithic tools.

By 1872 an extended human antiquity is undisputed by serious researchers, but the time-scale involved is quite obscure. Lyell, Lartet, Huxley, Lubbock and the rest have crossed the threshold of a new time-perspective in which they have no firm bearings, knowing only that the times involved are inconceivably vaster than those envisaged by even their most speculative predecessors. (See Albritton, 1980; Toulmin and Goodfield, 1965, for accounts of the time-scale issue.)

A final source of data affecting the perception of human evolution was cross-cultural evidence, now rendered more significant by the victory of monogenism. The superiority of western civilisation was obvious to nineteenth-century Europeans, and was attributed to the European's more advanced character, now construed as their having achieved a higher stage of evolution. They have emerged from the savagery in which others are yet mired, and from which it is the imperial task and destiny to deliver

TABLE 2.2 *Mortillet's classification of the Stone Age*[1]

Version	*Sequence (earliest numbered 1)*
1869	4 Epoque de la Madelaine
	3 Epoque d'Aurignac
	2 Epoque de Solutré
	1 Epoque de Moustier
1883	6 Robenhausian
	5 Magdalenian
	4 Solutrean
	3 Mousterian
	2 Chellean
	1 Thenaisian

[1] From G. Daniel (1950).

TABLE 2.3 *MacCurdy's 1924 classification of the Stone Age industry sequence*[1]

Phase	*Industry*	*Sub-classifications*
Neolithic	Carnacian	
	Robenhausian	
	Campignian	
	Maglemosean	
	Azilian-Tardenoisian	
Palaeolithic	Magdalenian	I–VI
	Solutrean	Lower, Middle, Upper
	Aurignacian	Lower, Middle, Upper
	Transition	
	Mousterian	Lower, Upper
	Acheulian	Lower, Upper
	Chellean	Lower, Upper
	Pre-Chellean	(transition)
Eolithic		

[1] From G. Daniel (1967), adapted.

them. The condition of Europe's ancestors was, then, displayed in the life of the contemporary savage.

Darwin had met 'savages' of a most pitiful kind during his voyage in the *Beagle*, the now extinct 'Fuegians' of Tierra del Fuego. Yet compared to his cousin Galton he was disinclined to attribute much intrinsic biological significance to cultural differences. Darwin was no racist, except insofar as he unquestioningly accepted some of the racist conventional wisdoms of the day. It is perhaps anachronistic to use the term 'racist' for nineteenth-century European (if not US) thinkers, since many of its present connotations are inapplicable. At the same time, in the absence of direct evidence, the 'savage' life-styles and customs served as a major data-source in speculations on the more recent phases of human evolution and, for example, the use of stone artefacts. Lubbock's work is a classic example. The favourite 'savages', of the late nineteenth century were always Australian aborigines, 'Hottentots', 'Bushmen' and Fuegians. Of the Australian aborigines the Tasmanians, rapidly becoming extinct even as Darwin wrote, were the most popular. Though remaining important to this day, the anthropological perspective is no longer one in which those whom it studies are considered as unevolved or less evolved 'savage' specimens lingering on from an earlier phase of humanity.

4 *Darwin's contribution*

The data available to Darwin when writing *The Descent of Man* thus comprised, firstly, the findings from comparative anatomy and embryology showing how close we were anatomically to chimpanzees and gorillas, and how our foetal development matched that of other mammals, being virtually indistinguishable from them until quite late in pregnancy. Secondly, there was the evidence of human antiquity drawn from archaeology and palaeontology, albeit very little of the latter. To these sources Darwin was able to add more material of his own, particularly his further investigations into the nature of inheritance and variation (Darwin, 1866) to which he constantly alludes. He also drew on data and ideas he had long been garnering in his 'M' and 'N' notebooks (Gruber and Barrett, 1974).

To appreciate the subsequent course of human evolution research it is necessary to have at least an outline picture of what Darwin achieved in *The Descent of Man* and *The Expression of the Emotions in Man and Animals*. The first appeared in two volumes in February 1871 (and in a revised 2nd edition in September 1874). Its title is something of a misnomer, since over half the work is devoted to a middle section on sexual selection in animals, a mechanism which Darwin considered highly relevant to human variation, as we will see.

The book has two aims; firstly, to establish that the physical and behavioural gap between humans and animals is only a matter of degree. Secondly, he wants to explain the differences between human races (invoking sexual selection). To achieve the first was, with regard to the physical aspect, fairly easy, given the groundwork of Huxley, Haeckel and others, and in Chapter 2 Darwin discusses a number of features such as human hairlessness, tooth-form, tail-lessness and erect posture, showing how each may be explained by natural selection, and how similar trends and features are evident in higher primates. To prove that humans were psychologically and behaviourally contiguous with 'lower' animals was more challenging and controversial. Chapters 3–5 present a thorough review of the 'Mental Powers' of men and animals. To the modern student of behavioural evolution these are the book's most significant chapters, for in them are provided the first evolutionary formulations of virtually the entire area. 'My object' he says, 'is to shew that there is no fundamental difference between man and the higher mammals in their mental faculties' (p. 99). Wonder, curiosity, imitation, attention, memory, imagination, can all, he believes, be shown in rudimentary form in animals. Candidates for 'human uniqueness' which others have proposed are systematically demolished; ability to learn, tool-use, self-consciousness, even language, 'sense of beauty' and religion, are all, he shows, either present in animals or less than universal in humans. His evidence on these matters is similar, if less copious and detailed, to that which would be invoked today, e.g. primate behaviour. Whereas we have the field studies of a Goodall or Schaller he must rely on zoo-keepers' anecdotes, but his conclusions are very similar.

As well as individual psychological functions he tackles 'social instincts' and the roots of morality, still areas of great controversy, particularly in regard to the evolution of altruism and self-sacrificing behaviour. It is interesting then to see that Darwin was fully aware of the difficulties such phenomena apparently presented for a natural selection theory, and he tackles them at length in Chapters 4 and 5. The existence of animal sympathy and sociability is demonstrated and the advantages to the social group of noble social virtues expounded. Darwin is sure that human morality is rooted in such social instincts and arises from the operation of natural selection on them. Yet his account has a number of weaknesses, stemming in turn from his lack of genetic knowledge. In the decade since *The Origin of Species* he had felt himself obliged to back-track on the matter of Lamarckian inheritance, and suggests that ' . . . some intelligent actions, after being performed during several generations, become converted into instincts and are inherited, as when birds on oceanic islands learn to avoid man' (p. 102), although he is sceptical about the transmission of 'virtuous tendencies'. In deriving morality from social instincts he proposes a model of what would now be termed 'group selection'; the advantages of self-sacrifice, etc. accrue not to the individual but to the group. But this raises the difficulty of understanding how, since such individuals will reproduce at a *lower* rate than their less noble companions, it is *their* behaviour which is selected. The bravest 'would on average perish in larger numbers than other men' (p. 200). Clearly, natural selection cannot straightforwardly explain this and Darwin proceeds to explain altruism by: (a) the evolution of improved reasoning leading our ancestors to see that it is advantageous to help their fellows, such habits becoming inherited over many generations (i.e. 'intelligence into instinct'); and (b) love of praise and fear of shame. Tribes in which noble behaviour was high would come to dominate those in which it was low, thus group selection would favour tribes of brave and self-sacrificing individuals over the selfish and cowardly, whose indiscipline and lack of moral fibre would result in their succumbing should the groups come into conflict. (See Chapter 5 below for the current state of this issue.)

The latter part of Chapter 5 also provided the basis for what Galton was later to develop as Eugenics, with its doubts about the effects of civilisation on the quality of the human stock '. . . the weaker members of civilised societies propagate their kind . . . this must be highly injurious to the race of man' (p. 206). The 'reckless and degraded' apparently increase more rapidly than the 'provident and generally virtuous', though their higher mortality rate to some extent checks the effects of this. With side-swipes at the deleterious effects of celibacy in the Catholic Church and the selective culling of the intelligent by the Spanish Inquisition, he brings the topic to a close. Whilst Darwin cannot be held responsible for the later excesses of social Darwinism, nor does he really share the associated terror of 'degeneration' which widely affected later European thought (e.g. Nordau, 1895), there is little doubt that in *The Descent of Man* an influential rationale for them can be discovered. Some modern writers have, in my opinion, been a little too eager to let their hero off this particular hook.

It was still, incidentally, necessary to spend a few pages proving that 'All Civilised Nations' were 'Once Barbarous', for such stubborn worthies as Archbishop Whateley and the Duke of Argyll even then maintained that our present state was a decline from a former 'golden age'. Chapter 6 deals in more detail with the 'Affinities and Genealogy' of Man, containing the oft-quoted prophetic passage that 'it is more probable that our earlier progenitors lived on the African continent than elsewhere' (p. 240). (In fairness to Lubbock, he said something very similar, for identical reasons, in *Prehistoric Times*, though specifying only 'tropical climes'.) Darwin also makes a vain effort to explain how we and apes share a common ancestor, we do not descend *from* them, but the Victorians were not to be robbed of the mirth-giving consequences of an Ape-ancestor model by such trifling qualifications.

On the race issue Darwin was a resolute monogenist, and Chapter 7 is devoted to demonstrating the untenability of alternative positions. But why were the different races so physically diverse, even accepting that ours is a 'polymorphic' species? Two possibilities presented themselves: (a) domesticated animals frequently show higher variation than wild ones, so was civilisation equivalent to domestication of the human species? (b) high variation also characterised animals with large geographical range, and in this respect humans were supreme. The former might account for within-group differences, but not racial differences, and Darwin tends to opt for the geographical-range explanation. This is only the beginning of the story, however, for he was struck by the fact that the features in which variation occurs are not particularly significant for survival. Variations devoid of either positive or negative survival value tend to be maintained, but the systematic nature of racial physical differences still puzzles him. To solve this riddle he digresses into the extensive study of sexual selection in animals which constitutes the book's central section.

The phenomenon of 'sexual selection' augments 'natural selection' of the usual type. It refers to the fact that animals select mates according to physical attractiveness. This amplifies sexual dimorphism as each sex evolves those traits found most alluring by the other, and Darwin shows how at all levels, from reptile and insect to bird and mammal, this generates an extraordinary range of skin colouring, plumage, horny excrescences and the like which serve to attract the opposite sex. Sometimes males compete for females, sometimes vice versa. In humans Darwin believes it is males who have the monopoly of choice! *Why* a feature should be found sexually attractive is often mysterious, though there might be a link with survival-relevant characteristics, thus large males might be more attractive than small ones. Nevertheless there is a high arbitrary component. It is this which Darwin uses to explain racial differences when he returns to Man in Chapters 19–20. (See T. H. Clutton-Brock (1983) for an important review of the current status of sexual selection in evolutionary theory.)

Again these pages contain pioneer forays into issues still under debate; the nature of sex differences, effects of different marriage customs and even such topics as the origins of music and clothes, plus a theme which recurs throughout the book and clearly fascinated the author – hair. He concludes that the greater size, strength and pugnacity of males compared to females has arisen from contests between rival males for possession of females in 'primeval times' and been 'subsequently augmented', while racial differences are due to sexual selection of the 'secondary sexual characteristics' (features not functionally involved in reproduction but found sexually attractive). There is no universal standard of beauty with respect to the human body, and possibly tastes become inherited, each race evolving its own innate ideal. Such phenomena as selective infanticide, relatively greater beauty of higher social classes and the adornment customs of savages are all incorporated into this discussion. Towards the end of the book he echoes the Huxley-Wilberforce debate by claiming that he would rather be descended from a heroic monkey than a vicious savage. He is not concerned with hopes or fears, only with the truth. Finally comes an oft-quoted passage in which he proclaims that for all his nobility, 'Man still bears in his bodily frame the indelible stamp of his lowly origin' (p. 947).

It has been said that while *The Descent of Man* looks at the human in animals, *The Expression of the Emotions in Man and Animals* explores the animal in humans. More specifically it addresses human emotions and their expression, especially facial expression. This is governed, according to Darwin by three principles: (a) 'associated habits' – behaviour habitually accompanying an action becomes a signal of emotions associated with that action; (b) 'antithesis' – behaviour opposite to that accompanying an action signals the opposite emotion to that associated with it (e.g. a dog's fawning is the opposite of its aggressive behaviour); and (c) 'direct action of the nervous system' – blushing, or trembling, for instance. The whole range of emotions, from anxiety and dejection to modesty, joy and pride are examined in the light of these, using a similar strategy of comparison with animals and attention to physiological detail as had characterised *The Descent of Man*. The roots of the smile can be traced to teeth-baring, of raised eyebrows in astonishment to a habit 'gained in order to see as quickly as possible all around us' (p. 296). The psychology of emotion had already been dealt with by Bain (1859) and Spencer (1855), but the former's approach was primarily classificatory and the latter's, though evolutionary, weak on hard data and less rigorous in conceptualisation. Darwin's book is a landmark in the study of emotion, remaining a classic in the history of Psychology. It served to bolster the thesis of *The Descent of Man* that human and animal behaviour were a continuum and that even our highest emotions were based upon common instinctive responses physically built into our neurological structure. It is the prototype too of a genre of studies familiar to us now in the works of ethologists such as Lorenz and Eibl-Eibesfeldt, where behavioural parallels between species are graphically displayed.

Darwin succeeded in convincing most, though not all, scientists that humanity had evolved, but his account of how it had done so was speculative and inferential, however plausible the general picture he had painted. His evidence was of three basic kinds: anatomical; analogy with animals; and anthropological evidence of the lives of 'savages'. The first established beyond doubt that humans belonged among the primates for the purposes of zoological classification. The second could provide innumerable analogies with human behaviour and physiological adaptations, as well as being a source of hypotheses as to the original adaptive significance of many of these. The last reminded his contemporaries of the sheer diversity of human behaviour and the inadmissibility of taking for granted the psychological character of western European 'Man' as typically human. But when all was said and done, the concrete evidence was missing. Neither the scanty fossils nor the copious flint tools shed much light on the actual evolution of our species. Time-perspective was still a major difficulty which would not make progress until further advances had been made in geology and palaeontology, and it would be nearly a century before techniques would be developed permitting any form of regular direct dating. Darwin's discussion of human evolution is, even so, extraordinarily broad and he clearly identified the key problem-areas of behavioural evolution: language, tool-making, social structure, the role of brain-organisation and altruism, for example.

All later discussion and research stems from Darwin's work. Having accepted his perspective, the hunt was now on among Europe's palaeontologists and archaeologists for that elusive creature who would link us to our simian progenitor. But it was a hunt conducted under the mocking eyes of the remaining anti-evolutionists; Richard Owen, Louis Agassiz and Rudolf Virchow. And it is arguable that no real advance was made at all during the subsequent period, 'the great confusion', to which I now turn.

B The great confusion, 1873–1913

These four decades, though in retrospect achieving little progress in the direct understanding of human evolution, were not sterile. On the contrary, the confusion was perhaps necessary in order that a basic groundwork eventually be laid in those disciplines like genetics upon which deeper understanding of evolution depended. It was a confusion arising in part from very genuine conceptual problems which only time and patience could unravel, but it also arose from the fact that European culture at large was preoccupied with trying to construe afresh human nature and the human situation. It was in the context of this cultural concern that the human evolution debate was being conducted, and from which it derived much of its *raison d'être*. In this section I will attempt to impose some order on a very chaotic scene by discussing it under the following headings: 1 Genetics, 2 Social Darwinism, 3 Fossil evidence, 4 Archaeology and 5 Psychology and Anthropology.

1 *Genetics*

An important difficulty arose almost immediately for the credibility of the central tenet of Darwin's theory – 'natural selection from spontaneous variations'. The problem lay in the nature of inheritance. Darwin (1868) had proposed a theory he called 'pangenesis' in which 'gemmules' from each part of the body circulated in the bloodstream and became stored and multiplied in the reproductive organs of each parent. Each gemmule was responsible for the development of the part of the body whence it originated. Parental contributions mingled in a fairly straightforward fashion. Gemmules could be somewhat affected by the parent organism's current state, permitting a degree of Lamarckian inheritance. But empirical evidence from hybridisation studies – which had been going on with varying degrees of methodological sophistication since the end of the eighteenth century – suggested (a) that variation was *not* continuous as Darwinian theory required, and, equally seriously, (b) that a single variant trait, however adaptive, could not be passed on for more than a few generations since it rapidly became swamped again, 'diluted' through successive matings of offspring with those lacking it. Galton had also identified the statistical phenomenon of 'regression to the mean' which exacerbated the point, the very tall, for example, do not have equally tall offspring, but ones nearer the population mean. Another notion which entered the arena, though in the end proving false, raised other difficulties for natural selection, 'orthogenesis'. This doctrine held that certain evolutionary trends were unstoppable and might end up dooming the species. The antlers of the extinct Irish Elk were cited; they just kept enlarging until the animal, devoting a totally disproportionate amount of energy to their growth and haulage, met its demise. (See Gould (1977) on this theory and its refutation.)

Lamarckism received a near-fatal blow in the 1880s when Weismann (1834–1914) identified the locus of heredity in the germ-cell's nucleus, the 'germ-plasm', (he also introduced the term 'chromosome') (Weismann, 1885, 1893; Sturtevant, 1966). This was followed by the work of a number of experimentalists, the foremost being W. Bateson (1861–1926), De Vries (1848–1935) and T. H. Morgan (1866–1945), which confirmed the discontinuity of variation and raised even more problems for the Darwinian model. In the 1860s Mendel (1822–1884) had performed his famous hybridisation experiments, while a Frenchman, Naudin, and Galton too, were doing similar work. Such research suggested to some that evolution proceeded by hybridisation, not by natural selection, thus right at the heart of genetics lay an anti-natural selection argument. It was Mendel's mathematical analysis of the nature of hybridisation, and his concepts of dominance and recessiveness to explain this, that led to the eager reception of his 'rediscovery' in 1900. Modern work (Olby 1966; Brannigan, 1979; Meijer, 1983) has largely discredited the myth of Mendel's neglect, explaining his re-emergence as a way of resolving a priority dispute between three researchers (De Vries, Correns and Tschermak) who had each independently arrived at Mendel's formula, but from within the orthodox evolutionary framework. As D. L. Hull (1982) points out, strictly speaking Mendel was not a 'Mendelian' as the term was subsequently used. In recent years the Mendelian 'revolution' has received much attention from historians of science and many points of contention persist regarding its interpretation.

Bateson and the new Mendelians turned increasingly against natural selection. By 1913 it was difficult to see how the proven importance of 'Mendelian' inheritance was reconcilable with the Darwinian account; everything in an offspring generation was received from the parent generation, changes in phenotype resulted from hybridisation, and the 'unmasking' or disinhibiting of traits already in the genetic material as potential. Nothing could, it seemed, be added to this. Occasional mutations (a term by then in use) could not account for evolution, being invariably sterile and monstrous. Bateson in fact washed his hands of the whole evolution issue, saying the time was not yet ready to understand it (Bateson, 1915). Though the evolutionary perspective was well established, acceptance of natural selection as its fundamental mechanism was if anything waning. In the 1920s, as we will see, Mendelism and natural selection were reconciled, a reconciliation becoming the new orthodoxy by the 1940s (J. Huxley, 1942). Even within the geneticists' ranks controversy was so rife that Mendelists and their opponents almost ceased speaking to each other. The latter were most strongly represented by the 'Biometricians', led by Karl Pearson and Galton, who tended to defend the Darwinian orthodoxy and adopted a basically statistical approach. (Olby (1984) has pointed out a major inconsistency in Galton's position, his Eugenic evangelism being pro-natural selection and his actual studies of heredity contradicting it.) Already the split between 'gradualists' and 'saltationists' is discernible, the former believing in continuous gradual change and the latter, including the Mendelists, that evolution occurred in qualitative jumps. Theorising about human evolution could make little headway against this background of controversy over the nature of evolution in general.

2 *Social Darwinism*

In spite of its name, this movement would be better called 'Social Spencerism', as Herbert Spencer was its most enthusiastic advocate. The British gentry, from whose ranks the likes of Darwin, Galton and Lubbock were drawn, were nothing if not enthusiastic stock-breeders. Darwin's message was rapidly interpreted as having portentous significance for the quality of the human stock, a view from which he in no way dissented. By allowing the unfit, or less fit, to breed rather than letting them be culled by natural selection as happened among animals and savages we were letting the human stock degenerate. 'Degeneration', a term introduced by a French psychiatrist, Morel (1857), became an obsession in the period under consideration, its eventual apotheosis coming under the Nazis. Late nineteenth-century European culture was haunted by the vision of society being undermined by promiscuously breeding degenerates passing on to their offspring their vicious, weak, immoral or otherwise tainted characters as hereditary criminals, idiots, prostitutes, lunatics, drunkards or mere masturbators. Ray Lankester (1880) invoked degeneration as a warning against national decline which would ensue if Britain failed to fund the sciences as the Germans were doing.

Darwinian theory was not entirely without its reassuring aspects though, for it implied that the higher social classes owed their position to natural fitness. Competitive nationalism and capitalism were, alike, 'natural' as mechanisms ensuring the survival of the most fit and robust nations, races, products, and individuals. To no ruling class did this appeal more strongly than to the new self-made millionaires of the United States, who attended avidly to Spencer's message.

For human evolution theorists degenerates could be seen both as a weakening of the stock and as atavistic throwbacks to more primitive types of humanity, and for most (such as Romanes, one of Darwin's later collaborators and writer on the evolution of intelligence) they join apes, idiots, babies and savages as representatives of the lower rungs of a ladder atop which middle-class white men are, somewhat precariously, perched. In the forefront of this was Francis Galton (1822–1911) whose Eugenics movement devoted itself to the problem of the human stock's declining quality, and exploring possible countermeasures. In Italy Lombroso was engaged in a similar crusade. Eugenics was not exclusively aligned with right-wing political ideology; by the early years of this century it was being incorporated in most social philosophies, socialist as well as capitalist. In the eyes of such reformers as the Webbs, socialist planning would naturally include Eugenic control over marriages (Freeden, 1979; Paul, 1984). Social Darwinism did, however, become increasingly right-wing ideologically. Darwinism showed that unbridled competition was somehow 'natural' and like the 'Invisible Hand' of Adam Smith's economic philosophy a century earlier, ultimately guaranteed the best possible outcome for the species. Into this scene the image of an ancestral ape-man fitted perfectly. Our primitive ancestor would be brutal, savage, immoral, similar in appearance to microcephalics or Downs syndrome sufferers (long known as 'Mongols'). The Neanderthal remains provided ideal 'projective material' in which to see this character. But a twist was to come. So unacceptable did this ancestry become that, as we will see shortly, on the eve of the Great War it became possible to evict him from our lineage altogether in favour of a better candidate.

3 *Fossil evidence*

The human fossil record had been expanding. L. Lartet (son of E. Lartet) had found Cro-Magnon Man as early as 1868, thereby putting 'modern' Man on stage alongside the Neanderthals. This set the scene for the widely accepted account of how Cro-Magnons (our true ancestors) ousted the primitive Neanderthals (from whom they may or may not have evolved even earlier). Whatever its truth such a drama surely reflected contemporary anxieties about the threat of degeneracy. Neanderthals were *not*, in any sense, degenerate but they perhaps symbolised to the late nineteenth-century mind that brutal 'throw back' aspect of it they so dreaded. In constructing a story in which they had once been defeated in a distant ancestral past, the current situation could be seen as a re-emergence of a perennial struggle between opposing forces of civilisation and savagery, a theme to erupt disastrously in the ensuing decades.

Table 2.1 shows how the number of Neanderthal finds increased considerably in these years, notably the Spy and Krapina remains, and then, in 1908, comes the 'Old Man of Chapelle-aux-Saints', a complete skeleton. The damping effect of Virchow and others on genuine scientific study of such finds has been noted already. However, this continual upturning of Neanderthals was for a long time unmatched by finds of any other sort. The precise relationship they bore to *Homo sapiens* came increasingly into contention during the first decade of the twentieth century (as it has remained ever since).

In the 1890s a major new discovery came to light, one predicted by Haeckel who christened it 'Pithecanthropus' (i.e. ape-man) in advance. Remains fitting this prophesied hominid were found at Trinil in Java by Eugene Dubois (1858–1940) (though he initially called it Anthropopithecus). 'Java Man', *Pithecanthropus erectus*, radically altered the picture of human evolution, for though clearly far older and smaller-brained than the Neanderthals, its gait was apparently bipedal. It fitted the 'missing link' category excellently with its blend of simian and hominid features, as well as pushing the evolution story back to a far earlier stage. Unfortunately, the stir created by the find, along with Virchow's customary derision, led Dubois to deny anyone further access to the originals, and he even kept quiet altogether about some of the fossil fragments. They stayed beneath his floor until 1926.

An intriguing aspect of this find was its exotic provenance. Hitherto, all finds had been European, and while Darwin and Lubbock had believed our progenitors to have dwelt further south, this was the first major non-European discovery. It gave a strong credence to views now beginning to circulate (notably in the work of H. F. Osborn) that human origins lay in the Far East or Central Asia. This eager jumping to conclusions was a little naive, for clearly Dubois's discovery was in a sense due to chance, for had he chosen to go elsewhere there is no guarantee he would have returned empty-handed. The non-European world was virgin territory for human palaeontologists. The discovery justified only the view that *some* of our ancestors had lived in the Far East; it did not justify the conclusion that they *only* dwelt there, and even less the dogmatic insistence that they *originated* and evolved there (or in Central Asia) that now set in until the 1930s.

1907 saw the uncovering at Mauer, near Heidelberg, of a jaw-bone which became known as the Mauer Jaw, or 'Heidelberg Man'. This bore interesting similarities to Java Man, but differences too, and even now the classification of this jaw as *Homo erectus* (where Java Man is now placed) is speculative and an 'Archaic *Homo sapiens*' allocation is equally common. Had the fossil record yielded only these finds, progress might have been slow, but it would have been in the right direction. Unfortunately two other events misdirected the field until after the Second World War (although its orientation began to rectify itself in the 1930s). The first of these was the reconstruction of the 'Old Man of Chapelle-aux-Saints' by Marcellin Boule (1861–1942); the second was C. Dawson's 'discovery' of Piltdown Man, *Eoanthropus dawsoni*, 'Dawson's dawn-man'.

Boule was a palaeontologist of the highest calibre who was sceptical about Neanderthal's place in the direct *H.sapiens* lineage. The fossil in question was of an old man who had suffered from severe osteo-arthritis, but in his reconstruction Boule failed to take into account the implications of this, and while correctly concluding that the man had been stooped, generalised from this to the Neanderthal species as a whole. They were not fully erect but shambling, stooping, arm-swinging beings. Such creatures could not be among our close forebears, the time between them and Cro-Magnon Man was far too short for the latter to have evolved from them. Thus Boule 'expelled' them (Hammond, 1982). Bowler (pers. comm.) considers Hammond to have somewhat exaggerated Boule's position on 'expulsion', and argues that Boule's interpretation was more that Neanderthals were a 'frozen relic', i.e. a distant ancestral form from which the main lineage parted far earlier in time than the period when the European Neanderthals were alive. Another possibility, suggested later by Keith, and more in line with present interpretations, was that they were a specialised side-branch, different from the last common ancestor shared with modern *H.sapiens*.

The irony is that Boule's theoretical preference for a multi-branched, rather than unilinear, model of human evolution was basically sound. Such was Boule's authority that his conclusions regarding Neanderthal morphology remained substantially unchallenged and unre-examined until the 1930s. The reasons for his error, examined in Hammond's important monograph, are complex, involving theoretical presuppositions, the social context of his research, and broader psychological factors.

If Neanderthal Man was not our ancestor, who was? In *The Descent of Man*, Darwin had astutely written that as our ancestors began to use tools, their jaws and teeth could diminish and 'as the jaws and teeth in man's progenitors gradually become reduced in size, the adult skull would have come to resemble more and more that of existing man' (p. 80). In other words, cranial enlargement would tend to *follow* reduction in jaw-size and the evolution of the human 'face'. There would be ancestors with smallish brains (though bigger perhaps than in other primates) and a more *Homo sapiens*-like jaw and tooth morphology. This would be because a reduction of the musculature attached to the cranium was necessary before cranial enlargement could occur. But at the juncture we are now considering, in the light of Boule's expulsion of the Neanderthals, the search is on for a radically different candidate, predating the point at which Neanderthals branched away from our lineage to their dead end. The successful candidate will have no heavy brow-ridges, no retreating forehead, and a more simian jaw (because earlier in the evolutionary line). In 1913 Dawson, an amateur archaeologist, and Smith Woodward of the British Museum of Natural History, reported a find Dawson had made the previous year at Piltdown in Surrey, and this Piltdown Man fitted the bill perfectly. Before pursuing this story further we need to consider briefly what had happened to Darwin's 'face-first' model.

A persistent psychological need to re-elevate humans above other animals had, over the latter nineteenth century, led to a concentration on the human brain as the key factor in our evolution. 'Man' was not to be deposed *that* easily after all. The up-and-coming young palaeontologist G. Elliot Smith eagerly espoused the combination of large-brain and ape-face as that most likely for a 'missing link', while Sollas had been predicting it for some time. The situation was made more complex since some workers such as Smith Woodward, Osborn and Wood Jones were proposing that apes too had evolved since the ape and *Homo* lineages split, developing their large jaws subsequently, and Smith Woodward considered them to be a degenerate side-branch (Bowler, pers. comm.). Nevertheless, the view that the human brain was the key factor, and would have had priority over other morphological changes was being generally espoused, and advocates of face-, or hands-, first models were few on the ground, although among them was Marx's collaborator Engels (Gould, S. J., 1973, Ch. 26). It was this big-brained, ape-jawed physiognomy that characterised Dawson's find.

This was not surprising since it was indeed a combination of orang-utan jaw and human skull, a fact not discovered until 1953 (Weiner et al., 1953). *Eoanthropus dawsoni* was a hoax. Whoever perpetrated it probably did not intend it to go undiscovered for quite so long, but presumably reputations and loss of face meant that soon too much was at stake. The Piltdown forgery literature is now extensive and culprits proposed are Dawson (Weiner, 1955), G. Elliot Smith (Millar, 1972), Chardin (Gould, 1983a), Sollas (Halstead, 1978) and, the latest, Conan Doyle (Winslow and Mayer, 1983). Louis Leakey (1903–1972) wrote that one reason why it went so long undiscovered was that Bather, Curator of the Natural History Museum between the wars, treated it with such extraordinary reverence, only allowing researchers access to plaster casts (L. S. B. Leakey, 1974). Leakey claimed that he was always suspicious of Piltdown Man.

In some respects the forgery was rather crude, with the teeth on the mandible visibly filed down out of alignment, for example. Leading authorities such as Elliot Smith, Chardin, Keith, Hrdlička and Sollas all eventually accepted it as genuine. Darwin's logic, which was quite sound, had been obscured by the wish to find a separate, non-Neanderthal, ancestral line compatible with the 'brain first' expectation of eminent opinion, an ancestor to redeem Man's self-respect. And it was British. Piltdown Man did not exist, so necessity mothered his invention.

As if these developments were not confusing enough, a German, Reck, reported the first fossil hominid from Olduvai Gorge, now classified O.H.1, which erroneously appeared to be associated with quite early beds. In fact it was a fairly modern *Homo sapiens*. This supposed antiquity of 'Olduvai Man' added a gratuitous final twist to the situation.

4 *Archaeology*

In archaeology attention was divided between the continuing task of unearthing and analysing stone tools, and the extraordinary cave-art discoveries in France and Spain. These tended to merge with Ethnology so that by the end of our period there was little distinction to be made between the study of prehistoric European artefacts and those of extant primitive peoples. An influential and compendious work which appeared in 1911 (with further editions in 1915 and 1924) was W. J. Sollas's *Ancient Hunters and Their Modern Representatives*. (Sollas was Professor of Geology and Palaeontology at Oxford.) The archaeological material covered in this work illustrates the immense amount of new data brought to light in the previous decades.

By the time Sollas was writing well over 100 prehistoric sites yielding artefacts, often on many different levels, had been discovered in Europe and North Africa, mostly in France and Northern Spain. Since Boucher de Perthes archaeologists had succeeded in classifying the European stone-culture sequence from the Palaeolithic to the Neolithic and the terms they introduced (usually derived from site-names) have largely persisted down to the present, although with important modifications and additions. Following the main architect of this scheme, Mortillet, referred to earlier, came important figures such as Cartailhac, Peyrony, and, above all, l'Abbé Breuil. (See Table 2.4 for important archaeological discoveries down to World War I.)

Sollas is able to review the main features of each of the principle phases from the Lower Palaeolithic to the Early Neolithic, and flint-working techniques of even Lower Palaeolithic Acheulean and Mousterian cultures were analysable in some detail as work sites had been identified (Plate 1), and as early as 1880 an English archaeologist, F. C. J. Spurrell had managed to reassemble a worked flint from its scattered fragments. The association between the Mousterian industry and Neanderthal fossils was

TABLE 2.4 *Major European prehistoric archaeological sites discovered or explored prior to 1914*[1]

Date[2]	*Name of site*	*Discoverer or principal researcher*[3]	*Nature of site/find*
1797	Hoxne, Suffolk	J. H. Frere	Flint tools
1803	Salisbury Plain	W. Cunningham and R. Colt Hoare	Barrows
1823	Paviland Cave	Dean Buckland	Flint tools found with 'Red Lady'
1828[4]	Grotte de Bize	M. Tournal	Pottery, with bones
1829	Kent's Cavern	J. MacEnery	Flint tools with extinct faunal remains
1834?	Chaffaud	Brouillet	Engraved reindeer bones
1834	Gristhorpe, Yorks	Beswick and Alexander	Oak-tree coffin with grave goods
1837	Danish shell mounds	Steenstrup, Forchhammer, Worsaae	Flint tools, food refuse, various artefacts
1837	Abbeville	Boucher de Perthes	Flint tools
1839	Lagore, Co. Meath	W. R. Wilde and G. Petrie	A crannog – a form of artificial island, 46 known by 1857
1848	Lake Onega, Karelia, (USSR)	n.a.	Rock carvings
'1850's'	Cresswell Crags, Derbyshire	Boyd Dawkins	Flint tools
1852	Aurignac	E. Lartet	Flint and ivory tools
1853	St Acheul	Dr Rigollot	Flint tools, 'Acheulean' type-site
1853–4	Obermeier, L. Zurich	F. Keller	Lake dwellings built on piles (over 200 known by 1875)
1858–9	Windmill Hill Cave	W. Pengelly	Flint tools
1860	Massat, Ariege	E. Lartet	Flint tools, bone and ivory artefacts
1861	Les Eyzies and La Madelaine	E. Lartet	Various artefacts – Les Eyzies continued to be explored into present century. Major cave art site. La Madelaine; type-site for 'Magdalenian' culture

Date[2]	*Name of site*	*Discoverer or principal researcher*[3]	*Nature of site/find*
1863	Le Moustier	E. Lartet and H. Christy	Various artefacts, 'Mousterian' type-site
1864	Niaux Cave	Garrigou	Cave art (importance unrecognised)
1868	Cro-Magnon	L. Lartet	Artefacts found with 'Cro-Magnon' fossil, including shell ornaments
1868	Ightham, Kent	B. Harrison	Flint tools including supposed 'eoliths'
1871	Grimes Graves, Norfolk	Canon Greenwell	Major flint mine still being studied (G. de G. Sieveking et al. 1973)
1872	Grimaldi, Monaco	E. Riviere	Numerous artefacts accompanying skeleton as grave goods
1872	Laugerie-Basse	Massenat	Numerous artefacts inc. engravings, with skeleton
1878	Le Chabot	M. Chiron	Cave art, no interest aroused
1879	Altamira	Marquis de Sautuola & daughter	Cave art, initially thought to be fake
1879	Fere-en-Tardenois	Vielle	Microlithic flint tools; 'Tardenoisian' type-site
1880	Predmost, Moravia	Szombathy, Obermeier	Engraved rocks as well as human remains, this date is the beginning of excavations which went on for many years
1881	La Cotte de St Brelade, Jersey	C. Burdo, R. R. Marett	Serious work did not begin until 1894. Research still continues, see McBurney and Callow (1971). Mousterian and Levalloisian artefacts, plus Neanderthal teeth
1880s	Cranborne Chase	Gen. A. Pitt Rivers	Pioneering excavations of extended site introducing many methodological innovations
1886	Campigny	Salmon	Numerous artefacts, 'Campignian' type-site
1887	Mas d'Azil	Piette	Coloured pebbles, 'Azilian' type-site
1895	La Mouthe	Riviere	Cave art, part of Les Eyzies complex

Date[2]	*Name of site*	*Discoverer or principal researcher*[3]	*Nature of site/find*
1896	Pair-non-Pair	Daleau	Cave art
1898	Clacton-on-Sea	Kenworthy	Flint artefacts, later studied by S. Hazzledine Warren, Oakley and Mary Leakey. 'Clactonian' type-site
1901	Les Combarelles, Font-de-Gaume	Capitan, Breuil	Cave art, part of Les Eyzies complex
1903	Calapata, Spain	J. C. Aguilo	Cave art – Spanish Levantine style
1904	Wolvercote Channel, Oxford	A. M. Bell	Flint tools
1906	Hornos de la Peña, Spain	Breuil	Cave art
1906	Castillo (and other Cantabrian sites)	Del Rio and Sierra	Cave art
1906	Gargas	F. Regnault	Cave art inc. hand silhouettes
1908	Le Portel, Ariège	Dr. Jeannel	Cave art
1909	Willendorf	Szombathy	'Willendorf Venus'
1909	Laussel, Dordogne	Dr. Lalanne	Carvings, cave art, inc. a bas-relief 'Venus' with horn
1912	La Pasiega	Obermaier and Werner	Cave art
1912	Les Trois Frères	Comte de Bergouen & sons	Cave art
1913	Swanscombe, Kent	R. A. Smith and H. Dewey	Flint tools. Known for some time but Smith & Dewey's work especially important. Later source of Swanscombe skull
1914	Tuc d'Audoubert	Comte de Bergouen, Breuil	Cave art, inc. bison sculptures

1 During the latter part of the nineteenth century the number of sites increases dramatically; in 1915 (Sollas, 1915) Sollas maps 23 Lower Palaeolithic, 32 Mousterian, 26 Aurignacian, 10 Solutrean and over 30 Magdalenian sites of major importance in Europe. By and large I have confined the list to Palaeolithic sites although some later ones are included where they had special significance for promoting the 'antiquity of man' perspective. Inclusion of African, Asian, American and Australian sites would have made the list too unwieldy and at this time they were not having any great influence in any case. D. J. Mulvaney (1971) provides an interesting chronology of Australian finds. African prehistoric studies were in their infancy, J. D. Clark (1959) references two pre-1914 works: L. Peringuey (1911) and J. P. Johnson (1907), while Tongue's work on rock painting (1909) is mentioned in the main text. Rock art of North Africa had been known since General Cavaignac's expedition of 1847 against the Ksour tribes, and was further explored by Barth, in the 1850s, Duveyrier and Nachtigel in the 1860s and Foureau in the 1890s (Lhote, 1961). North American work on ancient Indian mounds, etc.

was abundant but threw little light on the evolution story, being too recent, while the Near Eastern work was dominated by ancient and classic civilisations. Of Indian, Chinese and South East Asian prehistory virtually nothing was known by 1914.

[2] These dates refer to the date at which the site first became known, or at which work on it began. Often the most important archaeological research came much later e.g. Swanscombe, Clacton-on-Sea, Les Eyzies and La Cotte de St Brelade, Jersey. Occasionally the date is that of the first publication on the site rather than its discovery.

[3] This refers only to the initial discovery phase, with a few exceptions. Nearly all the French cave-art sites, for example, were extensively studied by Breuil and/or Cartailhac at some stage, although they figure in the initial discoveries of relatively few.

[4] Date of publication.

This has been compiled primarily from M. Boule (1923), H. Breuil and L. Berger-Kirchner (1961), G. Daniel (1950, 1967), Lubbock (1865, rep. 1912), Lyell (1863), D. Roe (1981), G. de G. Sieveking et al., (1973), W. J. Sollas (1911, rep. 1915).

well established and until fairly recently the assumption that the Mousterian was uniquely Neanderthal (and vice versa) was unquestioned (see Ch. 6 below). The principal types of artefact had also been classified, though the function of some remains obscure (see Chapter 3, Fig. 3.8). The Upper Palaeolithic was demarcated by a radical change in working techniques, and the appearance of 'art', distinguishing the Aurignacian from the previous Mousterian (a more transitional Chattelperonian phase is now identified at this point).

The great cave-art discoveries of Altamira (1879) and Niaux (1861) were long unappreciated and even, in the case of Altamira, assumed to be of recent origin. At the turn of the century Cartailhac, the most eminent sceptic, retracted his rejection of Altamira (Cartailhac, 1902) and a burst of cave-art studies ensued. Much of this new material was still being examined as Sollas wrote. But his book was not merely descriptive; he was advocating a quite definite thesis based on comparisons between prehistoric artefacts and those of contemporary primitive cultures and bolstered with comparative anatomical data. This thesis, briefly, was that the early Upper Palaeolithic Aurignacian culture, associated with a fossil assemblage known as 'Grimaldi Man' from Monaco, was produced by people ancestral to the contemporary 'Bushmen' of South Africa, while the somewhat later Magdalenians, as represented by the 'Chancelade Skull', had followed the retreating ice at the end of the Ice Age to become modern Eskimos. Neither of these theories is now accepted. Without actually equating Australian aborigines with Neanderthals, he also devotes considerable space to them, suggesting that there was at least a link between the two stocks.

This synthesis of anthropological, ethnographic and archaeological evidence had been under way for some time; J. W. Dawson (1880) had used American Indians in the same way. By Sollas's time it was possible to offer a detailed elaboration of his highly plausible theory on the basis of the newly available data. *Ancient Hunters* remains a valuable source of information. Given the richness of the archaeological data, and the fact that he was Professor of Geology, it is surprising that his dating was so inaccurate, by a factor of 10 in some cases. This arose from reliance on his own inaccurate calculations of deposition and erosion rates, though it is somewhat odd, that while Geikie had already calculated the antiquity of the Palaeolithic as 100,000–240,000 years (and Croll and Stone had given similar figures in 1864 on the basis of calculations involving the eccentricity of the Earth's orbit), Sollas crams everything after the Chellean, the earliest Lower Palaeolithic, into a mere 27,000 years. In the US Boltwood was already exploring the use of uranium dating of rocks which would bear fruit half a century later for researchers on human origins. More puzzling, Keith (1925) considers Sollas's figures as overestimates!

A further archaeological issue Sollas discusses is the 'raging vortex', as he calls it, of the Eolith controversy. To identify whether or not a stone has been worked by human hands is very difficult at the lowest levels of artifice, and during the early years of the century a number of finds were made (at Cromer and Ipswich particularly) of flints which were extremely ambiguous in this respect. These were termed 'Eoliths' (dawn-stones) by their advocates, the foremost of whom were Reid Moir and Sir Ray Lankester in Britain and Rutot and Verworn on the continent. They believed they had found the earliest crude products of human-like tool-making skill. Their opponents, including Sollas, were convinced that eoliths were produced naturally by water action or fortuitous pressure-flaking. *Eoanthropus* was obvious candidate for their maker. Sollas's doubts have generally been accepted, though Miles Burkitt was giving eoliths a sympathetic treatment into the 1960s (Burkitt, 1963).

In spite of misinterpretations and lacunae, archaeology in 1913 had changed beyond recognition since the 1870s. Many of the great Palaeolithic masterpieces – the Altamira Caves, the Willendorf Venus and the Les Eyzies complex – had been discovered and their ancient origins accepted. The close inter-relationship which had developed between Archaeology and Anthropology was soon to be weakened as a result of theoretical shifts in anthropology, notably those brought about by Franz Boas in the US (Boas, 1911). If there seems to be more rigour and coherence in archaeology than in human palaeontology at this time it is primarily because the data for the latter were still so minimal. Advances in archaeology were nonetheless only possible as a result of the methodological changes introduced by archaeologists such as Pitt Rivers, Montelius and Petrie whom we can only mention in passing here.

5 *Psychology and Anthropology*

A fifth strand requiring attention is that of the impact of evolutionary thought on Psychology and Anthropology (see Oldroyd, 1980). Darwin's own contributions had helped found several psychological research traditions such as the study of emotion and child development (Darwin, 1877). The origins of Psychology as a discipline in this period have been discussed in numerous histories of Psychology (see Hearnshaw, 1964 for the British roots). Pride of place necessarily goes to Galton, whose psychological writings are unified by an underlying preoccupation with the natural variation of the human race upon which natural selection operates. He also bridges the gap between Psychology and Genetics, though as already noted his positions on these issues are not entirely consistent. Behind his diverse contributions to Psychology, such as the initiation of the study of individual differences, the launching of the nature–nurture controversy and research on imagery, he is nonetheless haunted by the spectre of degeneracy and the need for Eugenic control, a preoccupation not all psychologists shared.

Haeckel's recapitulationist model inspired G. S. Hall's monumental *Adolescence* (Hall, 1904) which depicted children as recapitulating the stages of social evolution from savagery to civilisation as they grew up. The behaviour of animals naturally received enormous impetus from evolutionary thought, with Lubbock, Romanes and Spalding among the most important contributors (Hearnshaw, 1964; B. Singer, 1981). Animal behaviour studies finally came of age, however, in the work of C. Lloyd Morgan (1852–1936), with his fierce strictures against anthropomorphism.

The general import of evolutionary theory for Psychology was that behaviour was largely instinctive, built-in either at the overt behavioural reflex level or, more profoundly, at the motivational level. After the 1890s clear divisions begin to emerge between those whose thinking is becoming more environmentalist and those who base their theories on instincts. In a way the former represent a retreat from the Darwinian perspective, in which the human/animal boundary was abandoned, to re-establishing the uniqueness of our species, a uniqueness lying in our extraordinary adaptiveness and learning capacity. This is deeply ironic since it was the last thing someone like J. B. Watson intended. Thorndike and J. B. Watson, founder of Behaviourism, are the most eminent in this school, while European psychologists tended to espouse instinct accounts (e.g. McDougall and Freud). McDougall's most successful work, *Introduction to Social Psychology* (1908), systematically equates emotions and instincts. Freud, in 1913, published his famous foray into anthropology and evolution, *Totem and Taboo*, a theme continued just after our period in Trotter's *Instincts of the Herd in Peace and War* (1916). Freud's relationship to Darwinian thought is complex (Sulloway, 1979); certainly he sees us governed by deep-rooted instinctive processes which determine our motivation and maturation, but he also has a Lamarckian streak and his focusing on sexual instinct as the source of all others was unique. Sulloway additionally argues that Haeckel's recapitulationist 'Biogenetic Law' was a major influence on Freud's stage model of psychological maturation.

In France the 'degeneration' perspective widely affected psychiatric and social psychological thought, being evident in the work of Charcot and, more influentially perhaps, in Le Bon's best-seller *The Crowd* (1896). It would not be an exaggeration I think to suggest that European (if not US) psychological thought at this time was becoming increasingly preoccupied with, and giving expression to, those cultural anxieties referred to earlier; obsession with primitiveness, lurking savagery and degeneracy, and, in Psychology, fascination with hypnotism, multiple personality, spiritualism (that late-nineteenth-century form of ancestor cult) and Freud's murderous, lustful, unconscious. Current neurological models, such as Hughlings Jackson's, often supplement evolutionary theory, especially in psychiatric thought (e.g. Charcot and Freud). It is difficult in retrospect not to see in all this the psychological prelude to the catastrophe of the Great War. In considering contemporary revolutionary developments in the arts one cannot help but feel that Europe is being racked by a genuine social psychological crisis revolving around the relationship between 'Civilisation' and the primitive, ancestral, savage, irrational aspects of human nature which it sought, like Boule, to expel, like Freud to master or like Picasso joyfully to sublimate.

Things were calmer across the Atlantic, where technological and social optimism reigned. Here the basis was being laid for somewhat less fraught reconstructions of the human sciences, even if Behaviorism now seems extraordinarily naive and patently rooted in Watson's own repressions (D. Cohen, 1979; Creelan, 1974). Not that the USA was immune to fears of degeneration; on the contrary some of the most enthusiastic Eugenics measures were implemented there, and the attitude to the 'Negro' showed no signs of amelioration. Boas's anthropology nevertheless withdrew that discipline's support from the racialist consensus.

In Europe the dominant anthropological traditions remained those of the late nineteenth century; in Britain this meant the evolutionary orientations of Tylor and Haddon along with the Olympian mythological studies of James Frazer, whose encyclopedic survey of myth, magic and religion, *The Golden Bough* (1890–1915) was already exerting enormous influence. In France Lévy-Bruhl, Durkheim and Mauss were studying 'primitive thought'. Earlier, the evolutionary perspective had been adopted in America by L. H. Morgan (1877) who put forward a seven-stage model of social evolution (Table 2.5) which remained an important reference point, but his influence in the US was waning dramatically. New theories of 'diffusionism' were being put forward by writers such as G. Elliot Smith, who ascribed civilisation entirely to a diffusion of ideas and techniques from Egypt. This was incompatible of course with the notion of regular universal stages of evolutionary progress.

TABLE 2.5 *L. H. Morgan's seven stages of social evolution (1877) (p. 12)*

Periods	*Conditions*
I Older Period of Savagery	Lower Status of Savagery
II Middle Period of Savagery	Middle Status of Savagery
III Later Period of Savagery	Upper Status of Savagery
IV Older Period of Barbarism	Lower Status of Barbarism
V Middle Period of Barbarism	Middle Status of Barbarism
VI Later Period of Barbarism	Upper Status of Barbarism
VII Status of Civilisation	
I Lower Status of Savagery	From the infancy of the human race to the commencement of the next period
II Middle Status of Savagery	From the acquisition of a fish subsistence and a knowledge of the use of fire, to, etc. (*sic*)
III Upper Status of Savagery	From the invention of the bow and arrow to etc.
IV Lower Status of Barbarism	From the invention of the art of pottery to etc.
V Middle Status of Barbarism	From the domestication of animals on the eastern hemisphere, and in the western from the cultivation of maize and plants by irrigation, with the use of adobe-brick and stone, to etc.
VI Upper Status of Barbarism	From the invention of the process of smelting iron ore, with the use of iron tools, to etc.
VII Status of Civilisation	From the invention of a phonetic alphabet, with the use of writing, to the present time

A split was in fact occurring in Anthropology between the physical and cultural branches. The former, rooted in the old craniometric tradition, was now being led by Ales Hrdlička in the US and, in Britain, was represented by Duckworth and Elliot Smith. This branch remained intimately involved with palaeontology and many of its leading figures span the two disciplines (the term 'palaeoanthropology' being used today for this overlap area). Cultural anthropology having emerged from the older ethnographic tradition was now, under Boas in the US, becoming increasingly environmentalist and the French figures just cited were, though theoretically very different, moving away from evolution towards, in their case, a structuralist approach. Thus a breach was rapidly opening up between cultural anthropology and evolution research.

To summarise the situation in 1913, at the height of the Great Confusion:

(i) Genetics is rendering the entire Darwinian conception of evolution by natural selection problematical, with influential figures like W. Bateson demanding a virtual moratorium on the subject pending further progress in genetics. Internal squabbles among geneticists had become in some instances more constrictive than constructive.

(ii) Darwinian theory had become bound up with cultural, socio-psychological, preoccupations with degeneracy and the threat to society of the prolifically breeding unfit. By 1913, even *within* evolutionary theory, there was a division between the older view of a 'ladder' from apes via idiots, savages and children to European Man, and a newer 'bush' conception of numerous branchings in the course of evolution making ancestral lines more difficult to trace. To this school Boule belonged.

(iii) Within palaeontology the combination of Boule's misdescription of the Neanderthal 'Old Man of Chapelle-aux-Saints' with the complementary Piltdown hoax had succeeded in totally misleading all concerned. The major finds of the period, *Pithecanthropus* and the Mauer Jaw, could not be fruitfully analysed in such a context, and Pilgrim's discovery of *Sivapithecus* in India (Pilgrim, 1915) was a single flicker from the Miocene darkness. Even the first Olduvai find was misleading.

(iv) After a faltering start prior to 1900 prehistoric archaeology had, by contrast, made great progress, even though time-perspectives were still too compressed and methodologies of site-excavation and analysis, while advancing, yet left much to be desired. The close link with anthropology led writers like Sollas to assimilate prehistoric cultures rather too comprehensively into the moulds of contemporary 'primitive' cultures, even suggesting direct links between some of them.

(v) Within Psychology the impact of the evolutionary perspective had at first been considerable but by 1913 a split was opening between Darwinian instinct-oriented theories (largely European) and environmentalist approaches (mainly in the US). Succeeding decades saw the latter come to dominate academic psychology and human evolution was largely ignored by the discipline until the 1950s.

Clearly then, caught by the vicissitudes of theoretical developments in closely related disciplines and by virtue of its mythic aspect caught up in Europe's collective fantasy-life in the run-up to its greatest catastrophe, human evolution research had made little headway since Darwin. Even within palaeontology a misguided 'brain-first' model of hominisation held sway. Franco-Spanish prehistoric archaeology was flourishing, but almost autonomously from evolutionary thought generally, and the importance of the few major fossil finds could not as yet be gauged. After forty years hunting for missing links, it was not even clear anymore what the chain itself looked like.

C Finding the 'missing links'

The story of events between 1914 and 1960 is a combination of success and error as the search for human origins expands from Europe to the Asian and African continents. Many 'missing links' had indeed come to light by 1960, but theoretical understanding of human evolution often remained inadequate. But bones are not the whole story, we need to pursue further the developments in the related disciplines before tackling the dramatic events in palaeontology which occurred over this period.

1 *The modern synthesis*

We have seen the negative initial impact of genetics on Darwinism. Change was held by the Mendelians to arise from recombinations of genes (or 'factors' as Bateson called them) in hybridisation, or occasional mutations which were invariably sterile. Continuous spontaneous variation, central to Darwin's model, was thus a myth, and the evidence for it illusory. As research continued the operation of Mendelian inheritance was shown to be far more complex than it had seemed when just dealing with trait-pairs (like green/yellow in peas). The reign of the fruit fly, *Drosophila melanogaster*, in genetics research had begun. T. H. Morgan (1866–1945) and E. B. Wilson (1856–1939) rendered the effects of selection amenable to experimental study with a rigour hitherto unknown. In 1930 R. A. Fisher published *The Genetical Theory of Natural Selection*, which clarified, in mathematical form, the role of small-scale mutations, which occurred on a scale the pioneers had not appreciated. Julian Huxley (1887–1975), grandson of T. H., finally consolidated the reincorporation of natural selection in his book *Evolution: The Modern Synthesis* (1942). In fact it was now reinstated on a firmer basis than it had ever enjoyed in the nineteenth century, when rival mechanisms abounded even in the theories of those who broadly accepted the evolutionary position. Mutation, recombination and selective pressure from the environment were shown to account quite adequately for the evolutionary process, particularly when allied to the more sophisticated analyses of plant and animal ecology and demography which were now available.

The discovery of the structure of DNA in 1953 by Crick, Watson and, at one remove, Franklin, confirmed Weismann's conclusion that no direct environmental effect was possible on the genetic material itself – the central dogma of modern evolutionary theory. Popular Lamarckism was finally scotched after a brief, damaging resurgence in the work of the Soviet charlatan Lysenko, and being clung to desperately by McDougall at the end of his psychological career, in North Carolina. The DNA discovery facilitated explorations of more holistic conceptions of the way in which genetic and environmental factors interacted by such figures as Dobzhansky, Lerner, Waddington and Mayr (Allen, 1983). Not that all is sweetness and light, as we will see in the next chapter. The period between 1914 and 1960 thus saw an integration of genetics and evolutionary theory resulting by the 1940s in a triumphant comeback for the Natural Selection model, albeit in updated form. Genetic recombination and mutation could indeed provide the raw material on which natural selection operated.

For human evolution however there was a second, more problematical, development introduced by, among others, J. Huxley: the notion of 'cultural' or 'social' evolution, a process which was supposed, in the case of humans, to have taken over from orthodox evolution by natural selection. Once again, humans are special. I examine elsewhere the present state of this debate (Chapter 5 below), but it is at this period that it starts emerging as a major issue, since it seems to provide a way of rescuing us from the impersonal tyranny of blind natural selection. In a sense cultural evolution had figured in Darwin's discussion of sexual selection, while kindred rescue operations had perennially featured in popular responses to the topic as a whole. From now on, though, it acquires a more focused and technical character as the relationship between genes and culture is analysed.

2 *Animal behaviour studies*

The study of animal behaviour was, we have seen, boosted by Darwin's theory and in *The Descent of Man* he marshalled an impressive array of anecdotal evidence from explorers who had been pelted with nuts by orang-utans, and zoo-keepers who had witnessed the heroism of baboons. Only with Lloyd Morgan at the end of the century had its scientific study begun to make appreciable headway. But this was initially from a direction which failed to shed much light on human evolution at all, since after Lloyd Morgan the principal research was done, as noted earlier, by US psychologists studying learning behaviour within an environmentalist framework. Although Thorndike (1898) had used cats and chicks, research soon concentrated on the white rat. There was an unnoticed circularity in the logic of this choice, for the reason why the species was such a convenient research animal was because it was highly adaptable to different environments and thus easy to maintain and breed. But the underlying thrust of the research for which they were used was to establish that behaviour was environmentally determined. It is not surprising therefore that this assumption appeared to be vindicated, given that it was that very fact that its behaviour *was* so environmentally adaptive that led to the rat being selected for such experiments in the first place! (See Boakes, 1984, on the history of animal behaviour research.)

During the Great War the German scientist Köhler (1887–1967), interned by the British on the island of Tenerife, used the opportunity to explore learning in chimpanzees. Although R. M. Yerkes in the US had also been studying apes, it was Köhler's subsequent book *The Mentality of Apes* (1925) which caught the public's imagination and placed primate studies on the map. But Köhler was a leader of the German 'Gestalt School' of Psychology who were opposed in many fundamental respects to the prevailing US tradition of experimental rat-learning. For Psychology the major problem Köhler's research posed was the nature of 'insight learning'. His chimpanzees had shown themselves capable of solving such problems as 'how to reach a bunch of bananas' by devices like piling boxes on one another or dragging them into their cage with sticks. These solutions appeared suddenly, apparently after some rumination, and not simply as a result of 'trial and error' – chance rewarding of random behaviour – as favoured by the US behaviourists. In historical retrospect Köhler's work is important for more than these problems it posed for academic learning theory. In the book he is concerned from the outset 'to ascertain the degree of relationship between anthropoid apes and man in a field which seems to us particularly important' (Köhler, 1925 (rep. 1957), p. 9). He concludes that 'chimpanzees manifest intelligent behaviour of the general kind familiar in human beings'.

In an appendix Köhler reports observations of chimpanzee social behaviour and other non-cognitive aspects of performance, while much of the work concerns the use and making of implements. It is in fact the first of the major studies on primate behaviour to provide solid evidence of the behavioural affinities between our species, thereby providing a rationale, from the psychological direction, for giving such comparative behavioural studies a central role in the study of human evolution. Köhler himself spells out this evolutionary context, especially in regard to cognition:

> At the present time it is impossible to decide whether the processes which have become mechanical, and appear to us the easiest, have *originally* evolved most easily and, therefore, earliest. We can only judge what is originally easy and originally difficult, by means of

> experimental tests with anthropoids and perhaps other apes, with children and primitives (for more advanced problems), and perhaps also with imbeciles and mental defectives. (1925, p. 63)

This passage is an odd mixture of late-nineteenth-century presuppositions of an evolutionary ladder through apes to savages with a more prescient appreciation of the role which comparative behavioural data should play. He is halfway between laboratory and field, with some of the book being distinctly 'ethological' in flavour.

Between the two world wars primate studies in the field were begun by Nissen (1931) on chimpanzees, Bingham (1932) on gorillas, and, though in zoo confines, Zuckerman (1932) on baboons. Yerkes (1916) had published experimental research, comparable to Köhler's in some respects, primarily with orang-utans, and Kohts (1935) reported Russian work on chimpanzee intelligence. The earliest primate field study, on baboons, was done by the South African, Marais, in the early years of the century but remained unpublished until 1969 (Marais, 1969). Carpenter (1940) reported field research on the more distantly related gibbons.

By the 1930s Lorenz and Tinbergen had begun to open up the discipline of Ethology and were developing methodological techniques for scientific study of animal behaviour in the wild, concentrating initially on birds. New conceptual terms such as 'imprinting', 'fixed action pattern' and 'attachment' appeared, soon to become the prevailing coin of ethological discourse. Further major primate studies did not appear until after 1960, but the basis for incorporation of this field into human evolution research had been laid.

3 *Anthropology*

As far as cultural anthropology was concerned, developments over this period were of little use to human evolution research. The discipline was dealing with a number of theoretical and practical tasks of greater urgency. In the US the Boas tradition was being continued by such workers as Margaret Mead and Ruth Benedict who stressed the environmental and cultural relativity of human behaviour and social organisation. British anthropologists were concerned with important methodological problems of both data gathering and analysis, rooted in the work of Radcliffe Brown and developed by Evans-Pritchard, Meyer Fortes and G. Bateson. Another leading figure was the psychoanalytically influenced Bruno Malinowski whose concerns included a fuller articulation of the relationship between the anthropologist and the people being studied. In France the move towards structuralism, to become identified with Lévi-Strauss, continued, remaining centred on the nature of 'primitive thought'.

There was, though, a continuing tradition of ethnography, in which L. S. B. Leakey himself was engaged for much of this period. The father-figure of this was the Cambridge anthropologist Haddon. Bridging the disciplines of Anthropology, Psychology and Psychiatry, W. H. R. Rivers wrote a number of works on such topics as magic, politics, and the unconscious which seem to have fitted rather awkwardly into the intellectual framework of the time and been unjustly neglected since. Frazer meanwhile, who never saw a 'savage' in his life, continued to produce additional volumes of *The Golden Bough* (see Downie (1940) for more on Frazer).

Only towards the end of the period under discussion did these various strands of Anthropology begin to bear fruit for human evolution studies with the emerging analyses of primitive life-styles and economies, notably the 'hunter-gatherer' life-style assumed to have been typical of our ancestors. Where Galton had seen people of lower intelligence than his dog, these anthropologists discovered adaptive strategies and skills comparable to any in the human repertoire. But in the absence of substantial palaeontological and archaeological evidence, it remained unclear, at least before the 1950s, what role cultural anthropology could play in the study of evolution.

A development to be noted here, a corollary of the growing environmentalism of cultural anthropology, was the gradual demise of the concept of 'race' as a scientific term, even though it persisted in the literature in a loose descriptive way. Physical anthropologists, notably Carleton S. Coon (Coon, 1962) did continue to use the term uninhibitedly, but its mystique was waning fast. The arbitrariness of the criteria for defining 'race' was becoming widely apparent – skin colour, blood-groups, hair colour and, ear-wax, for example gave quite different pictures. Population geneticists had replaced it with the neutral notion of the 'gene-pool': the pool of genes contained in an in-breeding population. On occasion anthropology was positively in the forefront of anti-racism (Kuper, 1973), while the majority of serious scientists had abandoned the old idea of an evolutionary hierarchy of races.

4 *Archaeology*

Prehistoric archaeology is a sub-field within Archaeology as a whole, which during these decades saw tremendous advances in the study of early phases of human civilisation in Egypt, the Middle East and the Indus Valley. These, though, are peripheral to the present concern and for prehistory the important developments may be dealt with under three broad headings.

(a) Continued expansion of the data-base

In Europe the discovery of new sites continually outstripped the availability of trained researchers and funds, while self-supporting gifted amateurs of the older kind were fast disappearing. Among the most important finds were the caves at Lascaux, discovered in 1940, which immediately took their place among the most magnificent Upper Palaeolithic cave-art sites. Within nine days of their discovery in September the doyens of French Palaeoarchaeology were on the spot: Breuil himself, Cheynier and Bouyssonnie, with Peyrony close behind (Laming, 1959). Lascaux's red bulls and horses have entered the popular iconography of 'Stone Age Art', endlessly reproduced. But Lascaux was more than just an addition to the existing corpus of cave art, it required a major revision of existing models of the development of the genre. Hitherto a simple 'progressive' model had been adopted, of a steady improvement in quality from cave art's beginnings around (it was then believed) 20,000 BC to the culminating Magdalenian art of Alta Mira of about 10,000 BC. Lascaux, halfway between, was indisputably as fine as any other cave art known. So prehistory, too, had seen the rise and fall of cultures. (See Chapter 6 below for the function of cave art, etc.) In general the new European archaeological data obtained in this period served to bring the existing picture of the palaeolithic cultural sequences into finer focus without radically altering it, though dating remained obscure until the 1950s.

Among Breuil's child visitors in the 1920s was Mary Douglas Nicol, later Mary Leakey. Her interest in prehistory became over-riding and her drawings of flint artefacts impressed Louis Leakey as the best he had seen. Her illustrations to his work *Stone Age Africa* (L. S. B. Leakey, 1936) began a collaboration which lasted until his death, and a career which still continues. Her first major field-work was on the palaeolithic Clactonian industry in collaboration with Kenneth Oakley (Oakley and M. D. Leakey, 1937).

It was in the area of African prehistory that the most dramatic palaeoarchaeological advances were being made, advances in which the Leakeys played a pioneering role. The discovery of Gamble's Caves in Kenya in 1926 was a major landmark in this, though overshadowed by much older sites found since (such as Olorgesailie in the 1940s). As well as the Leakeys we may note other early contributors: Dorothea Bleek (1930) and Helen Tongue (1909) who published on rock and cave art, E. J. Wayland who did extensive work in Uganda after the Great War, and, later J. Desmond Clark and Sonia Cole (e.g. Clark, 1950; Cole, 1954). The African evidence produced a much clearer picture of the Lower Palaeolithic, at Olorgesailie and Olduvai in particular, than Europe could offer. The sites were both richer and older than the European ones which in the case of Britain were often on the fringes of our ancestor's range and occupied but intermittently (Roe, 1981). The term 'Oldowan' (the adjective from Olduvai) was added to the bottom end of the industry-classification scheme. The major report on the early cultural sequences in Olduvai was published in 1951 (L. S. B. Leakey, 1951), and on Olorgesailie's the following year (L. S. B. Leakey and Cole (eds), 1952). In the former he identified eleven stages in the development of the 'Great Handaxe Culture'. As well as adding a new dimension of antiquity, African finds also proved that stone tools were not only made from flint, as dogmatically taught in some European universities, but could include any rock with the requisite properties, such as basalt and obsidian. There is an anecdote of Leakey's promptly discovering a stone tool on arrival at Olduvai with Reck in 1931, contrary to the latter's scepticism. Reck's failure to find such artefacts on his earlier expeditions was due in part to his expectation that all tools would be of flint. The handaxe in question (actually found by a Kikuyu aide) was of lava. Finds in China, at Choukoutien, associated with Peking Man (see below) suggested that the Lower Palaeolithic industries had been fairly homogeneous in character throughout the ancestral range, although some stylistic differences were observed indicating a separation between the African and Chinese traditions (Shapiro, 1976).

An expedition led by Dorothy Garrod in the early 1930s opened up a major new site at Mount Carmel (now in Israel) (Garrod and Bate, 1937). As well as important fossil material (see below p. 62), this was also notable for the thoroughness and innovativeness of much of its methodology, involving comprehensive analyses of faunal evidence and palynology. It yielded unbroken sequences of Upper Palaeolithic industries of remarkable length.

These short notes hopefully indicate how widely the data-base of prehistoric archaeology was now spreading.

(b) Theoretical controversy

There were a number of internal theoretical debates going on within archaeology throughout this period, many concerned with interdisciplinary relationships (Daniel, 1950). Of more relevance to us, however, are certain internal controversies such as the nature of the transition from Upper Palaeolithic to Neolithic. This concerned the status of the 'Mesolithic' phase. Though long available, many archaeologists had shied away from the term, and V. Gordon Childe (1892–1957) in particular rejected its utility. The point at issue was how the phase from about 10,000 BC to 5,000 BC was to be conceptualised, which had important implications for how social and economic evolution and change were to be interpreted. Childe and other Marxists (notably the Russian, Marr) were inclined to view it as occurring in periodic revolutionary jumps of a radical, qualitative kind, interspersed with periods of stasis. The transition from Upper Palaeolithic to Neolithic was seen as a single step from a rather degraded version of the former, declined from its cave-art producing peak, to a radically new life-style altogether, emanating from the Middle Eastern 'cradle of civilisation' where settled agricultural communities were being established. Of Childe, G. Clark (Clark, 1980) says 'The impression his readers were likely to form was that of survivors from the Ice Age living on at a miserably low level of culture against the time of enlightenment issuing from revolutionary change in south-west Asia' (p. 7). Clark himself had begun to criticise this position in the 1930s and subsequent work has confirmed that a distinctive intermediate Mesolithic stage of cultural development can be regularly identified, not only in Europe but in America and Asia too. V. G. Childe was nevertheless one of the present century's greatest prehistorians, writing several classics on the late prehistory of Europe and the origins of civilisation (Childe, 1925, 1942, 1958), as well as providing a charismatic leadership for the discipline in this country during his period as Director of the Institute of Archaeology (Green, 1981).

A further arena of theoretical controversy of a more technical kind concerned the classification of stone-tool industries. This arose partly because the current nomenclature was so monopolised by French place-names, but more seriously because even the 'three ages' themselves were becoming problematical. What did these names mean? To what did they refer? Did they imply chronological sequence? If so then difficulties arose in accounting for the apparent co-existence of different industries. Were they cultural? This seemed to rule out the possibility of a single culture using a diversity of techniques, or different ones the same techniques. Although handy labels, the status and role of these taxonomic categories, from Chellean to Azilian and Palaeolithic to 'Iron Age' became increasingly hard to unravel, and writers throughout this period were frequently proposing revisions and amendments. This poses difficulties for any beginner naively assuming the continuing validity of a scheme encountered in an early text. Yet for all the heat the traditional taxonomic scheme survived remarkably intact for European material, though far more leeway was created for introducing novel taxonomies for sequences at new sites which did not clearly correspond to the classic French ones. Elsewhere, in Africa for example, the European classification after the Acheulean became useless (and in 1965 was formally and radically revised). As far as possible the current taxonomy is treated as a theory-free pragmatic frame of reference, but no taxonomy can be entirely theory-free and in due course we will see that controversy remains alive over the nature, if not the naming, of stone-tool industries.

(c) Methodology

The greatest advances over this period were probably methodological, ranging from aerial photography (Crawford, 1923) to pollen analysis, palynology, first used by Lennar van Post in 1916. However, any modern review of archaeological data is peppered with laments over the innocent delinquencies of earlier generations of diggers (Roe, 1981). The dating techniques reviewed in the next chapter nearly all originated during this period. Behind the steady development of more adequate methods of dating, site-analysis and excavation lay the increasing appreciation that archaeological research should be guided by specifically formulated aims and hypotheses; it was not simply a matter of digging to see what was there. It had therefore to be conducted in a way that maximised the amount of genuinely important information extracted, being after all a process of controlled destruction. Once ruined, a site is gone forever. Whatever its status as a 'science', archaeology was now conducted at the confluence of many different scientific disciplines, from physics to biology. This reappraisal of the task of archaeology, its interpretation as a hypothesis-testing enterprise trying to reconstruct vanished economic and social behaviour, was taken furthest perhaps, after 1960, by L. Binford, some of whose work will concern us in later chapters.

5 *Errors unmasked*

The final exposure of the Piltdown hoax (Weiner et al., 1953) came as a relief to the palaeontological world generally, for whom the object had become rather an embarrassment. Even in the 1925 scheme of such a devoted believer as Keith, it is shunted out onto a cul-de-sac side branch, away from the main line (Keith, 1925). Scepticism had been growing since the 1930s as African and Oriental finds pointed to a quite different sequence of morphological changes to that implied by the Piltdown remains. Brain-size increase followed upright posture and reduction of the jaw. Weidenreich had diagnosed the true situation with penetrating acumen in the early 1940s (Weidenreich, 1943) while Marston, finder of Swanscombe Man (see below), argued that jaw and cranium were from different animals. As early as 1949 Oakley's analyses of fluorine content had indicated that the remains were of fairly recent origin, but he was as yet unable to detect differences in fluorine content between the two parts. Oakley (Oakley and Randall Hoskins, 1950) then speculated that *Eoanthropus* was perhaps 'a late specialised hominid which had evolved in comparative isolation'.

In 1953 J. P. Weiner and LeGros Clark re-examined the object with particular attention to tooth-wear. Its credibility now collapsed rapidly – on closer examination, the molars proved to have been artificially abraded while the teeth were not even properly aligned. Artificial staining was identified and new, more sensitive, fluorine tests proved that Marston was right; jaw and cranium were of different ages. *Eoanthropus* was nobody's ancestor. By 1953 few, aside from the ageing Keith, were giving it much prominence in their accounts anyway, simply because it had become too much of an oddity. Genuine or not, how could such a being have had a role in modern *Homo sapiens*'s evolution?

The same non-European finds which had subverted Piltdown Man's credibility also raised problems for the Boule version of Neanderthal Man. *Australopithecus* and *Sinanthropus pekinensis* (discussed below) were both demonstrably upright and bipedal. Boule's Neanderthal, a known toolmaker of much more recent date, was a shambling stooper. His 'dead-end' theory that Neanderthals were an extinct side-branch had the effect of rendering new Neanderthal finds relatively uninteresting and hence under-examined. This, in a circular fashion, postponed the falsification of his reconstruction. Garrod's discoveries at Mount Carmel raised the possibilities of a closer link between Neanderthal's and *Homo sapiens*. Hrdlička (1930) and Weidenreich (1943) were inclined to believe that they were directly ancestral to moderns and the whole question was gradually re-opened. Neanderthal bipedalism with an upright posture was soon accepted as other remains came under scrutiny. Boule's own errors were not fully unravelled until the 'Old Man' was re-examined by Arambourg and Patte in the early 1950s (Arambourg, 1955; Patte, 1955). Both reached the same conclusion: he had suffered from grossly deforming osteoarthritis. This, plus lesser errors, had distorted Boule's image of Neanderthals totally. Again the formal exposure of the error confirmed what was already generally accepted by palaeontologists. In any case the social psychological framework of the pre-1914 era within which Piltdown Man and Boule's Neanderthals derived their meaning had long evaporated.

The riddle of Reck's O.H.1 was also settled. After some to-and-froing due to difficulties in analysing the overlying rock, it was eventually confirmed that Olduvai Man was, as suspected, an intrusive burial and not contemporaneous with the rock in which it had rested (Boswell, 1932; L. S. B. Leakey et al., 1933).

With these errors out of the way palaeontology was, by the end of the 1950s, free to pursue a less treacherous course.

6 *Palaeontological discoveries*

Between 1914 and 1960 the fossil picture changed beyond recognition. Viewed romantically, it might be termed the subject's 'Heroic Age', with its practitioners vying in charisma with their finds. The events in which they got embroiled and the lengths to which they went in pursuit of their osseous quarry contribute much to this aura, epitomised by the early Olduvai expeditions and the tragic tale of Peking Man. All the same, human palaeontology continued to suffer from erroneous theoretical assumptions, bad luck, and occasional plain stupidity. Certain finds (and groups of finds) stand out as landmarks but, though restricting attention to them, I hope I can still indicate the tortuous route of palaeoanthropological understanding during these years. They fall handily into six groups according to geographical provenance: Chinese, South African, East African, Javanese, Palestinian and European.

(a) China: '*Sinanthropus pekinensisa' or Peking Man*

Peking Man's story is a dramatic and ultimately tragic one (Shapiro, 1976; Reader, 1981) ending in the loss of all the original material, bar three teeth, in unclear circumstances in November 1941 during the Japanese invasion of China.

For many years Europeans had been aware of the fossil nature of the 'dragon bones' ground up in traditional Chinese medicines and sold by apothecaries. In 1921 a Swede, Andersson, having tracked one source of these to Choukoutien visited the site with Zdansky (with whom he later quarrelled) and W. Granger, a visiting US palaeontologist. Numerous fossils were found including two hominid teeth, the first fossil evidence of hominids in China. Zdansky, the finder, chose to keep quiet about them, not even telling Andersson, fearing, as he later explained, the premature publicity their announcement would bring. (Shapiro and Reader differ slightly on this; Shapiro has Zdansky finding them in 1923.) Zdansky returned in 1923 but with little success in the way of hominid fossils. He finally chose to reveal the teeth for the Crown Prince and Princess of Sweden (sponsor nation for the research) during a state visit to China in 1926, prompted by Andersson's request for something exciting to impress them. The only other oriental hominid remains known at this time were Dubois' *Pithecanthropus* discoveries in Java in the 1890s – which, as mentioned earlier, were not available for study. The 'Oriental origin' theory they had inspired persisted unabated, however, and between 1921 and 1928 a series of major central Asian expeditions under R. C. Andrews had sought, in vain, for evidence of this, although their efforts yielded many important earlier fossils including a clutch of dinosaur eggs, the auctioning of one of which brought the whole enterprise to an end – but thats another story . . . Pilgrim's 1911 *Sivapithecus* and subsequent finds of the Miocene anthropoid ape, *Dryopithecus* (10–16 myr) by Gregory in the same Siwalik foothills of the Himalayas reinforced the theory.

Andersson, as director of the research, passed the teeth over to Davidson Black, a dedicated Canadian palaeontologist obsessed with human evolution. In 1926 they were reported in *Nature* (Black, 1926), but classified only colloquially as 'Peking Man'. Zdansky went to Cairo in 1927, Black having taken over from Andersson. The funding shifted to the Rockefeller Foundation. 1927's season yielded a further tooth in fine condition and the classification *Sinanthropus pekinensis* was adopted. Nearly every year from then until 1937 the site was excavated by a team including Bohlin, Pei, and intermittently, Chardin. After Black's premature death in 1934, Weidenreich took over as director. An almost complete skull discovered in 1929 enabled comparisons with Java Man to be undertaken, at least as the latter was described in the original Dubois report. Peking Man was felt to be the later, more progressive and generalised of the two. By the time local unrest halted work at least 40 individuals were included in the sample, the largest single collection of fossil hominids then known. Years of scientific analysis lay ahead, and the cessation of field-work itself would not have mattered seriously at this time. Fortunately Weidenreich took casts, on which basic morphometric analyses are still possible. Only Zdansky's original two teeth, plus a third he found while re-sorting the material, are now safe, in Uppsala, Sweden.

Sinanthropus, attracting publicity from the start, was rapidly accepted, due partly, at least to start with, to the neatness with which it confirmed theoretical expectations. In addition to the fossils there were indications that fire had been used and cannibalism practised (inferences widely accepted until recently, L. Binford (Binford and C. K. Ho, 1985) pouring cold water on the whole subject as discussed in Chapter 6 (L. Binford, 1981, pp. 291–93). Lower Palaeolithic tools in sandstone and quartz were also found, principally classifiable as 'choppers'. To summarise the find's contemporary significance:

(i) It supported the 'Asian origin' theories then in favour.
(ii) The apparent evidence of fire-use and cannibalism gave at least a glimmer of the ancestral life-style.
(iii) As we will see, it overshadowed finds from South Africa which were more at odds with theoretical expectations.
(iv) Gould (1980) suggests that it also lent support to notions of white supremacy since while Piltdown Man, of apparently similar age, had a *Homo sapiens*-size brain capacity, Peking Man could only muster two-thirds of it. If modern Chinese are descended from *Sinanthropus* and modern Englishmen from Piltdown Man . . . . Personally I think the admirable Gould is stretching it a bit here: (a) Piltdown Man was already being placed off the main ancestral line (see above); (b) it would be current brain capacity not ancestral brain capacity that would indicate current relative status (accepting the logic of looking at things this way in the first place); (c) at the time Peking Man was in its heyday, the 1930s, fashionable opinion was surely swinging away from racism in the face of the rise of Nazism. Is there any evidence that current palaeontologists, aside from Elliot Smith perhaps and

Coon, were racist and/or were interpreting Peking Man as a direct ancestor of the modern Chinese?

(v) It inaugurated a major shift in attention away from European material.

Its current place in the evolutionary picture is somewhat different. We now see it as a major landmark in the morphological sequence from *Australopithecus* to *Homo sapiens sapiens*, classified along with Java Man as *Homo erectus*.

For the moment, in the 1930s, oriental ancestry held sway. Slightly earlier a much-acclaimed US tooth find, prematurely christened 'Hesperopithecus' (Sunset Man) was greeted as the first all-American fossil anthropoid ape. It turned out to belong to an extinct variety of pig. Peking Man now hogged the limelight until World War II, but equally important developments were taking place in the South African shadows.

(b) South Africa: the australopithecines

In 1924 Raymond Dart, a young Australian and ex-student of G. Elliot Smith, was struggling to come to terms with a grudgingly given professorship in the then back-water of the University of Witwatersrand, Johannesburg. His sole woman student happened to show him a fossilised baboon obtained from a nearby lime-quarry at Taung. Accounts of what happened next are somewhat confused, but the upshot was the appearance of the 'Taung Skull', including a cranial endocast. This infant skull was odd in a number of respects; the foramen magnum's forward location indicated an upright posture, the brain size was only slightly bigger than would be expected in an ape, though apparently different in shape, and the teeth were more hominid than pongid. It all added up, in Dart's view, to something very peculiar and contrary to the 'brain first' assumptions still dominating European palaeontology, not least Elliot Smith's own thinking. Dart dubbed his find *Australopithecus* (Southern ape, but one cannot help wondering if the fact that he was Australian did not have something to do with it).

He then proceeded to play his cards all wrong. Its announcement in 1925 came in a fanfare of publicity in the popular press rather than the customary dead-pan note in *Nature* (though this followed, Dart, 1925). Furthermore Dart drew inferences about the species' life-style that just could not be substantiated on the basis of his evidence. The facilities available to him hampered his preparation of a full scientific paper, though he did his best. A more cautious approach, involving earlier consultation with leading authorities, would have helped. As it was, their noses were well put out of joint, the great Sir Arthur Keith having to rely on what he could see of a plaster cast behind glass at the South African stand in the British Empire Exhibition. He and Elliot Smith were unimpressed and dismissive, consigning it to chimpanzee status and frowning on Dart's breaches of scientific etiquette. When he finally managed to get to England and present the object at a scientific meeting it was a disaster, there was simply no longer any interest in it and Dart apparently dried up. He returned home totally demoralised by the whole business. *Australopithecus* fell from sight, the Taung limestone was converted into cement, the small-brained biped nearly forgotten. This at any rate is the story as now generally related. How far did the palaeontological establishment really neglect *Australopithecus*? In 1934 G. Elliot Smith gives a full-page picture of the skull and writes

> . . . the specimen is of exceptional interest and importance, not only because (it) . . . is more nearly complete than any other known fossil of an Anthropoid Ape, but also because it presents a closer affinity to Man even than *Dryopithecus*. (G. Elliot Smith, 1934, p. 65)

Has a certain degree of myth-making taken place about the rejection of *Australopithecus*?

At this point Robert Broom enters the story. This already ageing Scottish-born eccentric doctor and palaeontologist (the greatest ever in his own opinion) had retired from medicine to devote himself full-time to fossils. In 1936, having come to some financial arrangement with the foreman of his local limestone quarry, Sterkfontein, Broom found more evidence of *Australopithecus*. By the time of his death in 1951 he had totally vindicated Dart. In addition to Sterkfontein he discovered australopithecine sites at Kromdraai (1938) and Swartkrans (1948), while Dart himself returned to the field and found Makapansgat in 1947. In the course of this Broom recovered a virtually complete spine and pelvis, long bones, a further skull and numerous teeth and other fragments. Classification of the species was in continuous flux (and remains so), but at least two forms, a 'robust' and a 'gracile' were clearly identifiable. Other, East African, finds later supplemented the South African ones. Broom's major work on the australopithecines appeared in 1946 (Broom and Schepers, 1946).

Keith capitulated in 1947 and by Broom's death it was accepted that the species represented a bipedal, small-brained, hominoidal creature of very early date. Dart, ever speculative, continued to ponder on its life-style and character. The nature of bone-damage to other associated fossils suggested to him an 'osteodontokeratic' culture, which meant literally 'bone, tooth and horn-based'. *Australopithecus* was depicted as a carnivorous hunter and scavenger (Dart, 1957). This too ran counter to the conventional wisdom which held the earliest people to have been gentle herb and fruitivores. Dart's model has since been severely undermined, but the diet of protohominids remains a controversial focus for current theorising (see Chapter 4 below).

But when did these australopiths live? The geological character of the sites rendered calculations on this impossible: they were in limestone and not amenable to the techniques dependent on radioactive decay or geomagnetism which began to be widely applied towards the end of our period, nor were they clearly stratified, having been buried in caves and crevices subject to various types of erosion and collapse which made stratigraphical analysis a nightmare.

By 1947 attention was beginning to shift towards Africa as a possible location for *Homo sapiens*'s birthplace. Broom's work, and the activities of the Leakeys were mainly responsible for the change and the first Pan-African Congress on Prehistory organised in Nairobi by Louis Leakey instituted a new era in palaeoanthropology. But harmony did not reign. Louis Leakey and the South Africans did not see eye to eye over *Australopithecus*. For Leakey the creature was a side branch, the key ancestor would be a tool-maker. Leakey had a life-long preference for an extended *Homo sapiens* ancestry which affected his interpretations of all fossil hominid finds. Much current controversy in the area has its roots here.

(c) East Africa: *Zinjanthropus* and *Homo habilis*

Initially Louis Leakey's Olduvai work had yielded artefacts galore but only one fossil hominid, O.H.2, represented by two meagre skull fragments. The 1930's expeditions in any case amounted to preliminary reconnoitres and surveys; for practical as well as methodological reasons little else was possible. In fact Leakey's own first fossil hominids were found near Lake Victoria after the 1931 Olduvai expedition with Reck. These were the Kanam Skull and Kanjera Jaw. Leakey's already considerable reputation enabled him to persuade many leading palaeontologists that these proved the antiquity of Reck's O.H.1, of which, after earlier doubts, he was now convinced. He invited a sceptical geologist, Boswell, to accompany him back there on the Fourth East African Archaeological Expedition. It was a fiasco. Everything went wrong for him: there was no accurate map of where they had been found, iron markers had been removed by local people for their own purposes and the original photographs were spoiled. Boswell returned, scepticism undiminished and the antiquity of O.H.1 was refuted soon after, as previously described.

Leakey's own life now changed direction, his divorce and remarriage were still sufficiently unacceptable to British society to render a return to his native Kenya prudent. In the late 1930s he was commissioned to write a thorough study of the Kikuyu with whom he was raised, whose adolescent initiations he had undergone, and whose language he dreamed in and spoke, he claimed better than English. (Only now is this study actually 'in press'.) The war took him into counter-espionage and amateur detective work (L. S. B. Leakey, 1974). His, and Mary's, archaeology continued at sites such as Gambles Caves and Olorgesailie, with periodic forays into Olduvai, in neighbouring Tanganyika (Tanzania). After 1947 his fossil-hunting continued at Rusinga Island, an important site yielding the Miocene hominoid which he named 'Proconsul'. The 1947 Pan-African Congress, already referred to, was called by Leakey partly to celebrate the promise and glories of Olduvai Gorge, Olorgesailie and the other prehistoric sites in his domain. Few academic meetings can have been quite so eventful, it nearly ended with the cream of the world's prehistorians being stranded in floods after visiting Olduvai, where their adventures had already included a rhinoceros eating the vegetable supplies.

In 1951, supported by funding from Charles Boise, the Olduvai Gorge work began in earnest. By 1955 only a single fossil hominid find had been made: two australopith milk teeth (O.H.3). The 1959/60 season saw the breakthrough. Six finds added two new species to the fossil hominid catalogue (an increase of 50 per cent over the two pithecanthropus and two australopithecine species then known!). The first to come to light, O.H.5, was a skull discovered by Mary and associated with Oldowan artefacts. This was as embarrassing theoretically as it was exciting palaeontologically, for it bore all the hallmarks of an australopithecine. But Louis Leakey did not, as we noted, believe they could make tools. The dilemma was solved by denying it australopithecus status and calling it *Zinjanthropus boisei* (Zinj = East Africa). Morphological differences from the South African finds could, according to Leakey, be readily identified, although a leading South African authority, John Robinson, disagreed. The find was announced at the 1959 Pan African Congress on Prehistory in Kinshasa (Congo), where it was nicknamed 'Nutcracker Man' from its massive jaw. O.H.5 also received the honour of being the first fossil hominid to be Potassium-Argon (K-Ar) dated (see Chapter 3). This gave an age of 1.75 myr.

Embarrassment over Zinj's tool-making ability was short-lived. The Leakeys' son Jonathan soon discovered further fossil fragments catalogued O.H.7 and O.H.8 (with O.H.4 and O.H.6 also now seeming to belong to this new species). These represented a second new hominid, *Homo habilis*, so named for its assumed ability to make tools. *Zinjanthropus* and *H.habilis* had lived contemporaneously, and the former had not made the tools after all. But how different in fact are *H.habilis* and the robust australopiths (where *Zinjanthropus* was later reallocated as *A. robustus boisei*)? Morphologically the differences are far from being as substantial as their different scientific names suggest, the primary factor being *H.habilis*'s apparently greater brain size. The matter is still controversial and it is even suggested that among the original *H. habilis* fossil fragments are some australopithecine intruders. The O.H. number has now risen to over 50, but Olduvai has produced no more new species, although as expected, *Pithecanthropus*, now renamed *Homo erectus*, has been found.

Many other East and South African sites have since been opened up, often by associates of the Leakey team. For all the wealth of new material, it must be kept in mind that fossil hominids invariably form but a small percentage of a total fossil assembly, at East Turkana (Koobi Fora), for example, eleven years' work yielded 5,000 fossils, only 200 are hominid and only 59 at all substantial, including a mere 9 reasonably complete skulls (Reader, 1981).

The appearance of 'Zinj' and *H.habilis* in the 1959/60 season established the primacy of Africa generally in hominid palaeontology, and East Africa in particular. The majority of later developments have been elaborations from the basic ancestral picture revealed here.

(d) Java: *Pithecanthropus* and *Meganthropus*

Although first China and then Africa have dominated the scene, Java has never been entirely off-stage. In 1936 G. R. von Koenigswald finally persuaded Dubois to release his original finds. After Peking Man the hunt was on for further Javanese fossils, a hunt headed by von Koenigswald (von Koenigswald, 1956). His first success was in 1931 at Ngandong on the Solo River. This fossil, now classified as *H.soloensis*, was earlier thought to be comparable to Europe's Neanderthals. In 1936 a small child's skull was found at Modjokerto by Cosijn, stimulating further investigation of comparable fossil beds, and during 1937–9 von Koenigswald explored the Sangiran region of the upper Solo River. By 1952 he and other later researchers had managed to find remains of six more early hominid individuals. Of the total now known (another came to light in 1969) six may be termed *H.erectus* and two *Meganthropus*. Classification of Javanese finds has remained difficult. Dubois himself took issue with von Koenigswald's claim to have found more 'Pithecanthropus' remains, although the similarities were clear to everyone else. Javanese *H.erectus* specimens are more robust in some respects than Chinese ones, and their precise relationship to African finds is difficult to establish, particularly as geological dating of Javanese fossil beds is as yet unclear, though recent estimates give 700,000 years as the oldest possible date for Dubois' Trinil finds and 100,000 at most for *H.soloensis* (Bartstra, 1983, p. 426). To make matters worse, the Trinil II femur's association with the skull-cap and other *H.erectus* remains is now in serious doubt (Kennedy, 1983). *Meganthropus*, related perhaps to a similar colossal early anthropoid found in China, *Gigantopithecus*, is now widely believed to have some kinship to robust australopiths. A substantial hand-axe culture is known to have existed, notably at Patjitan, but these tools were found (1933) in their secondary internment and no tools associated with pre-*sapiens* fossils have been recovered. These tools are now thought to be younger than 50,000 years and their classification is, frankly, confused (Bartstra, 1983, pp. 426–8).

The Javanese finds are important primarily for the evidence they provide of the range, both geographical and morphological, of the *H.erectus* grade, off-setting biases likely to enter the picture if the African evidence alone were available.

(e) Palestine: Neanderthals and *sapiens*

Dorothy Garrod's Mt Carmel expedition of the 1930s has been referred to earlier in connection with archaeological methodology. The immediate consequence of the research, however, was to extend the data-base for examining the final *H.sapiens sapiens* emergence. Of the two sites uncovered the older, Tabūn, was clearly Neanderthal, though the specimens were not identical to western European ones, and were classified first as *Palaeoanthropus palestinus* (McCown and Keith, 1939). The finds at the second site, Skhūl, were also initially classified this way and the two sites were thought to be contemporaneous. Current thinking has shifted as we will see in due course (Chapter 6 below). A further major Neanderthal find in the Near East was at Shanidar, Iraq, in 1953 (Solecki, 1955). The existence of a somewhat less robust Near Eastern Neanderthal population in apparent contiguity with more *sapiens*-like successors was a major factor in subsequent theorising regarding the transition between the two, and the origins of both. The Garrod expedition well and truly re-opened the Neanderthal question in the late 1930s, and in fact the Tabūn and Skhūl skulls remain problematic.

(f) Europe: Archaic *Homo sapiens*

Although the most significant hominid fossils were no longer being found in Europe, it did not entirely cease producing interesting remains. The number of Neanderthal sites was still increasing almost annually both in western and eastern Europe, though a major tragedy occurred in 1945 when the Mikulov Castle in Czechoslovakia was burnt, with the loss of material from at least three major sites: Mladec, Predmosti and Šipka.

Britain at last managed to claim a *bona fide* hominid of its own: Swanscombe Man (probably female in fact (Wolpoff, 1971)), unearthed by Marston in north Kent in 1935 and 1936 (with a further fragment of the skull being found, amazingly, in 1955). Germany produced the Steinheim skull (Berckhemer, 1933), and after the war the Fontéchevade skull was found in France (Vallois, 1949). This group of skulls and skull fragments may be considered along with the earlier Heidelberg Man (or Mauer Jaw) and the Petralona skull from Greece (Kokkoros and Kanellis, 1960; Poulianos, 1982) as the most enigmatic hominid fossils facing European palaeontology. Their dates are agreed to be pre-Neanderthal and they would appear to be transitional between *H.erectus* and *H.sapiens*. This of course raises problems for the place of the Neanderthals; were these an ancestral population which split, with one group becoming *H.s.sapiens* and the other *H.s.neanderthalensis*? As we will see later there are many open questions remaining with regard to this group, provisionally labelled 'Archaic *Homo sapiens*'.

Summary of Part C 6

The state of play in 1960 was then utterly different from that in 1914. Throughout the period researchers had been juggling to no great effect with genealogical trees of varying degrees of bushiness or unilinearity, Elliot Smith, Hrdlička, Keith and Leakey to name but a few. Meanwhile the fossil-tracking of human evolution had been pushed back to nearly two million years, a time-span for hominid life that would in itself have astonished pre-Great War palaeontologists. The '*dramatis hominidae*' included at least two varieties of bipedal australopithecines and a somewhat bigger-brained contemporary of their later days, *Homo habilis*. 'Pithecanthropus' now stretched from China to Java and probably to Europe and Africa also, while a mixture of Neanderthals and enigmatic unclassifiables covered the last 100,000 years or so prior to the appearance, a mere 30,000 odd years back, of modern humans. Some hefty but doomed cousins, *Zinjanthropus*, *Meganthropus* and *Gigantopithecus* looked on. Far back lay *Dryopithecus*, *Proconsul*, *Sivapithecus* and *Ramapithecus*. The exact ancestral relationships and kinship affiliations of these beings, their degrees of hominisation and life-styles, were a matter of guesswork, though it was fairly obvious that some had coexisted with others and that no simple single lineage united them. K-Ar dating at the very end of the period, along with better stratigraphical techniques and knowledge of other areas of faunal evolution, was beginning to make an impression on the ever-frustrating problem of dating. The 1959 centenary of *The Origin of Species* could be greeted optimistically as a new level of interdisciplinary involvement and interest in human evolution began to crystallise (see Tax, 1960). LeGros Clark's classic textbook *The Antecedents of Man* appeared (1959). Genetics, Ethology, Taphonomy, Cultural Anthropology, even Psychology, could all have something relevant to offer now that fossils themselves were to hand as a concrete focal point. Missing links had been discovered, but how did they join up? Although events since 1960 have not answered this definitively, they have continued to expand our knowledge of our ancestors.

D Conclusion

The study of human evolution has had a rougher ride than most areas of scientific endeavour over the same period. Its problems have been severalfold.

Firstly, Palaeontology is not an experimental science, and in relation to humans has had to depend on whatever happens to turn up. Since new finds are relatively rare they tend to stimulate speculation and conclusion-jumping out of all proportion to the physical evidence. Theories cannot be tested in a controlled way and though each existing theory is in a sense tested against each new item of fossil evidence, that evidence itself is invariably ambiguous and controversial. The very perception of what it *is* will be determined by the preconceptions of those studying it.

Secondly, leading on from this, we can see time and again how such preconceptions, often rooted in extra-scientific socio-psychological factors, affect the reception of new evidence. Piltdown Man and Peking Man are eagerly seized on and accepted, Neanderthal Man is at first rejected, then distorted; Java Man receives such a mixed reception that its finder retreats into a thirty-year sulk, reinterring the finds beneath his floorboards; *Australopithecus* was, we are told, dismissed out of hand. Such assumptions as the 'brain-first' and 'Asian origins' models were in fact based largely on *à prioristic* reasoning and seriously misled researchers for decades.

Thirdly, human evolution theories are dependent on current understanding of evolution and inheritance. Theoretical changes and phases of confusion at this level have often left human evolution researchers rather at sea, and they have sometimes gone on their own sweet way, paying scant attention to evolution theory in their own theorising. This problem has largely abated, being most marked at the turn of the century, but the dependency relationship persists.

Fourthly, echoing our first chapter, the charisma, the psychological potency, of the topic has presented a perennial problem. Theories of human evolution and current notions of 'human nature' are always interwoven, and the latter can often set limits on the kind of account the world is prepared to accept. Nor is this an 'imposition' from outside, for theorists themselves operate within the framework of such current ideas and may even strongly identify with some of them. There is clearly much interesting work to be done in the history of science field concerning the nature of theory-generation and evaluation in human evolution research.

Fifthly, it must be said that the field has had more than its quota of temperament problems. Virchow's smug and stultifying dogmatism, Dubois and Dart's inability to stand their ground under criticism, Smith Woodward's initial gullibility regarding Piltdown Man (assuming he wasn't the culprit), Zdansky's odd behaviour with Peking Man's teeth, even Louis Leakey's earlier stubborness about the importance of his pre-War finds and hostility to *Australopithecus*, all betoken a vulnerability on the part of hominid palaeontology and palaeoanthropology which can be held in check a little better in more experimentally oriented and perhaps institution- and laboratory-bound regions of science. This would make it a duller world too and I am not offering a criticism of the discipline as such, only an observation on it in the light of its history so far. The personnel are after all self-selected and though you don't have to be mad to be a fossil hunter, it probably helps.

I have tried to introduce the reader to the principal figures and developments in the study of human evolution from the early nineteenth century to 1960. I have scanned, often too superficially, a wide range of disciplines beside the central ones of Palaeontology and Archaeology. Psychology and Sociology hover in the wings, more affected by the debate than affecting it, although their perspectives are felt from time to time. Initially indeed, in *The Descent of Man*, the psychological and sociological content is considerable (though not labelled as such), as it had to be in the absence of fossil data. For a recent critical account of the major fossil discoveries the reader is referred to Eldredge and Tattersall (1982), while Reader (1981) has pulled together an enormous range of information in his account of the historical high-points. Daniel (1967) is probably the best source on the history of Archaeology. Wendt (1972) is also useful as it contains discussion of early material hard to find elsewhere. Nevertheless it remains true, as I wrote at the outset, that no thorough history of the whole topic is available. Bowler (1984) has covered the late-nineteenth-century anti-Darwinian phase of the history of evolution theory. F. Spencer (ed.) (1982) covers the recent course of US physical anthropology.

In the most recent period of rapid expansion in human evolution research the problems I have just described appear to have diminished in force, but it is in the nature of such things that they only become visible in retrospect. Controversy is as rife as ever, though the issues involved have altered. During the 1960s and early 1970s the field saw dramatic developments on several fronts. The fossil data from East Africa, the Fayum Depression in Egypt, the Siwalik Hills in India and Pakistan and from Lufeng in China (as well as intermittent European finds) has increased enormously, ranging from the Miocene protohominids to Neanderthals, though interestingly only one new species of hominid, *Australopithecus afarensis*, has come to light. The cross-fertilisation of Archaeology, Anthropology and Taphonomy has moved studies of life-style to an entirely new plane of sophistication. Finally the advent of Molecular Biology has entirely altered the way in which inter-species relationships are being tackled. At the level of theory, such developments as the rise of Sociobiology, cladistic analysis and feminist perspectives are all involved. We can now move on to the current state of play.

CHAPTER 3
Methods and data-bases

Introduction

Before examining the principal current pictures of hominid behavioural evolution and the theoretical issues involved, it is necessary to review the nature of the evidence on which knowledge of hominid evolution is based, although even this is pervaded by theoretical controversy. In this chapter therefore I will be describing the principal methods and data-bases of modern human evolution research. In addition I have felt it advisable to include a section on the present state of evolutionary theory as a whole since this forms the backdrop against which many issues concerning human evolution have to be understood. There are major shifts afoot in this field and it is important that the reader be alert to them. I have, throughout, pitched the discussion at an introductory level sufficient to guide the reader into the more technical literature. This inevitably means that a price has been paid in terms of oversimplification and omission.

The chapter is in eight parts: A a preliminary discussion of dating methods, B the geological context of hominid evolution, C the hominid fossil data, D artefacts, E molecular biological approaches, F cross-cultural research, G primate studies, H evolutionary theory.

A Dating methods

Some indication has already been given of the perennial problems presented by the task of dating the various kinds of evidence on which knowledge of our evolution is based. This was long exacerbated by mistaken presuppositions about the time-scales involved. The original Swiss and Danish datings were calculated on the basis of sedimentation rates, rates of marsh-formation and the like. Today the range of techniques available has expanded enormously, but the field remains fraught with difficulties. Even for the Pleistocene epoch, the most recent geological period covering virtually all the Palaeolithic phase, a recent writer (Morrison, 1980) feels it necessary to present two alternative chronologies for the glaciation sequence, a 'long' and a 'short', on the first of which the initial major glaciation of the 'Ice Age' (Guna I) is over a million years ago, while on the latter it is about half a million. For events prior to about 200,000 years ago the two chronologies are seriously discrepant (see Fig. 3.1a). As we will see, numerous though they are, each dating technique has its own limitations and may often be usable only if certain fairly specific geological conditions are met. As far as Absolute dating (see below) is concerned in particular, a great many hurdles remain to be overcome.

Before tackling the techniques themselves, it is useful to mention the various types of dating which are logically possible, a topic to which Oakley (Oakley, 1969) provides a good introduction. The first distinction to be made is between Absolute (or Chronometric) dating and Relative dating. The first refers to the actual age of the material, usually, for pre-Mesolithic finds, in years before the present, the 'present' being by convention 1950. Absolute dating is extremely difficult and recent decades have seen a hunt for reliable 'clocks', especially in phenomena associated with radioactive decay. (Several abbreviations are used in dating: bp/BP for 'Before Present', mya/MYA and myr for 'millions of years ago', kyr/KYR for 'thousands of years ago'. When lower case is used this means the date is based on radioactive decay dating; capitals refer to ordinary solar years.) Relative dating refers primarily to sequencing: the age of a specimen vis-à-vis the deposits in which it was found, or whether associated finds are actually contemporaneous. An obvious starting point is in stratigraphical relationships.

A second distinction made is between Direct and Indirect dating. Direct dating refers to dating of material from internal evidence (e.g. Carbon-14 dating of a piece of bone), and Indirect dating to dating from context, at one or more removes. This distinction is more of a continuum than a simple dichotomy. Analysing the types of dating offered by these distinctions, Oakley identified four types of Absolute and four types of Relative dating. This hints at how complex the dating issue can become. Although most readers of this book will not require detailed technical knowledge of current dating methods, some acquaintance with them is necessary to render the literature in the field comprehensible. For further details of most techniques discussed see Fleming (1976).

1 *Carbon 14 (C-14), or 'Radiocarbon', dating*

C-14 is a radioactive isotope of carbon generated ultimately by cosmic ray interaction with the upper atmosphere. It is created continuously and is

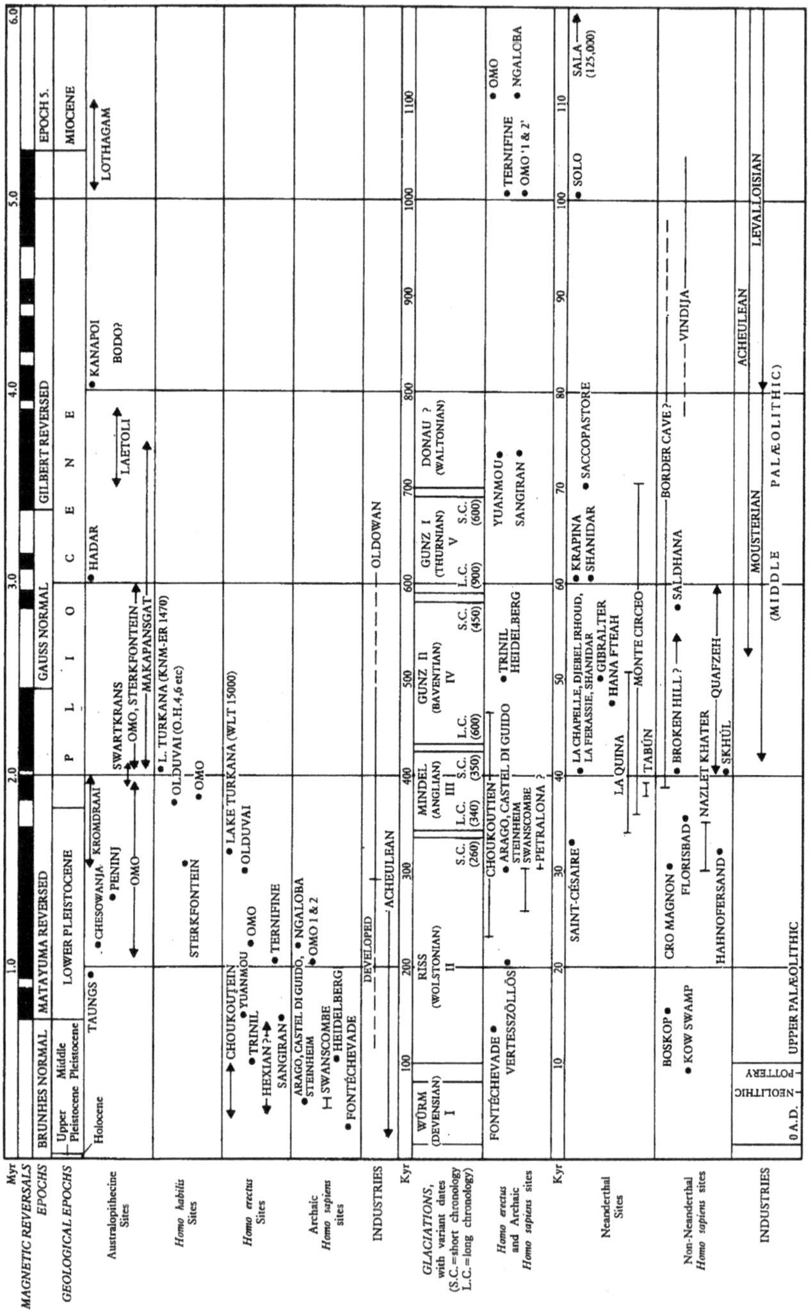

Figure 3.1a Time chart.

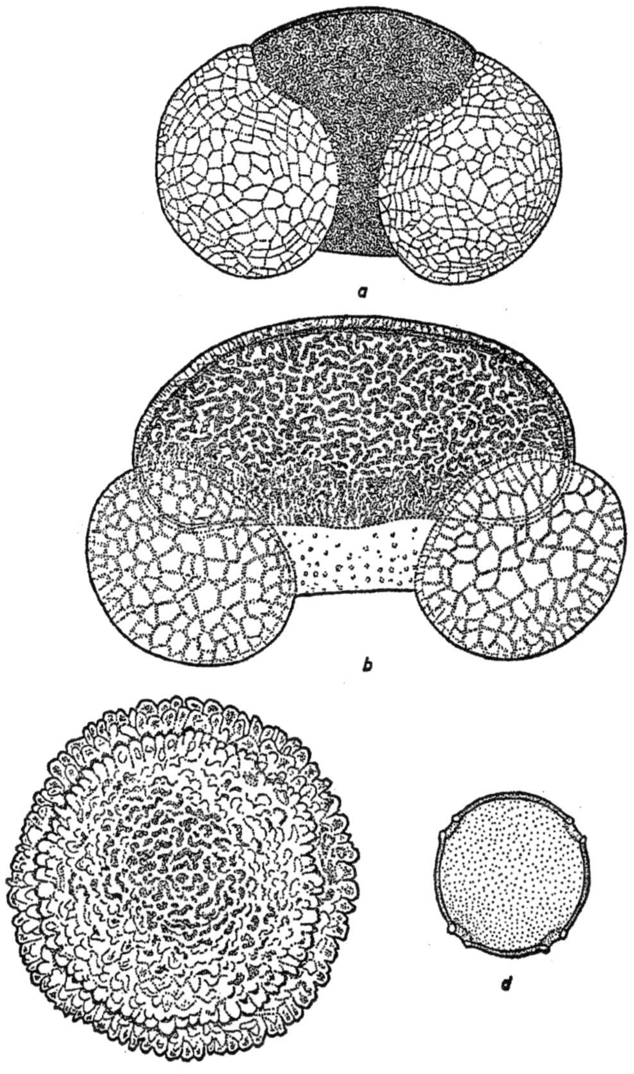

Figure 3.1b Palynologists use fossil pollen evidence to identify prevalent flora and, via that, climatic conditions (see p. 73).
From K. Oakley (1969) with permission of Weidenfeld & Nicolson.

assumed to constitute a constant, if minute, proportion of atmospheric carbon (in carbon dioxide). Living organisms continually ingest this via food and respiration but at death this ceases. A newly dead organism will thus contain a known level of C-14. As the C-14 decays the level reduces, giving an index of time elapsed since death. The half-life of C-14 is 5,568 years, so a specimen with half the base-level of C-14 activity for living organisms will be 5,568 years old. At twice this age the level of activity will be a quarter and so on, reaching an asymptote at around 70,000 years ago. In practice C-14 dating has been restricted to the last 50,000 years, due to increasing unreliability of measures of low activity. C-14 dates usually have a 1 standard deviation error figure given with them. Libby's introduction of this technique (1955) revolutionised the study of later prehistory, especially the Upper Palaeolithic transition. It has had its share of problems nevertheless, and the scales have required major revision on occasion. The assumption that C-14 has been at a constant atmospheric level was not entirely supported by subsequent research (notably on tree-ring dating which does give *both* a precise absolute dating *and* C-14 levels back to around 3–4,000 BC). It is of course inapplicable for pre-*Homo sapiens* periods. At Oxford Research Laboratory for Archaeology a new method of measuring C-14 directly from its mass rather than its radioactivity is being perfected using a radiocarbon accelerator. This will push the dating range back to the full 70,000 years and requires only a thousandth of the amount of material usually required (Gowlett, 1984).

C-14 based datings are conventionally expressed in lower case 'bp' for 'before present', as opposed to BP for solar years.

2 *Dendrochronology or tree-ring dating*

This need concern us only briefly, for it is limited to the last 6,000 years or so. Dendrochronology is based on tree-ring counting, matching tree-ring sequences from overlapping, but progressively older, specimens. As the annual growth rings vary in width according to climatic conditions, it is possible to build up lengthy sequences from specimens coming from the same geographical area. The oldest living tree is probably a Japanese cedar 5,200 years old, but the California bristle-cone pine has been used for C-14 calibration. The oldest of these is 4,600 years though a 4,900-year-old one was chain-sawed down in the mid-1960s (McWhirter, 1984).

3 *Fission-track dating*

Fission-track dating is a method of determining the proportion of Uranium 238 (U–238) which has decayed since a mineral specimen solidified. The procedure is complex, since it involves measuring the existing level of U-235 (of which U-238 is a constant proportion), and techniques for rendering visible the tracks left in the material by radioactive decay products. In theory the method could be used to date material of virtually any age, since the half-life of U-238 is in billions of years. Its use is, however, restricted to rather special circumstances where samples of vitreous or crystalline minerals such as obsidian are available which can be reliably given a relative dating contemporary with the specimen being studied. Its applications in geology and historic archaeology are extensive and it has served to provide a greater understanding of much of the East African geological background (e.g. Gleadow (1980) using fission-tracks in zircons).

4 *Potassium-Argon (K-Ar) dating*

The logic of this is somewhat similar to that of C-14 dating, but the range is far greater, since we are dealing with an isotope of potassium, K-40, with a half-life of 1,300 million years (1.3×10^9). As this decays, it generates, as a 'daughter product', an isotope of the inert gas argon, Ar-40. If we can determine the amount of potassium in a crystalline specimen and also measure the amount of Ar-40 in it we can, again, determine directly from their ratio the time elapsed since crystallisation. Minerals younger than 10,000 years cannot be dated in this way since the level of Ar-40 is too low.

This technique is more widely used than the previous one, potassium being commoner than uranium, although the geological circumstances in which it is feasible are broadly similar. Since receiving its debut in the dating of 'Zinjanthropus' in 1960, it has been widely used for dating hominid fossils and artefacts.

Numerous other dating techniques rooted in atomic physics will also be encountered – Uranium-Thorium, Protoactinium-Thorium, Electron-spin resonance (ESR) and Thermoluminescence (TL), the last useful for dating artefacts that have been fired or heated in manufacture, such as pottery and some flint tools.

5 *Amino acid racemisation*

The amino acids in living material are all in the L-isomer form, i.e. they spiral to the left. After death an increasing proportion of the molecules shift into D-isomer (right-spiral) form until an equilibrium is reached. This process is called 'racemisation' and has provided a method for direct dating of bone. Though promising, the procedure is difficult as rates of racemisation are temperature- and moisture-sensitive. Ideally specimens should derive from environments, such as caves, which have been fairly constant in both respects. It also of course requires the preservation of sufficient quantities of amino-acid-bearing material. The maximum theoretical range of this is about half a million years. A related method called 'aspartic acid racemisation' is going back to about 43,000 years only. These methods need to be interpreted with caution. In 1972 a skeleton called the Sunnyvale Skeleton was found near San Francisco and in 1975 an age of around 70,000 years was reported on the basis of aspartic acid racemisation, 30,000 years earlier than any known New World *Homo sapiens*. A book entitled *American Genesis* followed on the strength of it. In 1983, alas, C-14 and fission-track datings of the bones and source-bed material gave ages from 3,000–9,000 years ago, thereby effectively scotching the idea that we evolved in California (Taylor et al., 1983).

6 *Fluorine analysis*

An older method of *relative* dating of bone by chemical means is fluorine analysis. Put simply, buried bone absorbs fluorine from the soil at a rate very much dependant on local conditions. Bones contemporary with one another from the same site will nevertheless have similar fluorine levels. Comparing specimens will thus reveal their relative ages. It was by this method that the Piltdown hoax was detected (see Chapter 2 above).

7 *Geomagnetic dating*

This has been used in conjunction with K-Ar dating in volcanic areas of East Africa. Though unable in itself to provide specific dates, it helps geologists to identify, broadly, the epoch to which a given bed belongs. During the Earth's geological history there have been many reversals in the polarity (and indeed direction) of its geomagnetism (see Fig. 3.1). The present period, the Brünhes Normal Epoch (with north in the North) began around 0.7 myr and was preceded by the Matayuma Reversed Epoch to 2.5 myr. The period prior to this, the Gauss Normal was preceded about 3.4 myr by the Gilbert Reversed. The Matayuma Reversed Epoch was interrupted several times by short periods of 'normal' polarisation which can give a somewhat finer-grained picture. Tracking geomagnetic orientations through a stratigraphic sequence, and combining this with e.g. K-Ar dating, we can proceed to cross-reference and feed-back between the different absolute and relative dating methods to date beds otherwise unamenable to either. In theory very extended and absolutely datable sequences can be identified. This is an area in which great future progress may be expected.

8 *Palynology*

Palynology is the study of fossilised pollen grains. Because of the nature of their cell walls pollen grains tend to be preserved readily and their study can yield important data, often the only data, regarding vegetation and, *inter alia*, climate, at the time a deposit was laid down. There are 40–50 features of pollen grains which can be compared and actual procedures of pollen analysis are tedious and time-consuming. The outcome of this analysis however is a profile of the relative frequencies of pollens of the different species represented. In this way, especially in well-stratified beds of the Pleistocene, the nature of the flora can be tracked through whole cycles of climatic and ecological change. Though not in itself a dating technique, palynology has become an essential adjunct for many other dating procedures and archaeological techniques. (See Figure 3.1b.)

9 *Sedimentation analysis*

Earlier techniques of analysis of sedimentation rates can of course still be used, and are particularly useful where the annual climatic cycle produced an identifiably laminated stratigraphy. One example of this is the phenomenon of 'clay varves' characteristic of peri-glacial conditions, in which fine and coarse deposits alternate. During the summer when there is considerable water-flow the finer material fails to settle and coarser material is washed down from the melting glaciers and by normal erosion. In summer then the coarser deposits accumulate on the river-bed, but in winter the flow-rate slows, perhaps the surface freezes, and in these sluggish conditions finer sediments settle, giving a winter layer. In Scandinavia and Northern Europe especially extensive use has been made of this for dating purposes. Sea-bed cores are a further source of information since they tend to be well stratified and can provide, via micro-fossil evidence, a picture of temperature variations over very long periods which may in turn be correlated with other evidence of e.g. glacial phases and shifting sea-levels.

10 *Fossil fauna*

Palaeontologists have now built up a fairly clear picture of the succession of animals which have lived in Europe, much of Africa and Asia (as well as America) in the last few million years. Although regional variations in time of extinction or replacement of species are wide, the typical faunal assemblages in the fossil record at different periods are now widely identifiable, enabling the palaeontologist to locate finds in the general sequence. Since typical faunas are also associated with particular climatic conditions this data can also be used as palynological data is, providing a framework for wider palaeoecological research. Such techniques of using changes in fauna to track broader changes have been used for a long time by archaeologists (e.g. by Garrod and Bate, 1937). Much field palaeontology is concerned with precisely this kind of analysis. Relative dating of hominid fossils from associated fossil fauna is invariably a first research move. An interesting example of the use of fossil fauna for dating a Chinese hominoid site is the narrowing down of the Lufeng site to about 8 myr on the basis of the presence of fossils of three species of extinct rodent, known from Pakistan to have overlapped only briefly at around this time (Flynn and Guo-Qin, 1982).

For the modern researcher the task of dating is therefore a complex one involving weighing and integrating many different kinds of evidence. In the long run these techniques are complementary and by incorporating them all a much narrower absolute dating 'window' may be obtained than any single one alone can provide. It is important to bear in mind that all the statements about age and sequencing in this book are ultimately based on one or a combination of these procedures, and that not infrequently the most cautious as to their reliability are those actually doing the dating.

B Geological background

Prior to around 18 myr, before Africa and Asia came into contact, both continents are believed to have been thickly forested, providing the environment in which, in Asia, the Miocene apes evolved. Following the establishment of a land-bridge between the two continents there was an extensive irradiation of fauna from Asia into Africa, related also to climatic changes in Asia which attended the continuing rise of the Himalayas. A world-wide period of cooling now set in, lasting until about 8 myr. In India at this time the Siwalik foot-hills of the Himalayas, a rich fossil source, were being formed from the alluvium washed down from the mountains, which were, and still are, rising, under the impact of the Indian subcontinent on the Asian land-mass (see Bernor, 1983).

The African scene, especially in the east and north-east, changed drastically in the late Miocene as the Rift Valley opened up and mountain-building processes altered earlier continental drainage patterns. This created an entirely new mosaic environment in the north-east, where it is now believed likely the hominids finally evolved from Miocene apes of Asian origin. Continuous volcanic upheavals combined with climatic oscillations between moist and dry periods have created a very complex geological situation in this region. Elsewhere in Africa savannah vegetation spread in the wake of shrinking jungle and the current scene was established in broad terms by around the time of Lucy's appearance at Hadar circa 3 myr (see Appendix A). No single trend towards more dessicated conditions is detectable, as is sometimes claimed, but the African continent has seen major oscillations between desertification and afforestation ever since, with lakes such as Lake Turkana and, further west, Lake Chad, undergoing numerous rises and falls in water level (R. E. Leakey, 1981; E. Vrba, 1983; McIntosh and McIntosh, 1983).

Around 2½ myr it is clear that the creation of the Arctic ice sheet had consequences throughout the northern hemisphere (Shackleton and Opdyke, 1977) and this is marked in Africa by major floral and faunal changes, including the appearance of *H.habilis*. The extent of the Afro-Asian land-bridge varies as climatic changes affect sea level, the Mediterranean at times being entirely cut-off and a western land-bridge with Europe emerging.

The crucial phases of hominid evolution seem to have occurred in the north-east of Africa during the late Miocene upheavals. Against a backdrop of volcanic activity right down the Rift Valley, a variegated environment of mountains, lakes, gallery forests, sea-shores, arid plains and fertile valleys was created, savannah becoming more dominant to the south. The Omo Valley in Ethiopia and Olduvai Gorge in Tanzania form during the subsequent period.

Europe in its present form does not emerge until the end of the Miocene when the Tethys Sea – of which the Caspian and Black Seas are remnants – dried up. The date at which one starts the 'Ice Age' in Europe is, as mentioned above, a matter of some dispute, and the last Ice Age was certainly not the first the Earth has seen, although it might have been among the most extensive. It is normally subdivided into 5 or 6 major glacial epochs (see Fig. 3.1) named after the Alpine and German sites where they were first identified. The periods between them are called 'interstadials', and it is one of these in which we presently have the good fortune to dwell. There is now considerable variation in the naming of glacial phases in different countries, although the original sequence is used as a sort of reference point it is clear that conditions were not entirely homogeneous. The regular ecological sequence identifiable during the interstadials is proving increasingly helpful in dating of European Palaeolithic finds (see T. Nilsson, 1983, for a general overview).

Glaciations are accompanied by dramatic drops in sea level as water becomes locked in the polar ice-caps. Many important Ice Age occupation sites are undoubtedly under the North Sea and Channel. In the more northerly regions, Scandinavia and Scotland, the end of the Ice Age did not result in such a dramatic rise in sea level because the land itself, relieved of an ice-mantle thousands of feet thick, itself rose in what is called an 'eustatic' response. This actually outpaced rising sea levels in the further north, resulting in raised beaches appearing above current sea level. In the Far East lowered sea levels provided a land bridge across the Bering Straits around 30,000 years ago enabling humans to cross into North America, while a similar effect in South East Asia facilitated their migration to Australia and Oceania (though not entirely without sea travel).

The overall picture then, for most of the time we have been evolving, has been of continual geo-climatic fluctuations of a fairly cyclical nature, even in Africa where a picture of unidirectional change towards more arid conditions is sometimes given. In Europe, in the last million years or so, a cyclical round of dry glaciations and milder, moister, inter-stadials exerted, in the way some have described it, a sort of pumping action on the populations living there. There are gaps in the picture. Particularly we are unclear as to what was going on in central or southern Asia during the period of general *H.erectus* dominance, after its spread from Africa around 1½ myr.

C Hominid fossils

1 *The nature of fossilisation*

The term fossilisation generally refers to any preservation of organic material, or evidence of activity, in the geological record, thus even foot-prints and burrows may be described as fossilised ('trace fossils'). Technically it also has a narrower meaning, being a specific chemical process in which the protein (ossein) in bone, accounting for about a third of its weight, is replaced by non-organic material percolated into it by water. This is followed by mineralisation, eventually of the entire object, by molecular replacement of various kinds depending on the bone's original biochemistry. Since fossils represent the primary raw data from which our knowledge of our physical evolution is derived it is worth considering the nature of fossil evidence. For land species fossilisation is a most improbable fate for any given individual. The study of what happens to dead bodies is known as taphonomy, and certain very odd taphonomic processes have to occur to ensure fossilisation. Even then only the hard parts of a body are normally preserved, soft tissue rapidly vanishing.

First the putative fossil has to be buried in alkaline soil, or other deposits. In acid soils bone disappears rapidly. Marine life is more regularly fossilised both because the sea is non-acidic and because remains can sink to deep sea-beds where they stay undisturbed while being slowly buried in sediment. Chalk of course is really solid fossil. Although active burial may occur (e.g. by the action of sexton beetles) only modern humans deliberately inter their dead.

Once buried in alkaline soil, preferably moist if the fossil is not to become fragile, or in some other preserving medium such as a tar-pit, peat bog, heavy fall of volcanic ash or on the soaking stalagmite-forming floor of a limestone cave (most of these being very long shots indeed) the fossil has ideally to remain undisturbed until the palaeontologist comes to extract it carefully from its matrix. Naturally the odds are that any given fossil is too deeply embedded for anyone to know it is there, or that it long ago eroded out of its original fossiliferous bed to be washed away, or that it got quarried along with its limestone or gravel matrix and turned into cement. 'Eroded out' fossils are usually in a somewhat damaged or 'rolled' state which often renders them impossible to date or assess in many important respects. Most carcases are rapidly eaten by scavengers, their bones weathering to dust.

Taphonomic studies by Brain (Brain, 1976) and Ann Behrensmeyer (Behrensmeyer, 1978) have proved how the fossilisation chances of different bones also vary enormously. There are considerable inter-species differences, but among hominids teeth are the most likely to survive, providing the sole evidence for the existence of some species (as we saw initially in the case of Peking Man). Crania, mandibles and long limb bones are next likely to last, but smaller, thinner bones (e.g. the clavicle) are far rarer. Fossils are, in addition, often distorted by earth-movements during their long inhumation. Species living in some habitats or geological contexts may never enter the fossil record, and there is obviously a direct, if imperfect, relationship between size and chances of fossilisation. The more adventures a fossil has the less value it is to the palaeontologist. We have seen how some Chinese fossils nearly reached the apothecary's mortar, and thousands more were never thus rescued. So the chances of a given fossil, hominid or any other, ending up in the hands of a palaeontologist who has methodically extracted it from its primary resting place are millions to one against. Yet in the aeons of geological time billions of animals have died to become, if only partially, fossilised, so the chances of at least one member of a given species being found by a palaeontologist become reasonable, and for some epochs and habitats a near certainty. But even Louis Leakey personally found only around half a dozen substantial hominid fossils.

Hominid fossils prior to *H.sapiens* are rare. This is due to a number of factors: they were probably not numerous anyway (a maximum world population of 1 million at any one time is suggested for *H.erectus* and earlier African hominids would have been far fewer), the earliest were relatively small, their life-style may have led them to die in situations exposed to scavengers, and so on. Also regions of maximum fossil frequency might not actually be regions of maximum population so much as regions where conditions were favourable for fossilisation. As well as bones there are two other significant, extremely rare, forms of hominid 'trace fossil'; footprints and cranial endocasts, both of which are enormously important.

In the light of all this we begin to appreciate on just how narrow a basis palaeontologists have to work, and how tentative their conclusions really are, whatever the fervency with which they on occasion advocate them. The catalogues of hominid fossils published by the Natural History Museum may seem copious at first glance, but most items are mere weathered fragments of teeth and enigmatic bits of skull.

2 *The hominid fossil record*

This can be considered in three sections: the Miocene hominoid precursors of the hominid and pongid (great ape) lineages, the australopithecines (the earliest known hominids), and the subsequent *Homo* fossils. It is assumed that australopithecines evolved from a Miocene ancestor during the Pliocene period, but there is a virtual fossil void between 10 and 3.5 myr, due primarily to erosion outpacing deposition in the areas of proto-hominid habitation (See Plates 2–8 for skulls of principle hominid taxa, and Appendices A and B.)

(a) Miocene hominoids

Although several families of ape-like creatures have been identified during this period, three in particular are considered of importance in connection with human evolution. The smallest and earliest in appearance in the fossil record (20 myr) is *Proconsul*, until recently considered to be an African sub-species of *Dryopithecus*. In 1983 and 1984 the taxonomic status of the two species was to some degree reversed, with *Proconsul* being reinstated as a species in its own right (as Leakey had originally preferred) and the Dryopithecines being considered to be a later family of European and Asian Miocene apes – ranging from Spain to China, via Turkey and the Siwalik Hills of India. Many species of *Proconsul* are found in East Africa, at the Fort Ternan and Rusinga Island sites especially. *P.africanus* is best known from the specimen KNM-RU 7290, discovered by Mary Leakey in 1948, which has recently been subject to reconstruction (A. Walker et al., 1983) arising from the location of two crucial missing cranial fragments among material collected at the site in 1947, and hitherto unnoticed. This has somewhat altered the profile of the specimen, and demonstrates perhaps the insecurity of palaeontological speculation. Its cranial capacity, at around 167.3 cc, represents an encephalisation quotient (EQ, see below, Chapter 4) of 48.8 per cent, slightly more than modern monkeys of similar size, though this interpretation has not gone unchallenged (Leutenegger, 1984). A somewhat larger cousin of this is *P.nyanzae*. Whether *Proconsul* walked along the top of branches, like modern monkeys, or swung beneath them like modern apes, is not entirely clear. Fleagle (1983) suggests that *P.africanus* was a relatively arboreal quadruped, 'the impression one gets is of an arboreal quadruped with some suspensory and climbing abilities but no . . . brachiation' (p. 315). *P.nyanzae*, by contrast, he sees as more terrestrial.

The second group, including *Sivapithecus* and *Ramapithecus*, discovered by Pilgrim in the Siwaliks, is larger than *Proconsul* from which it is assumed to have evolved, and has a more characteristically hominid dental arch (see Appendix B.) Long known from teeth and jaw fragments alone, recent excavations at Lufeng in China under the direction, primarily, of Wu Rukang (Yu-kang in some translations), have brought to light more or less complete skulls ascribed to these species, as well as nearly a thousand teeth (Wu and Oxnard, 1983).

Finally there is a third group, less immediately involved in human evolution but fascinating nonetheless, which is classified as *Gigantopithecus* and has been identified in China on the basis of huge teeth and mandibles. This seems to have survived, perhaps in the niche now occupied by pandas, until as late as ½ myr, long after the other Miocene hominoids had vanished. Or is one still with us? Current thinking is moving increasingly towards the view that orang-utans are descended from *Sivapithecus* (Andrews and Cronin, 1982; Smith and Pilbeam, 1980).

Classifications in this area remain in flux, as well as the changes regarding *Dryopithecus* and *Proconsul*, several authorities are wanting to sink the 'ramapithecines' into the genus *Sivapithecus* (Greenfield, 1979; Andrews and Cronin, 1982). It was long customary for textbooks to grant *Ramapithecus* the status of 'first hominid' (e.g. Pfeiffer, 1978) and this was the favoured hypothesis in R. Leakey and D. Johanson's two recent best-sellers (R. Leakey, 1981; Johanson and Edey, 1981). Although Wolpoff (1982) predicts that a ramapithecine proto-hominid from the fossil void will eventually come to light in Africa, this view of *Ramapithecus* is now rare, for two reasons. Firstly, there is increasing evidence for a late divergence date between hominids and pongids at around 5–6 myr. Traditionally a date twice as long ago was accepted. A leading factor in this change was the research on DNA similarities between humans, chimpanzees and gorillas initiated by Sarich and Wilson (1967) (and discussed later). Secondly, palaeontologists themselves have been reassessing the dental evidence on which traditional interpretations were based and have found their own reasons for expelling *Ramapithecus* from the hominids and accepting a late divergence date (Gantt, 1983; Corruccini and Ciochon, 1983). If the divergence date is revised to 5–6 myr then no

Miocene hominoid can be classified as a hominid. There is a fringe hypothesis that chimpanzees and gorillas descended from australopithecines (Gribbin and Cherfas, 1982), but most authorities now see the problem as identifying a common homo/pongid ancestor from the immediately preceding phase among the late Miocene hominoids. At any rate, the quartet of *Proconsul*, *Dryopithecus*, *Ramapithecus* and *Sivapithecus* disappear around 8 myr and *Australopithecus* appears around 4 myr. Going counter to the current consensus, Schwartz (1984) has argued for a closer relationship of hominids to *Sivapithecus* and their orang-utan descendants than to gorillas and chimpanzees.

The debate is bedevilled with conceptual problems rooted in evolutionary theory. Additionally there is a habitual assumption that *H.sapiens* features are always 'progressive' compared to those of other primates, which may be a subtly anthropocentric error. There may well be instances where *we* have the 'primitive' feature and the great apes the 'progressive', or more neutrally, 'derived' one (Stringer, per. comm.). This pertains particularly to the crucial technical issue of the significance of our thick tooth enamel (see Wolpoff, 1982; Gantt, 1983 for the *Ramapithecus* debate in this connection). It also raises difficulties in interpreting the degree of affinity between hominids and pongids and the timing of their divergence.

Interpretations of the Miocene and early Pliocene hominoid record at present shift virtually by the month. Ciochon and Corruccini (eds) (1983) provides the most comprehensive (and expensive) source for recent discussions of the issues, but even so the debate has since moved on in several areas.

(b) Australopithecines

Dart's discovery of *A.africanus* and Broom's subsequent researches have been described in Chapter 2. Our knowledge of this genus has continued to increase, the most dramatic event undoubtedly being the discoveries of Johanson and Taieb's Hadar (Ethiopia) expeditions from 1973–1975. A full report on the hominid fossils is in the *American Journal of Physical Anthropology*, vol. 57 (1982), while Johanson and Edey (1981) *Lucy – the Beginnings of Humankind*, provides a popular account of their recovery, as well as expounding Johanson's own view of human evolution. The fossil known as Lucy (AL-288 1) is the most complete (approximately 40 per cent of the skeleton) (see Appendix A). The age of the material is estimated at 3–3½ myr. Other non-South African *Australopithecus* fossils have been found at Omo, Laetoli, Baringo, Lake Natron, Olduvai and Lothagam (where a tooth and jaw fragment perhaps 5½ myr has been found, though little can as yet be gleaned from it). In 1977 a string of over 50 fossil hominid footprints extending in two trails for about 23.5 metres was discovered at Laetoli where Mary Leakey was working. These date from 3.6–3.75 myr and seem likely to have been made by an australopithecine species. Later excavation has extended the trail to over 70 footprints. Interpretation of this phenomenal data is controversial (see Chapter 4) but Johanson and Edey's book somewhat exaggerates the similarity to modern footprints. The Laetoli and Hadar finds were classified by Johanson as *A.afarensis*, and it was proposed that these be seen as ancestral to later South African and East African species.

Certainly by 2 myr we seem to be faced with two lineages, a robust and a gracile, with possibly two species of the former represented by *A.robustus* (*Paranthropus*) in South Africa and *A. boisei* (*Zinjanthropus*) in East Africa. The gracile variety, however, disappears while the robust – the earliest from around 2 myr – continues alongside *H.habilis* and early *H.erectus* to perhaps 1 myr. Some hominid fossils of this period are difficult to classify, combining features of both robust and gracile varieties (e.g. the Taungs skull itself). There is a move afoot at the time of writing to revive the term *Paranthropus* to cover all robust species, and to differentiate the South African species into two, re-instating '*A.crassidens*', the term originally given to the Swartkrans remains. (If accepted this would presumably result in the robust species being referred to as *P.boisei*, *P.robustus* and *P.crassidens*, with *Australopithecus* being confined to *A.afarensis* and *A.africanus*.)

The degree of controversy surrounding the genus *Australopithecus* is still considerable, tending to be heightened rather than resolved by successive finds. Participants in the debates now cover virtually the entire discipline of palaeoanthropology, with French, British, US and African workers in often heated contention. Major problems concern both classificatory and behavioural issues. The latter are dealt with elsewhere, but classificatory ones require some attention here. To the outsider many of the points under discussion hinge on highly technical questions about the interpretation of anatomical minutiae: the nature of venous drainage from the brain for example (Conroy, 1982; Falk and Conroy, 1983; Kimbel, 1984), or the shape of the humerus (Senut, 1980; 1981). But there are also underlying problems in what we might call the 'philosophy of classification', such as how species are to be defined, what degree of difference justifies segregation into different species, and the value of cladistic methods (discussed later). As far as *A.afarensis* is concerned, controversy arises because there are clearly two broad classes present at Hadar: a small (to which AL-288 1 belongs) and a larger one. Some morphological differences can be accounted for by size alone. Simple increases in size are necessarily associated with certain changes in morphology, or form, as well, a phenomenon known as 'allometry'. The best, and longest known, example is the increase of bone thickness as a cubic function of linear size (since the volume of an organism, which the bones support, is a cubic function of linear dimensions) (R. Gould, 1980; Harvey and Clutton-Brock, 1983). Many of the differences among *A.afarensis* specimens can be explained this way, indicating that they were a single, sexually dimorphic, species. Other differences, including tooth-shape and some anatomical features of the lower limbs (Stern and Susman, 1983) are not so readily explicable. There has thus been a controversy (unresolved at time of writing) in which Johanson, Kimbel and White are defending the original 'one-species' position against Olson and Senut.

A second matter, the degree of similarity between *A.afarensis* and *A.africanus* pits Johanson and his supporters against veteran South African palaeontologist Philip Tobias, an old associate of Louis Leakey (Tobias, 1980a). At present the former appear to be more successful in that the term *A.afarensis* is that normally used in the literature to refer to their finds. Though this might partly be for convenience rather than implying a genuine substantial distinction, cladometric studies tend to confirm that the fossils attributed to the two species can be differentiated (e.g. Corruccini and McHenry, 1980). The '*robustus*' = '*boisei*' controversy dates from Louis Leakey's original announcement of the latter species. Many palaeontologists, including Johanson, tend to view Leakey as having been a rather excessive 'splitter', creating new species for new fossils on the strength of what others considered only minor morphological differences. On this particular point the consensus now appears to be in Leakey's favour, the two being fairly closely related sub-species of robust australopith, but *A.boisei*'s tooth size considerably larger than that of *A.robustus* (Wood and Stack, 1980). As already mentioned, the taxonomy of this group is again in flux.

It is sometimes difficult to evaluate how far these controversies involve non-scientific psychological factors such as the tendency of palaeontologists to want to ascribe uniqueness to their own finds or long-standing professional rivalries (e.g. a certain level of animus towards Johanson for his general research style and jealousy of his finds). In the meantime, as with Miocene hominoids, the classifications within the australopithecine group must be considered as being as yet unsettled, and readers should be alert to ongoing changes if they are to keep track of the literature.

(c) *Homo habilis*

Although B. G. Campbell (1978), an opponent of the classification since 1964 (Day, 1977), proposed splitting this group between *A.africanus habilis* and *H.erectus habilis* according to whether they were early or late respectively, Eldredge and Tattersall (1982) believe in retaining the category on pragmatic grounds, and Campbell's suggestion has remained for the most part unadopted (although it is accepted by Tanner (1981)). The first *H.habilis* fossils, coming to light at Olduvai in the wake of *Zinjanthropus*, provided Louis Leakey with the opportunity to salvage his non-toolmaking view of australopithecine behaviour. The number of fossils ascribed to this species is small, and largely confined to a limited area of East Africa, although there are claims for finds at Omo (Ethiopia) and Sterkfontein (South Africa) too. The best known is KNM-ER 1470, at 2 myr (Plate 4). The key factor in favour of the classification was that these specimens had a greater brain size than the australopiths and hence, by inference, were the probable makers of the stone tools found in their source beds. For many writers, such as the late Kenneth Oakley, a proven ability to make tools has represented the defining characteristic of human status, the 'Rubicon' of hominisation (Oakley, 1972). In this context the desire to retain *H.habilis* as a separate species is given added weight beyond the strictly anatomical data. Yet connections between behavioural and morphological evolution are far from clearcut, and rarely perfectly correlated in a simple fashion.

Although the *H.habilis* debate is somewhat quiescent at present, Day's opinion that 'it would be premature even now to say that it is resolved' (Day, 1977, p. 151) remains valid. *H.habilis*'s fate probably rests on the eventual outcome of the theoretical battle between 'gradualist' and 'punctuationist' models of evolution (see below). The latter would tend to favour retaining it as an identifiable species, whilst for the former the tendency would be to see it as a transitional form between *Australopithecus* and *H.erectus*. The heyday of the species was around 1.8 myr – its cranial capacity, at an average of 659 cc, falls about halfway between the gracile australopiths' 451 cc and the early *H.erectus*'s 942 cc (Blumenberg, 1983, p. 590). Stringer (1986) has recently detailed the morphological grounds for a genuine *H.habilis* grade between australopiths and *H.erectus*, although some fossils often ascribed to it must be considered doubtful (KNM-ER 1805, KNM-ER 1813, O.H. 14 and O.H. 16).

(d) *Homo erectus*

These fossils span a great range, temporally, geographically, and to some extent morphologically too. A mandible from Sangiran in Java is perhaps 1.6 myr (although its classification is not entirely settled) and presumably it is with *H.erectus* that the global expansion of the hominids began, possibly even before 2 myr. As Stringer (1984b) stresses, however, the assumption that there is a simple descent from *A.africanus*, through *H.habilis* to *H.erectus* is probably far too naive a picture and hard to reconcile with all the fossil data. Furthermore, the origin of *H.erectus* could as well have been outside Africa as within it, even though the number of early fossils elsewhere is less, there is no evidence that *H.erectus* was in Africa earlier than in China or Java. The earliest fossil which may be ascribable to *H.erectus* is KNM-ER 1481 at around 2 myr (Day et al., 1975; Kennedy, 1983) though this is disputed, Trinkaus (1984) seeing no reason why it should not be ascribed to *H.habilis*. Stringer (1984b) in a wide-ranging review finds (*contra* Eldredge and Tattersall (1982)) that *H.erectus* is far from displaying long-term homogeneity and that continuous changes over its nearly 2 million years of existence can be demonstrated (including brain-size). The species begins to give way to more advanced forms at around ½ myr, finally disappearing at around 0.2 myr (Andrews, 1984). In October 1984 the most complete *H.erectus* skeleton yet found was announced by Richard Leakey, it having been discovered the previous year. This specimen, WLT 15,000 from Lake Turkana, has been dated at 1.6 myr and is of a boy about 1.6 metres in height, suggesting a mature height as great as 1.82 metres, or 6 foot. This is considerably taller than had hitherto been imagined for this species, which has conventionally been supposed to have fallen towards the bottom end of the current normal human range at around 5 foot, though robust. In due course this find may shed much light on such questions as *H.erectus*' gait and biomechanics.

There can be little doubt that *H.erectus* initiated the Acheulean tradition of stone tool manufacture. It has been suggested that they roamed in small hunter-gatherer bands of about 25 members, with as many as 40,000 such bands covering Europe, Asia and Africa (Pfeiffer, 1978), though in fact their presence in Europe is not entirely firmly established, at least as anything other than being very peripheral to their main areas of occupation. They certainly tamed fire to some extent, since the earliest fire-sites occur prior to the appearance of the transitional forms of 'Archaic' *H.sapiens*. Even so the use made of these, and whether they were made by *H.erectus* or opportunistically started from natural fires, is unknown. (See Chapter 6, Table 6.2.)

The principal mystery confronting those researching *H.erectus* is where and when it finally gave rise to *H.sapiens*. Andrews (1984) argues that Asian *H.erectus* (Peking Man) represents a lineage which became extinct, but others have maintained that the transition from *H.erectus* to *H.sapiens* occurred *in situ* across its range, a view bolstered by the widespread distribution of the transitional forms from Europe to Africa and the Far East, Thorne and Wolpoff (1981) making out this case for the Australian material. The two schools of thought, those believing in a 'single point of origin' and those espousing this widespread transition, can each invoke features of the evidence in their support – or at least produce arguments against supposed refutations. We return to this issue in Chapter 6. For the moment it should be noted that after ½ myr we meet fossils which are markedly different from *H.erectus* in some respects while not in others. One or some or all of these variants presumably led to *H.sapiens sapiens* and that or another to *H.s.neanderthalensis*. The number of sub-species allocated to the *H.erectus* group is greater than for any other hominid species; even leaving *H.e.habilis* aside, B. G. Campbell (1978) gives six, as opposed to five, *H.sapiens*.

(e) Archaic *Homo sapiens*

This term was introduced by Stringer (1974) to refer to the awkward clutch of fossils showing features transitional between *H.erectus* and anatomically modern humans, on the one hand, and Neanderthals on the other. It includes the famous Swanscombe skull, the Heidelberg jaw, the Greek Petralona skull, the Steinheim skull, the Vertesszöllos remains from Hungary and others from Africa (Broken Hill, Ngaloba and Bodo for example), and China (Da-li). Since they lack the specialist, derived features acquired by the Neanderthals, these skulls are in some respects closer to modern *H.sapiens sapiens* than are the latter, which indeed diverge more markedly from ours towards the end of their record (Stringer, 1974). It is unclear what the nature of this transitional material really is, whether it represents one main 'stem' evolving from *H.erectus*, or several running parallel; as just noted this is a central mystery in the whole human evolution topic (see Appendix C). The earliest anatomically modern human remains come from Border Cave, South Africa, though their dating varies considerably between 50,000 and 90,000 years ago. The fact is that in Europe and South West Asia, but not in Africa or the Far East, *H.s.sapiens* of a modern type appears after Neanderthal forms which most (though not all) authorities consider unlikely to have been its direct ancestors. The fate of Archaic *H.sapiens* forms is thus unclear and more fossils need to see the light of day before it becomes clearer. Poulianos (1982) considers the Petralona skull to be a *H.erectus* sub-species and uses the classification *Archanthropus europaeus petraloniensis*. This usage must be considered eccentric. The Chinese are claiming, on morphological grounds, that *in situ* evolution from *H.erectus* to *H.s.sapiens* occurred there (contrary to Andrews, 1984), lending support to the 'widespread transition' theory. This is discussed further in Chapter 6, as is the Neanderthal question.

(f) Neanderthals

I will here confine myself to summarising the fossil evidence. Neanderthal fossils are relatively common (about 350 individuals are known), although they are scarcer for the earlier phase of their dominance which lasted from around 100,000 years ago (the date of the Krapina and Saccopastore material) to their demise around 35,000 BP, when modern humans replaced them. Before 0.1 myr a few Neanderthal fossils are known, the rear half of a skull from Biache, in France and a mandible from La Chaise, also in France. The best-known later Neanderthal sites are La Chapelle-aux-Saints, La Ferassie, La Quina, Shanidar (Iraq) and Tabūn (Israel). The often neglected East European material has recently been reviewed by F. H. Smith (1982) and accessible summaries of the whole topic are a *Scientific American* article (now rapidly dating) by Trinkaus and Howells (1979) and one in *Natural History* by Stringer (1984a).

The anatomical difference between Neanderthals and 'Moderns' is greater than between any two living ethnic groups, while internally they exhibit relatively little morphological variation other than a degree of sexual dimorphism similar to that of modern human populations (Heim, 1982a; 1982b; 1983). Whether, as has been alleged, their appearance was sufficiently similar to ours for them to have passed unnoticed in a New York subway is a matter of debate. It might, anyway, say more about New York subways than it does about Neanderthals. Pfeiffer (1982), basing his picture primarily on work by Trinkaus, describes them as having 10–20 per cent heavier bones than us and a hand-grip 2–3 times stronger, the strongest being able to lift a ton. Distinctive muscular features enabled them to maintain fine control without loss of strength e.g. in tool-making, but health-wise they were vulnerable to arthritis and 30 per cent show signs of injuries. Pfeiffer also pictures them trekking 20 miles a day through snow, fully laden, and without snow-shoes.

Transitional Neanderthal-Modern forms have proved extremely elusive, reinforcing the view that Moderns replaced them rather than descended from them. Some Near Eastern fossils have been interpreted on occasion as transitional and even now, fifty years after their discovery the classification of the Tabūn and Skhūl finds is sometimes queried, though the consensus is that the former are Neanderthals and the latter *H.s.sapiens*. At one time it was widely suspected that western European Neanderthals were more distinctively robust and extreme in appearance than were oriental ones, as a physical adaptation to the peri-glacial Ice Age climate. This has not been borne out, the Shanidar Neanderthals from Iraq being as robust as any (Solecki, 1971). Stringer (1984b) does not rule out climatic adaptation as a factor in the Neanderthal morphology, however, and it is possible that during their most expansive phase they could have moved *out* of this colder region into warmer climes without losing their cold-adaptive features. The qualifiers 'early', 'Classic' and 'Late' are commonly used and contrary to Trinkaus and Howells' account (1979), Stringer (1982) continues to argue for the progressively increasing morphological divergence from non-Neanderthal forms mentioned earlier.

Without entering here into the various theories about the Neanderthal's rapid extinction, the advent of *H.s.sapiens*, and the relationship between them, it is noteworthy that the gradualist v. punctuationist issue is a major underlying factor. Both here, and in the earlier *H.erectus* to Archaic *H.sapiens* transition the fossil data is too equivocal for either the gradualist or punctuated equilibria schools to emerge victorious. Stringer at present prefers an African origin of *H.s.sapiens* with a progressive northward replacement of Neanderthals (per. comm.) with a limited degree of interbreeding (or gene-flow). (Little credence has yet been given by those in the field to Gooch's (1977) theory that Moderns emerged from a hybridisation of Neanderthals and another *H.sapiens* branch.) Given the morphological distance between Neanderthals and Moderns, the interfertility between them is likely to have been low; however, some preservation of Neanderthal genes may well have occurred.[1]

The association of Neanderthals with Mousterian stone industries is less clearcut than once thought. Post-Neanderthals have been found associated with Mousterian industries in the Near East, while at St Césaire a Neanderthal has been found in the context of a post-Mousterian, Chatelperronian, industry (Lévêque and Vandermeersch, 1981). Again, behaviour and biology are not precisely linked.

The whole question of the relationship between Archaic *H.sapiens*, early anatomically Modern *H.sapiens* and Neanderthals is technically exceedingly complex. At bottom there is a hetereogeneous set of more or less battered skulls and post-cranial bits and pieces which can only be analysed morphometrically to see which most resemble which and which are least alike. Lineages may be tentatively discerned and the fossil-pack shuffled in various permutations, but due to the small numbers involved at any data point such analyses are notoriously sensitive to additions of new finds and revisions of dating – evolutionary 'trends' are prone to disappear and regional differences evaporate. Even the major, clearly defined, Neanderthal fossils from places like Krapina, Shanidar and La Ferassie differ from one another to some extent. 'Neanderthal' does not refer to a uniform species but a variable group of fossils sharing certain diagnostic features which can most readily be explained as derived or specialised characteristics indicating descent from a common ancestor, and absent from others. Some of these though, such as the famous heavy brow-ridges, are differences in degree, and problems arise when specimens display 'intermediate' levels of such features. The same is even more true of the Archaic and Early anatomically modern groups: the skulls are few and of scattered provenance.

(g) *Homo sapiens sapiens*, 'Moderns'

The replacement of the Neanderthals by modern humans in Europe is marked by the appearance in the fossil record of 'Cro-Magnon Man'. Subsequent European *H.s.sapiens* groups show a degree of ethnic variation that led earlier writers into seeing them as directly ancestral to some modern groups (see Chapter 2). This is now generally considered to be mistaken. The question as to when modern ethnic groupings *did* originate is a difficult one, and bound up obviously with the aforementioned problem of whether the *H.erectus* to *H.sapiens* transition was focused or widespread; if the latter, then ethnic diversity has its roots far further back than it would in the former case. Some preliminary light is beginning to be shed by the use of studies of mitochondrial DNA and other biochemical techniques for gauging genetic affinities.

On balance, though not on all indices, Europeans appear closer to the

[1] Mitochondrial DNA research now suggests this to be highly unlikely.

Mongolian (oriental) group than either do to Africans, hence presumably dividing later. Not much weight can be given to this as yet and it must be stressed that the interpretation of the mitochondrial DNA evidence given in Gribbin and Cherfas (1982) is now considered unreliable, or at best premature. What we do know is that *H.s.sapiens* was in Australia and New Guinea by around 40,000 BP and in the New World by 20,000, and perhaps earlier. Even the remotest, most isolated, tribal societies are thus relatively recent arrivals in their present homelands when viewed on an evolutionary time-scale. The first 'civilisations', defining this by the existence of permanent cities, only appear around 7–8,000 BC at a number of sites in the Indus Valley, Turkey and the Middle East (Jericho). The time-lag between different regions in this respect is almost entirely explicable by environmental circumstances, although the mechanisms by which culture spreads are not uncontroversial, as we will see in due course.

'Modern' fossils are highly numerous and their value is somewhat different from that of earlier ones. They enable us to enter into 'palaeodemographic' and 'palaeogenetic' investigations of such factors as life-expectancy, health, sex-ratios, sexual dimorphism, genetic drift and migration. Cultural differences in treatment of the dead seriously affect this research – obviously a society practising cremation leaves fewer remains than one which mummifies its dead!

Although a few fossils with 'Modern' features have been found in sub-Saharan Africa predating their European debut, none are elsewhere known to be contemporary with Neanderthals prior to the 'Upper Palaeolithic transition' of *c.* 35,000 BP. Their exact route of descent from *H.erectus* thus remains in an obscurity that new finds might at any time dispel.

On this data our knowledge of the course of our physical evolution is ultimately based. Though expanding this information is by its nature unpredictable and grows slowly. Until the Neanderthals the quantity is small. The challenge is thus to devise methods of analysis which will maximise information yield. Dating has already been dealt with and, as far as fossil bone is concerned, direct dating is confined to relatively recent material amenable to amino acid racemisation and C-14 dating. Dating of pre-*Sapiens* hominid fossils is therefore invariably indirect. (Figures 3.2 and 3.3 show geographical distribution of principal hominid fossil sites.)

3 *Analysing fossils*

At this point various disciplines meet and the methodologies arising from their interaction are frequently both ingenious and illuminating. Fossil material being usually deficient both in quantity and quality, the task of

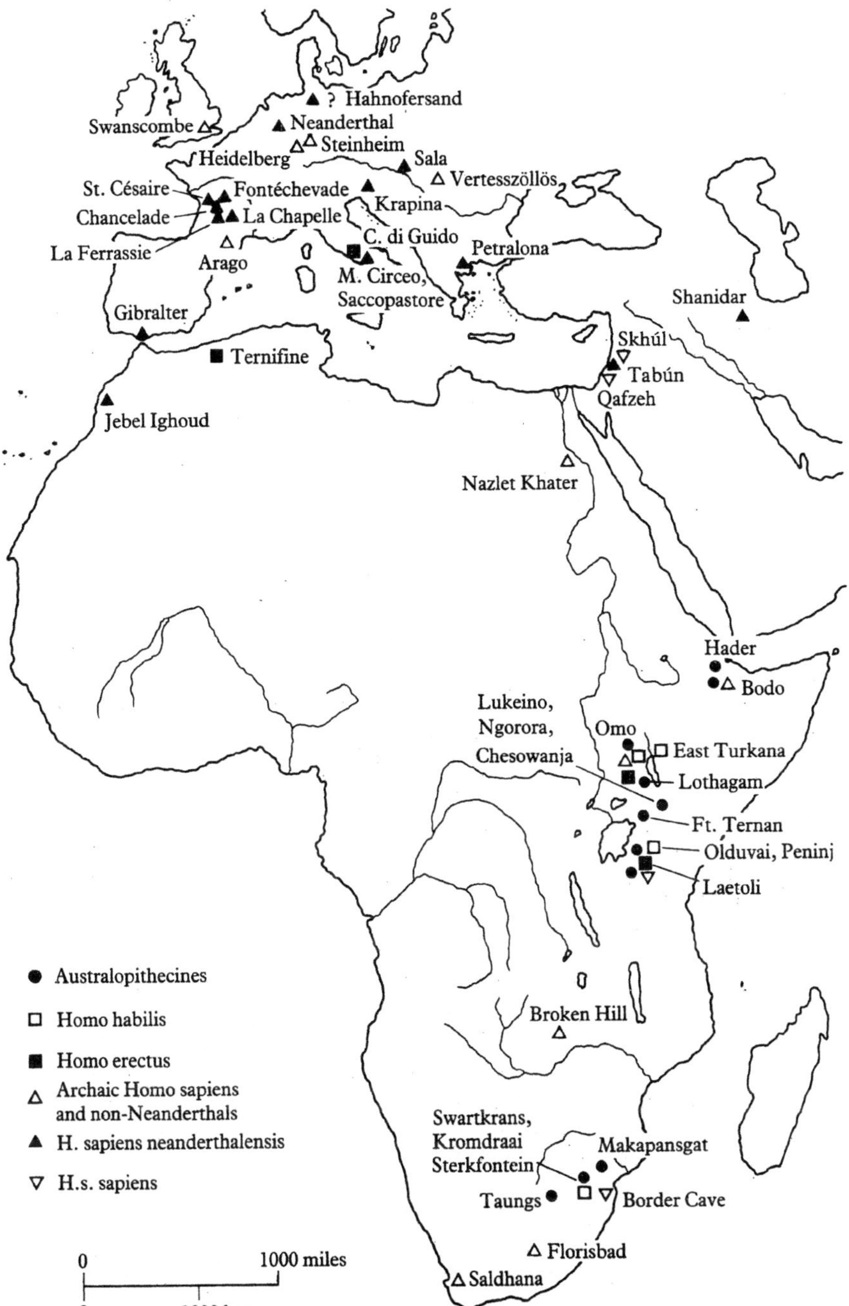

Figure 3.2 Hominid fossil sites of Africa, Europe and Near East.

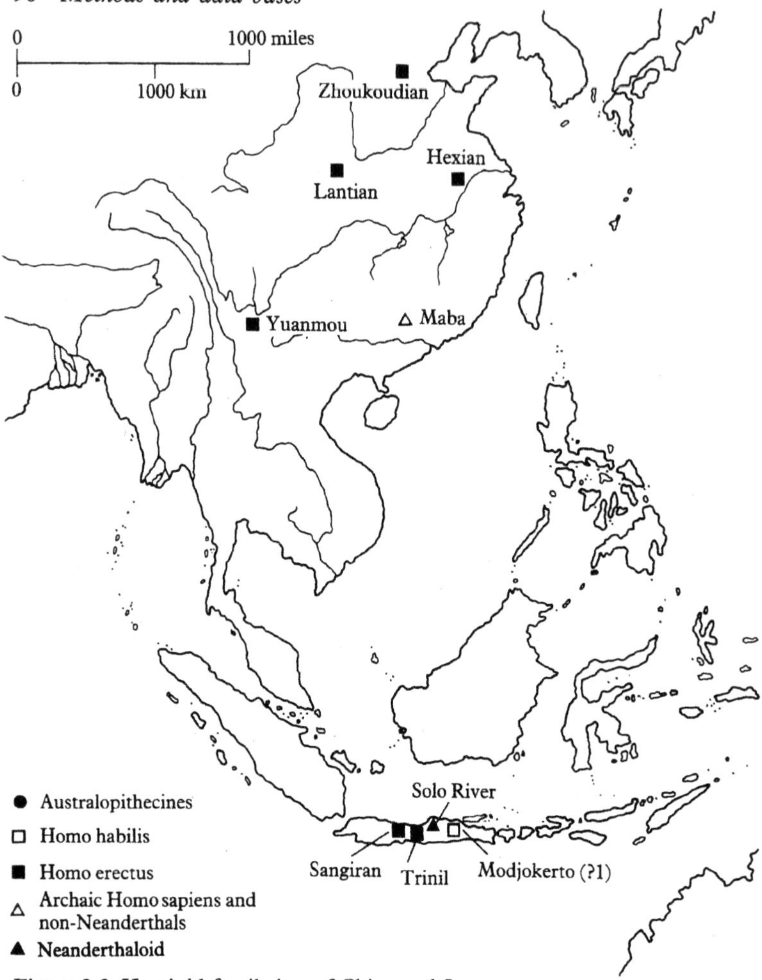

Figure 3.3 Hominid fossil sites of China and Java.

extracting the maximum information from it has long taxed palaeontologists. Until quite recently they had nevertheless to rely almost exclusively on readily quantifiable overt morphological features, combined with essentially qualitative interpretations of what these characteristics implied, derived from broader understanding of comparative anatomy. This is still the starting point of most fossil analyses and has reached a pitch of considerable sophistication, having been evolving since the nineteenth century. Of special interest are attempts to reconstruct gross anatomy and even appearance on the basis of very limited samples of skeletal material. It is, then, supposedly, an easy step to reconstruct behaviour from appearance. At their most extreme, palaeontologists will conjure up a whole species' life-style on the basis of a tooth or two. Although the orthodox study of fossils in terms of traditional comparative anatomy remains at the heart of contemporary research, there is increasing caution as to how far reliable behavioural inferences can be drawn. Furthermore, the more 'generalised' a fossil species is (and *Homo* is *very* generalised) the less clearcut the connections between physiology and specific behaviour. A snake can only wriggle across the ground but healthy humans can hop, skip, jump, walk, swing from branches, crawl, swim, slide down slopes on their bottoms *and*, at a pinch, wriggle across the ground.

Inevitably, theoretical as well as practical issues are involved in framing research procedures, since it is the researcher's theoretical position which provides the rationale for trying to answer the questions he or she is addressing. Method and theory are, as ever, bound up with one another. Let us turn briefly to some of the other methods now being adopted.

(a) Morphometrics and cladometrics

To clarify this terminology first: morphometrics refers to the use of advanced statistical techniques to analyse morphological measurements. Cladometrics is a particular use of this in the context of an approach to classification called 'cladistics', and most morphometric studies on human evolution are done in this context. Cladometrics is concerned with analysing the degree of similarity between different species. Modern computing techniques enable the analysis of far greater numbers of variables than was traditionally possible, revealing covert patterns of correlation which would once have remained unspotted.

The morphometric approach tends at present, then, to be combined with a theoretical (though not uncritical) commitment to some version of cladistics as a guiding classificatory principle. Simplifying somewhat, cladistics uses the overall level of similarity between species, rather than tracing lineages via particular inherited features, as a basis for classification and an index of affinity. This is an extremely controversial topic in theoretical biology and several schools of thought are in contention (see Ridley, 1983, for an introduction to the controversy). The object of the exercise is thus, as far as we are concerned here, to provide measures of mutual affinity or distance among fossil specimens both to identify the number of species present and to compare them. An important aspect of cladometrics is that it tends to endorse a punctuationist rather than gradualist evolutionary model (Wood, 1981). Some extremer cladists have eschewed evolutionary theory altogether, but these have not yet made themselves felt in the human evolution area. The aim of cladistics is to produce a theory-neutral system of classification to replace the traditional phylogenetic of evolutionary taxonomy which is derived directly from evolutionary theory. At this level it is dubious on philosophical grounds, but in practice cladistics has proved highly useful pragmatically in testing hypotheses about species affinities.

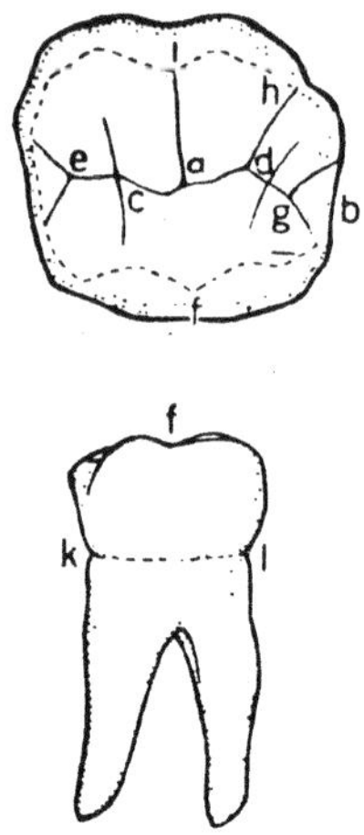

Figure 3.4 Identification of 11 'landmarks' as a basis for comparison between fossil teeth of different species. Teeth are the most commonly preserved part of the anatomy, the pressure to find ways of exploiting the dental evidence to the full is very great.

A right lower molar. (a) Central occlusal fovea; (b) distal marginal ridge opposite the distal fovea; (c) distal trigonid crest's intersection with longitudinal developmental groove; (d) distal talonid crest's intersection with the longitudinal groove; (e) mesial fovea; (f) metaconid-entoconid ridge intersection with the lingual transverse groove; (g) distal fovea; (h) hypoconid-hypoconulid ridge intersection with the distobuccal transverse groove; (i) protoconid-hypoconid ridge intersection with the mesiobuccal transverse groove; (k) and (l) mesial and distal-most points on the cervix (cemento-enamel neck).

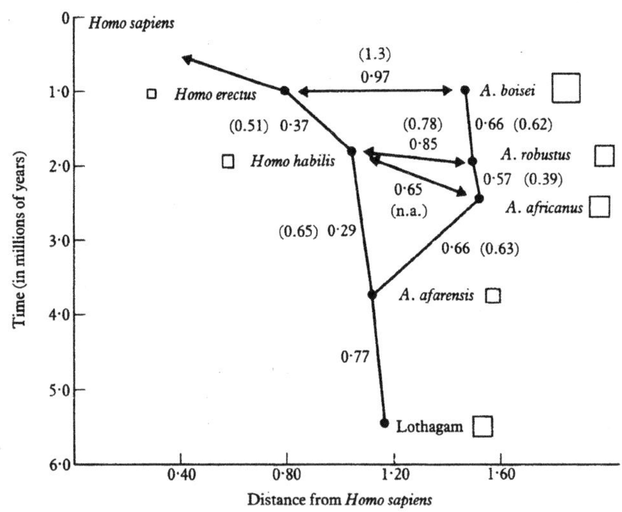

Figure 3.5 Hominid taxa distributed according to Penrose shape distance from *H. sapiens* (horizontal axis) and geological age (vertical axis) based on the first molar and mandible. The sides of the squares by each species are one-tenth of the Penrose size distance from modern man. Values in parentheses are those given in McHenry and Corruccini, 1980.

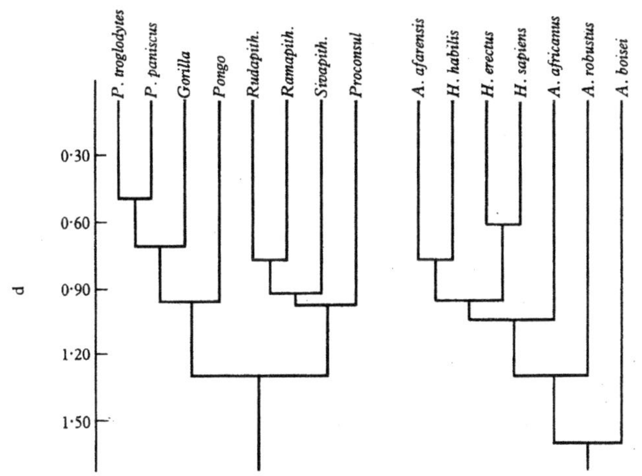

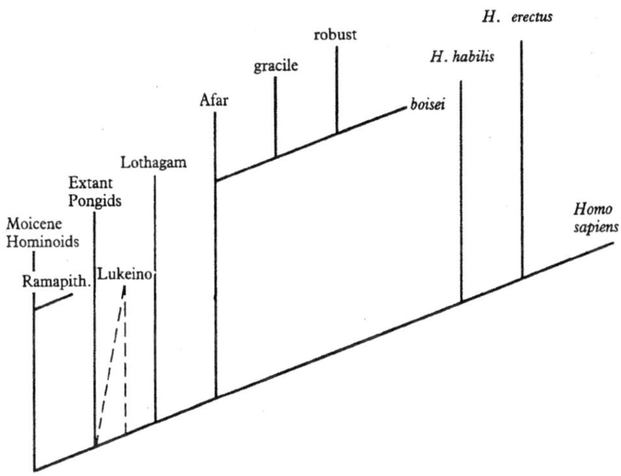

Figure 3.6 Phenogram based on Pythagorean distances among hominoids using the entire mandible. This phenogram, although presented in the form of a genealogical tree, is only a graphic representation of the outcome of a statistical analysis of the overall distances species are from one another, based on morphometric data from mandibles (including teeth). A different data-base might give a different picture. Thus such findings, although suggestive of the actual genealogical relationships, are not in themselves conclusive.

Figure 3.7 Cladogram of hominid relationships. From Corruccini and McHenry, 1980. A cladogram is based on the number of shared morphological features, the way these cluster being depicted, as here, in tree form. Again, this does not claim to be showing actual lines of descent.

A study by two leading exponents of morphometric technique, Corruccini and McHenry (1980) will illustrate the nature of this approach. In this paper attention is focused on three teeth. These are very special teeth because they come from that enigmatic Pliocene fossil void. They consist of a 10–11 myr upper molar from Ngorora (which though described in the introduction figures little in the remainder of the paper), a 7–5.4 myr lower molar from Lukeino, Kenya, and the two lower molars in the Lothagam mandible (5.5–5.0 myr) mentioned above, though only one of these is sufficiently complete for inclusion in the study.

Not very promising material on which to base hypotheses regarding the pre-*A.afarensis* course of protohominid evolution! Nevertheless, starting with 11 'landmarks' (see Fig. 3.4) the authors were able to obtain up to a dozen variables per tooth (the basis on which they were selected need not concern us here, but broadly they were traits known to be relevant to homo-pongid differentiation). They then measured a large comparison sample of *H.sapiens* and living primate molars, plus as many of the other known hominid molars as they could. Two figures from this paper may be taken as representative of the kinds of results which morphometric analyses provide as well as summarising the author's own conclusions. In Figure 3.6 we see that their general picture of hominoid material is that it falls into two discrete clusters of hominids and pongids respectively, with progressive internal differentiation. Figures on the vertical axis are a measure of distance between 'clades' termed the 'simple Pythagorean distance', thus 'gorilla' and 'pongo' are approximately 0.95 apart. Distance between the two clusters, being greater than 1.8, is beyond the figure's range. In Figure 3.5 the relationship between the Lothagam specimen and other hominids is depicted, illustrating the conclusion that the australopithecine and *Homo* lineages are probably separate after *A.afarensis*, with the Lothagam individual being ancestral to all, but closer (at just under 1.2 on the Penrose shape distance coefficient) to *H.sapiens* than to *A.africanus* (at about 1.4). On this measure (Penrose coefficient), too, though not shown here, chimpanzees differ more from the australopithecine line (1.7 from *A. boisei*) than they do from the *Homo* line (1.29 from 'early Homo' and 1.33 from *H.erectus*). The Lothagam specimen is in turn closer to both chimps and orang-utans than the australopiths are. The upshot of the paper is to conclude, among other things, that *A.afarensis* is probably ancestral to *Homo* (i.e. we did not descend via *A.africanus*) and that the pongids split prior to the Lothagam mandible, and possibly prior to the Lukeino specimen too (Fig. 3.7). The status of *Ramapithecus* as a hominid is also weakened.

None of these conclusions need be accepted on the basis of this paper alone; indeed the authors gave rather different distance measures in a version of Fig. 3.5, published a few months earlier (McHenry and Corruccini, 1980). It should perhaps be emphasised that the 'cladograms' with which much of the current literature abounds, especially that involving these cladometric statistical techniques, do *not* constitute hypothesised lines of descent as much as the nearest statistical descriptions of inter-specific affinities on the particular feature or set of features being studied, which might be anything from tooth morphology to skull shape or non-anatomical data such as immunological characteristics. A use of this approach comparing living primates only and the relationship of *Homo* to them is Oxnard (1981). Any conclusions as to actual descent are interpretations *from* this data, though of course the implicit assumption is that the closer the affinity, the closer the genealogical relationship. But this is complicated by the issue, raised before, of whether features are 'primitive' or 'derived'.

(b) Wear analysis

Still with teeth, our most commonly fossilised bones, we are learning to glean information about diet from the nature of tooth-wear itself. Similar techniques of micro-wear study are also shedding light on tool-use and we might for convenience discuss both here, after all many would view tools as initially substitute or supplementary teeth! Palaeontologists have long held that tooth-form and diet are directly correlated and they are obviously broadly right: the grinding teeth of browsing herbivores are quite distinct from the carnivore's ripping dental equipment. But at a finer level of analysis earlier easy inferences have been modified. The study of tooth-wear showed, for example, that the diet of the larger-toothed robust australopiths did *not* differ appreciably from that of smaller-toothed gracile species. Among leading researchers in this area are Alan Walker and Al Ryan. Their underlying assumption is that different diets produce characteristic forms of wear identifiable under the microscope. These include polishing, pitting, scratching, and micro-flaking. Ryan has also shown that different sorts of tooth-usage (grinding, nipping etc) produce characteristic wear patterns (Johanson and Edey, 1981, pp. 358–60), a phenomenon studied in some detail for earlier hominids by J. Wallace (1978). Caution is nevertheless needed as recent experimental work has shown fairly minor dietary variations to produce substantial changes in tooth-wear pattern: (Kay and Covert, 1983). Not all hominid tooth-wear is due to diet, the peculiar wear patterns of some Neanderthals (e.g. La Ferrassie) are probably due to the use of teeth for chewing animal hide, in order to soften it, as done by modern Eskimos.

Wear-analysis of tools is on somewhat sounder ground since it can be rigorously experimentally based. Usage of facsimile tools on different materials can be undertaken as extensively as necessary in order to establish a clear picture of the typical wear-pattern produced by e.g. meat-cutting, wood-working, bone-working, or skinning and hide-scraping. The principal researcher here has been L. H. Keeley (Keeley, 1974, 1980;

Newcomer and Keeley, 1979). Attention has focused primarily on stone tools but an important extension of these techniques has been the study of cut-marks on fossil bones which has produced some astonishing findings (see Chapter 4 below). Binford (1981) has attempted the ambitious task of analysing the nature of bone-wear and bone-assemblage composition in general with a view to placing site-classification on a firmer basis.

(c) Gait analysis

The discovery of the Laetoli footprints gave a great impetus to the study of how gait could be gauged from such data, although anatomical studies of the locomotor implications of hip and knee-joint morphology have long played a part in the study of the origins of human bipedalism. (See Jenkins (ed.), 1974 on primate locomotion.) An already classic paper by Day and Wickens (1980) compared the contours of modern human 'soft ground walking footprints' with those found at Laetoli and concluded that the makers of these 'transmitted their body weight and the forces of propulsion to the ground in a manner very similar to that of modern man' (p. 385). Stern and Susman (1983) have been more cautious, finding these footprints to lack the characteristic human 'ball' below the big toe and also signs of toe-curling during walking, unlike modern humans who splay their toes as they land. Such features as stride-length in relation to body-weight and height, and weight distribution on the foot are all currently under investigation. The leading authority on primate gait is probably Russell Tuttle of the University of Chicago. *Australopithecus* gait is at present a most heated area of debate, with the Stern and Susman Stony Brook (State University of New York) school opposed to Kent State University and Johanson, with the French also involved (Lewin, 1983).

It was not until the 1960s that the true antiquity of walking came to be widely appreciated, and even then Lucy's bipedalism was a considerable surprise. Traditional studies of fossil bones, combined with behavioural experimental studies and footprint analysis, might yet clarify outstanding enigmas regarding early hominid gait such as how far it resembles our own and how it evolved from prior modes of locomotion. *Why* it evolved though is still obscure (see Chapter 4 below).

(d) Endocast studies

Casts of the inside of the cranium have long been known to occur naturally, and the first Taung fossil included a very substantial cranial endocast leading Dart to speculate about the owner's brain structure. Such chance occurrences are rare, although their value as evidence for neurological evolution has been appreciated for many decades. The Russian palaeoneurologist Kochetkova provided a review of this material in her posthumously published treatise *Paleoneurology* (1978). To some extent progress in this field has had to await advances in our understanding of brain functioning which have only been made in the last decade or so.

Since around 1970 great technical improvements have been made in obtaining artificial endocasts of latex from hominid crania, Columbia University's Ralph Holloway leading this mode of research. Hitherto the study of brain evolution has focused primarily on size, standard methods for measuring this having long been available. Important though this is, providing us with the 'EQ' (encephalisation quotient) score, it sheds no light on changes in organisation or structure of the brain. By providing us with information about the surface features of the brain, insofar as the internal skull surface reflects them, we can begin to obtain at least a glimmer of insight into this. But further discussion of brain evolution can be postponed until the next chapter.

By its very nature fossil material presents scientists with a tantalising challenge. The response to this has been to elaborate a whole array of techniques, only a few of which I have sketched above, to maximise information retrieval from these battered remnants of our predecessors. Each sheds its own light on a narrow band of problems, but their sum total is managing to span an increasingly large portion of the spectrum.

D Artefacts

The second great source of data on our evolution are our ancestors' artefacts. In some respects they are more directly relevant to psychological evolution than most anatomical data, since they embody actual behaviour. For the greater part of human evolution stone implements alone survive, along with the wastage accompanying their manufacture. They are supplemented, though rarely prior to the Upper Palaeolithic, by evidence of occupation sites in the form of rocks arranged as walls or windbreaks, post-holes and hearth-sites. Towards the Upper Palaeolithic artefacts in other materials begin to creep into the archaeological record; horn, bone, shells, and, very rarely, wood, while fragments of ochre might imply activities involving colouration, perhaps body-painting. Only after 35,000 BP do cave art, painted stones and carvings appear, while by the Mesolithic and Neolithic a whole array of wooden and, in the Neolithic, pottery objects or their fragmentary remains, enter the picture. This means that for the majority of the period covered by our physiological evolution we are effectively reliant on stone artefacts for testimony as to our ancestor's behaviour, the rare additional evidence acquiring enormous importance. This absence of non-lithic artefacts prior to the Upper Palaeolithic is not of course due to their absence from our forebears' repertoire, but because organic material simply did not survive. Wood and other vegetable materials undoubtedly figured largely in the daily lives of both *H.erectus* and the australopithecines. Animal remains such as pelts, feathers, teeth and bones were probably utilised, we cannot know how. In later chapters we will see how evidence other than stone artefact data can also be incorporated into the study of behavioural evolution, but for the moment we will concentrate on the stones.

1 *Stone artefacts*

Not the least of the problems presented by the study of stone tools is knowing that what one has *is* indeed an artefact. Natural processes can produce objects indistinguishable from simple choppers and flake blades. When a number of such objects are found in close proximity, however, the chances of their all being freaks of nature rapidly drop to zero. The oldest known tools date from around 2½ myr but a recent paper claims the discovery of a 'crude unifacial hand-axe pebble tool' of 9–10 myr, and suggests it was made by *Ramapithecus* (Prasad, 1982). This raises the ghosts of the 'Eolith' controversy mentioned in Chapter 2. Regular manufacture of stone tools is ascribed usually to *H.habilis* and certainly from over 2½ million years ago to a few thousand our predecessors must have engaged in the virtually daily activity of fashioning them. The output of axes from the Grimes Graves site in Norfolk was probably several million, while the flint mines of north-west Europe reached an output of around 400,000 axes per annum (Coles et al., 1978). Even allowing for the fact that for most of the period total world population was well below a million, this is a lot of stone tools over 2½ million years, especially when one considers that a skilled flint-knapper could fashion a hand-axe in a couple of minutes (it requiring as few as 25 blows). While fossils are rare, stone artefacts are therefore numerous. Reporting from the Middle Awash Valley area of Ethiopia, researchers write: 'From Gargufia to Subalealo alone, a distance of 20 km, individual tool occurrences are virtually continuous and number in the hundreds, with artefacts in the tens of thousands' (Kalb et al., 1982, p. 28). Or again, the Mt Carmel expedition of 1929–34 reported 74,750 implements (Garrod and Bate, 1937), and Mary Leakey 2,470 artefacts from the early *H.habilis* occupation site FLK (M. Leakey, 1979). The great French archaeologist Bordes claimed not only to have examined 1½ million Neanderthal-style stone tools but to have personally manufactured over 100,000 to get the feel of it! (Pfeiffer, 1982). (Perhaps these figures should be taken with a pinch *du sel*, he would have had to examine 63 a day every day for 65 years to reach the former figure.)

The term 'artefact' refers to *all* products of tool-manufacture, including wastage (or 'debitage'), while 'implement' or 'tool' refers to the product itself. The ratio of tools to debitage can be very low at manufacture sites. The modern archaeologist is interested in all of this material since it is only by analysis of the total assemblage that manufacturing processes and the nature of the site can be reconstructed.

By the Mesolithic period, with its small flake blades (microliths) artefact numbers into five figures for a site are not uncommon. Although certain geological conditions can degrade flint, stone artefacts, once buried, will tend to be preserved rather than not. The major routes to oblivion are later erosion resulting in their exposure to weathering and water action, and human quarrying and mining activities. Although flint is preferred, other materials include limestone (e.g. in India (Paddaya, 1977)), lava, obsidian and quartzite. Identification of artefacts made of lava, for example, can be considerably more difficult than with flint as they do not flake cleanly.

Even until recently there has been great variation between tribal cultures in the extent to which stone implements are used; Kwakiutl Indian (US north-west) being far more wood-dependent, while Australian tribes make great use of stone. Environmental factors are obviously involved.

The three major aspects of this material to be dealt with here are: (a) method of manufacture; (b) classification; and (c) assembly analysis.

(a) Manufacture

Put at its most crude, there are two basic ways of making stone tools; you can take a rock and knock it into the shape you want, or you can take a rock and knock the shapes you want off of it. These are not entirely logically distinct, or mutually exclusive, but need to be differentiated as underlying orientations to the manufacturing task. The earliest tools were, as far as we know, made exclusively by the first technique, one end of a cobble being crudely chipped with two or three blows into a cutting edge. These 'Oldowan choppers' are the hallmark of the first identifiable industry, the Oldowan. Such tools can be broadly classified as 'core tools', as opposed to the 'flake tools' made by the second technique. This first method also produced the long-lived and eventually quite sophisticated 'hand-axe' industry known as the Archeulean. This tradition spread out from Africa around 0.4 myr and seems gradually to have pushed the Oldowan before it until versions of the latter survived only on the fringes, in South East Asia and one or two northern sites, the best known being at Clacton-on-Sea in Essex (the 'Clactonian' industry). One of the mysteries of human evolution is the extended period of co-existence of the two traditions in Africa itself, of which no good interpretation has yet appeared.

Eventually the second technique makes its appearance in the form of the Levalloisian flake industry, which involves preparing a 'core' by prior trimming in such a way that a flake of the desired form can be neatly removed from it. The manufacturing process takes four distinct stages, using well over 100 separate blows, and leaves a distinctive piece of debitage, the so-called 'tortoise-shell' core (see Fig. 3.8). Even in these early industries the 'waste' flakes were often used, appearing in reports as 'utilised flakes', nor are archaeologists quite so certain as they used to be that these were not purposely made. The Levalloisian still only produced one flake per core and was thus relatively inefficient. In the Middle Palaeolithic flake-industries proper emerge, the most important being the wide range termed 'Mousterian' and ascribed primarily to the Neanderthals. In its earliest phase it was still influenced by the Acheulean and hand-axes are still present (the 'Mousterian of Acheulean Tradition'). These decline in frequency and numerous flake tools appear: side-scrapers and various notched and 'denticulate' points. Relations between Acheulean, Levalloisian, Clactonian and Mousterian are not entirely clear, the Levalloisian appears to have been a result of contact between Acheulean and proto-Mousterian, while the Mousterian itself is seen as originating in the Clactonian, which had evolved from the Oldowan at some considerable remove into a crude flake-industry, and also been influenced by the Acheulean. It is evidently an area of some confusion, but the Mousterian emerges supreme and persists until the Upper Palaeolithic 'leptolithic' industries supersede it.

In the Upper Palaeolithic a move is made towards more finely worked blades and their manufacture from prepared cores. Techniques are invented enabling the striking of numerous blades, which need little or no retouching, from a single core. This trend peaks in the Mesolithic and Neolithic phases when the number of small microliths produced increases for use as arrow-heads, knives and possibly even such roles as saw-teeth. The final, Neolithic, phase sees the advent of a new, and final, innovation: the manufacture of axe-blades by grinding and polishing. The beautiful axe-heads produced by this method may be found in museums throughout Europe. Needless to say, many of these tools represent only a part of the finished implements, the remainder being wood or leather, long perished (Plate 9). With the advent of bronze, the 2½ million year evolution of lithic technology effectively ceases.

In addition to manufactured stone-tools we find other stone artefacts or utilised cobbles – hammer-stones, for example, identified by a much-pitted area on one surface, spheroid pebbles used as bolas weights or projectiles, and larger rocks placed to hold down animal hides to form primitive tents, or windbreaks.

Flaking techniques underwent a variety of progressive changes, from simply knocking one rock with another as a hammer-stone to highly delicate techniques involving 'soft' hammers of wood and bone, and 'pressure-flaking' – exerting a continuous pressure in a finely controlled fashion. Heating was also used sometimes to facilitate flaking and when this has occurred tools may be directly dated by thermoluminescence. A pound of flint yielded about 8 inches of cutting edge on a chopper or hand-axe, and 40 inches using the Levallois technique, but by the peak of the Mesolithic a skilled knapper could obtain up to 150 blades from a single nodule. The evolution of tool-manufacturing and usage constitutes a unique source of information about our behavioural evolution which is only beginning to be explored from a psychological perspective.

(b) Classification

Stone artefacts (indeed all artefacts!) require classification in two ways: first a classification of what they are, and second, of who made them or where they come from. Both present difficulties.

(i) TYPES OF TOOL

Although 'site-analysis' is to be discussed below, it must be stressed at the outset that the total tool-assembly of a site is the proper unit of analysis, and it is often only in this context that tool-functions can be identified or guessed at. The term 'tool-kit' refers to the total range of types of tool found at a site, or typical of a particular industry. The growth in size of the hominid tool-kit as a whole was initially very slow, increasing from 6 to 10 over a million years! (If it is meaningful to classify them at all at this stage, which is now doubtful.) In the Middle and Upper Palaeolithic the tempo of innovation increased along with changes in manufacturing methods. Prior to the Upper Palaeolithic there is relatively little regional variation within the basic traditions mentioned previously other than that due to differences in raw material. Even so, experts such as Roe and Wymer are able to differentiate regional variations and spot affinities, provided sufficient samples of material are available. During the Early and Middle Palaeolithic the main distinction is between the hand-axe-using industries (Acheulean, Early Mousterian) and the non-hand-axe-using ones (Oldowan, Clactonian, Levalloisian, Mousterian).

The obvious way to classify tools is by function, as an axe, awl or screwdriver. It is a source of perennial frustration for prehistoric archaeology that the functions of unearthed tools are often unknown, indeed unknowable. This problem is beginning to yield to micro-wear analysis but there are large numbers which remain a puzzle. Why, for example, do the large early Acheulean 'hand-axes' so often have a cutting edge all the way round? (Incidently there is no sign that they were hafted onto any other material.) The problem does not ease when the tool-kit increases if anything it worsens. Since the late nineteenth century when French archaeologists began to get to grips with the issue, the nomenclature of tool-types has been a mixture of description, guess-work and in one case, eponymy ('boucher' after Boucher de Perthes). Bordes has brought some

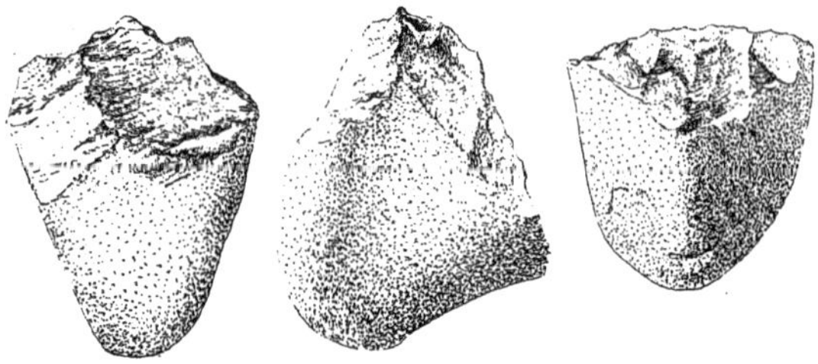

OLDOWAN

Two typical Oldowan pebble tools made of volcanic lava, from Bed 1 at Olduvai Gorge, Tanzania.

(See Figure 4.5 for further examples)

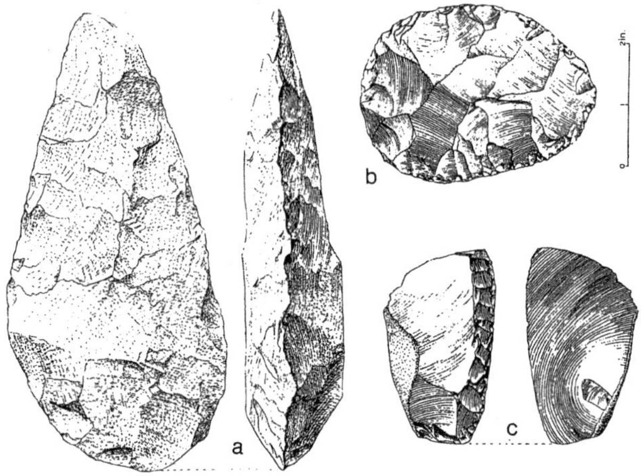

ACHEULEAN

(a) Lava hand-axe, Olorgesailie, Kenya
(b) Ovate hand-axe, South of Wady Sid, Israel
(c) Micoquian flake-tool (flint), Tabūn Cave, Mt Carmel, Israel

(See Plate 10 and Fig. 4.4 for further examples)

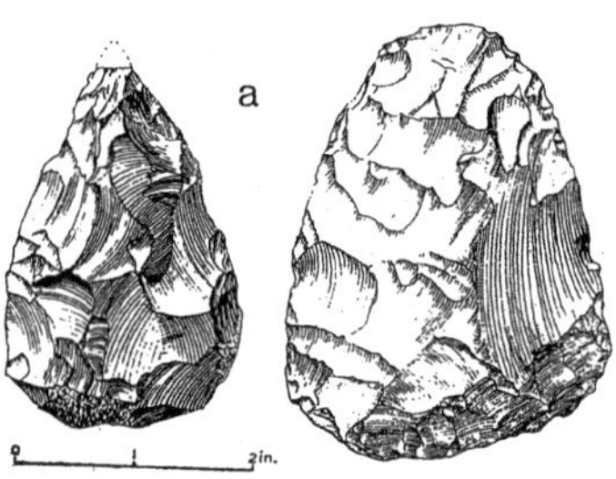

MOUSTERIAN

(a) Two 'Mousterian of Acheulean Tradition' hand-axes; *left*, from Le Moustier, *right*, from Kent's Cavern, Torquay
(b) and (c) Typical 'side-scrapers' or '*racloirs*'

The Mousterian is associated primarily, but not exclusively, with Neanderthals

Figure 3.8 The evolution of stone tools, examples from the major phases now identified. Selected from K. P. Oakley, 1975. *Reproduced by courtesy of the British Museum (Natural History).*

LEVALLOISIAN

(a')–(a''') Three views of the typical 'tortoise-core' by which Levalloisian industries are identified. This is from the Baker's Hole industry, Northfleet, Kent

(b) Flake with faceted striking-platform such as would have been struck from area 2 on (a'''), also from Northfleet

The Levalloisian flake industries overlap with the later Acheulean core-tool industries and the early Mousterian.

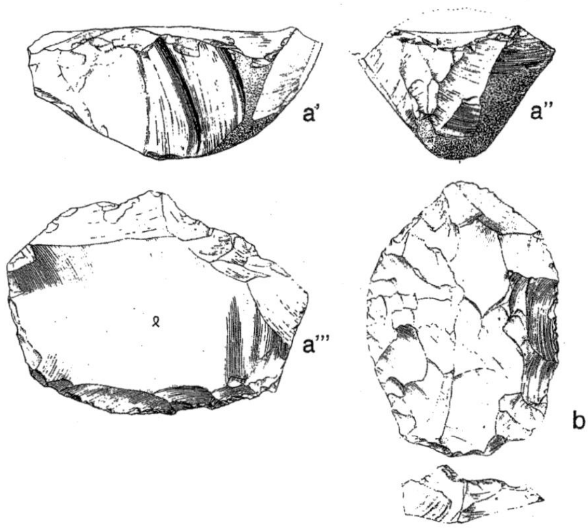

UPPER PALAEOLITHIC

A few of the numerous new types of tool which appear in the Upper Palaeolithic 'leptolithic' industries.

(a) Knife-point, from Laussel in the Dordogne ('Gravettian')
(b) End scraper or '*grattoir*' from the Welsh site of Cae Gwyn, Vale of Clwyd
(c) Nosed scraper, or 'push-plane', Aurignacian, Laugerie Haute
(d) Solutrean piercer or 'hand-drill' – Laugerie Haute

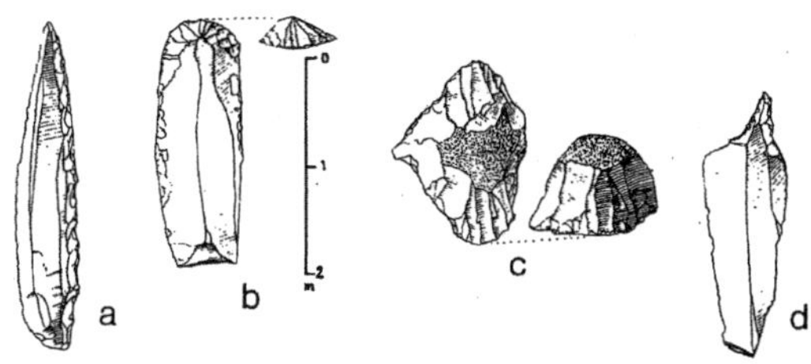

MICROLITHS

Microliths typical of late Palaeolithic, Mesolithic and Neolithic 'blade tool' industries

Microliths. (*a*) Capsian trapeze ('geometric'), Gafsa, Tunisia. *After Vaufrey.* (*b*) Azilian point ('non-geometric') Mas d'Azil. *After Piette.* (*c*) (*d*) Capsian lunate and triangle, both obsidian, Kenya. (*e*) Triangle of rock crystal. Bandarawela, Ceylon. *After Noone.* (*f*) Tardenoisian trapeze, Tardenois (Aisne). (*g*) Tardenoisian point, Ham Common, Surrey. (*h*) Maglemosian point, Kelling Heath, Norfolk. (*i*) Australian saw-knife (*laap*). *After de Pradenne*

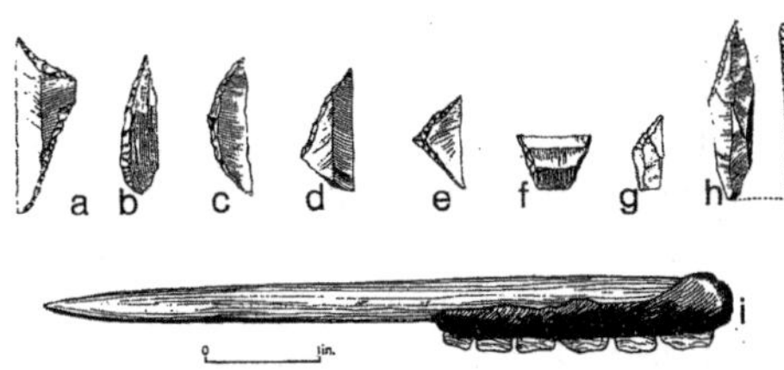

system to the Mousterian with a catalogue of 60 basic types. Much of the vocabulary in this area remains French (racloir, grattoir, pointe à cran, bout coupé hand-axe and percoir, for example), though there are standard English equivalents for most of them. The beginner in the field is likely to be quickly bewildered by burins, end-scrapers, side-scrapers, gravers, tanged points, cleavers and bouchers, with qualifying adjectives such as notched, cordate, shouldered, unifacial, bifacial and denticulate. This surprise is partly due to our not expecting such a technical vocabulary to have arisen in connection with what at first glance appear to be barely distinguishable objects, while we would expect it beforehand of a subject like medicine or chemistry. The general principle for identifying artefacts is to give them a sort of three-fold name incorporating a descriptor, a provenance term and a type-term; thus we can talk of blue Chinese vases, bronze Greek spears or plastic American beakers. The same applies to stone tools, except that the type-term and the descriptor may fuse; so the term 'point' occurs ubiquitously as a type-term. Function, physical appearance and stylistic affinity all enter into the classification system.

Earlier generations' hopes of a clearcut world-wide evolutionary sequence emerging, in which one type of industry succeeded another in orderly fashion, have evaporated except in very broad terms. Wymer (1982) writes that 'although there is no clear unilinear succession of stone industries throughout the world, there are obvious technological developments that proceed from one tool type to another' (p. 32). And he identifies four stages representing a series of 'progressive refinement and elaboration of material equipment, which is reflecting cultural advancement in a social and economic sense' (*ibid*). These are: (a) chopper-core industries, (b) hand-axe industries, (c) transitional industries and (d) leptolithic industries ('leptolithic' meaning 'light stone', i.e. usage of thin flakes and blades rather than robust tools).

Although the specific uses of most implements may remain obscure, the function of many is becoming clearer in the light of micro-wear analysis and archaeological excavation of butchery sites in which the tools used to skin, disarticulate and break bone can be seen *in situ* around the remains of the prey animal. Prominent among these are the Torralba and Ambrona elephant-kill sites in Spain (Howell, 1966) and three sites at Olduvai (Isaac and Crader, 1981). However, Binford counterbalances undue optimism in this area (Binford, 1981), as do Shipman and Rose (1983).

(ii) INDUSTRY CLASSIFICATION

Turning now to the classification of industries, or techno-complexes as some US writers are referring to them, in quest of theoretical neutrality, we are primarily tied to site names. In addition we also find ourselves faced with a major theoretical dispute, the latest turn in an issue which has been running since the 1920s and was mentioned in Chapter 2. To what do these industry names refer? Or, more directly, how do we explain the differences between tool assemblies? On this matter Bordes and Binford clashed head on. For the former, the differences were due to different tribes with idiosyncratic life-styles, each generating a characteristic distribution of tool-types in their tool-kit. Binford felt the differences were functional, each distribution pattern representing a particular activity. Using that old psychologist's friend, factor analysis, to analyse the rates of co-occurrence between different tool-types, he identified 5 basic Neanderthal tool-kits: a 'maintenance' kit used for repairs, mending, etc., a killing and butchery kit, a food-processing kit used mainly with meat, a shredding and cutting kit and, finally, a more specialised killing and butchery kit. This was based on a study of 5 assemblages drawn from Israel, Syria and France and including 4,000 tools. At Combe Grenal in France, Bordes's principal research site at which 55 layers between 100,000 and 40,000 BP have been identified in a section which reached 40 feet in depth, Binford found 14 tool-kits in an analysis of 19,000 tools. Bordes himself felt he could identify 4 only, which he ascribed to different tribal groups. (See Pfeiffer, 1982, for an account of this; the primary Binford references are L. R. Binford, 1972, 1973; L. R. and S. R. Binford, 1966.) The British archaeologist Mellars (1969) has produced a third account with elements of both Bordes's and Binford's models. This accepts a tribal picture for some regions such as the South of France, but a more fluid one in which temporal evolution of industries and function and group differences are interacting for others.

After the Middle Palaeolithic the nomenclature problem begins to get acute and the normal strategy is to identify an industry by the site-name where it was first identified or where it is especially well represented. Within sites we may use stratigraphic position and look for industries elsewhere in the region in a 'homotaxic' position. The problem of reconciling geography and chronology (or stratigraphy) has never been entirely satisfactorily resolved. Similar industries may be found at different sites of significantly different ages.

We have already mentioned the Mousterian industry, which is recognisable from its use of prepared cores to produce side-scrapers and other flake tools, leaving a typical 'discoidal' core among the debitage. But this term 'Mousterian' is applied to sites ranging from Germany to the Middle East and has in turn acquired a set of modifying labels, Early, Middle, Late and the aforementioned 'Mousterian of Acheulean Tradition'. The Upper Palaeolithic industries which succeed the Mousterian have been traditionally labelled according to classic French type-sites, (see Table 6.1). Stylistic criteria for differentiating industries become easier to identify during this period, the Solutrean people, for example, produced masterpieces of pressure-flaking in their beautiful 'willow' and 'laurel-leaf' points. Beyond Europe, however, the use of such terms was abandoned progressively from the inter-war period, and local type-site names are used. The African classification was totally revised in 1965 and the most recent reviews of this are J. D. Clark (1982), and D. W. Phillipson (1985) (see Table 3.1). Only the Lower Palaeolithic 'Oldowan' has received, like the French Acheulean and Mousterian, any universality of use.

TABLE 3.1 *Sub-Saharan African stone-tool industry sequence*[1]

Epoch	*Equatorial*	*South Africa*
Later Stone Age[2]	Capsian Nachikufan Pomongwan	Wiltonian, Smithfield
	Tshitolian, Tshangulan	Alexandersfontein Facies B Pietersburg
Middle Stone Age[3] (MSA)	Lupembo-Tshitolian (35–20 kyr) Bambatan, Stillbay (75–10 kyr)	Lesotho (45–35 kyr)
	Lupemban (75–35 kyr)	Howieson's Poort (end c. 65 kyr)
	Sangoan, Charaman (end. c. 70 kyr)	Early Pietersberg, Mossel Bay, Orangian (end c.80 kyr)
Earlier Stone Age[4]	Late Acheulean (= Fauresmith) (c. 1 myr–0.5 myr) Acheulean (1.5 myr – 0.2 myr) Developed Oldowan (inc. Hope Fountain, Tayacian) Oldowan (2 myr +), Karari (c. 1.4 myr) (Kafuan – now rejected as arising from natural breakage)	

[1] The African industries do not permit the relatively simple sequencing of European ones, for only at a few sites such as Kalambo Falls, Zambia, do extended sequences occur, and these cannot be easily generalised to other regions. This table represents a gross simplification of already simplified (and often tentative) tables and data in J. M. Coles and E. S. Higgs (1969), J. D. Clark (1982), G. Isaac (1982) and D. W. Phillipson (1982).
[2] The term Later Stone Age refers primarily to the advent of microlithic industries. This varies enormously from one site to another and cannot therefore be given a clear, general, starting date. This applies to all African epochs; they are defined by style of industry, they are not, strictly speaking, temporal, even though they are in fact a developmental sequence. Earliest Later Stone Age industries probably date from around 35,000 BP, corresponding with the European Upper Palaeolithic, but these can occur alongside Middle Stone Age industries.
[3] Earliest Middle Stone Age industries (e.g. Sangoan) evolved from Acheulean ones sometime between 100,000 and 200,000 years ago. J. D. Clark (1982) begins his tabulation at c. 110,000 BP, but several industries were already in existence at that point.
[4] Although this epoch begins before 2 myr, Acheulean industries persist to as late as around 57,000 bp at Kalambo Falls.

Stone artefacts gradually cease to be the main criteria for identifying cultures and industries after the Upper Palaeolithic. Shapes of barrows and megalithic structures come to prominence in Europe after about 7,000 BC while the range of non-lithic artefacts plays an ever greater role. From this phase too we can catch a last, more detailed, glimpse of the way stone artefacts were used, as traces of their handles or other organic matter used in making the complete tool may not yet have perished. One remarkable find in North Hannover in 1935 (*Proceedings of the Prehistoric Society*, 1937, p. 178) revealed all about a Neolithic dagger (Plate 9).

The problem remains that broad labels like Early, Middle and Upper Palaeolithic are chronological in character, while Aurignacian, Perigordian and Azilian, for example, are regional. The older striving for clear stratigraphic sequences was a vain quest since different industries expanded and contracted at different rates while co-existing with one another. There is no single continuous 'most advanced' culture in prehistory any more than in history. The very function of these culture/industry/techno-complex labels is not entirely settled – do they refer to actual cultures? To technological traditions? Ethnic groups? Economic functions? Evolutionary phases? In prehistoric times the human race was not partitioned in ways we are now familiar with: nationality, social class, language group, religion, perhaps not even ethnic group. Most authorities favour small family bands – but the interactions of such bands are a mystery. Certainly their social organisation was not identical with that of any animal species. By the Upper Palaeolithic we might assume that various levels of clan and ethnic division were operating, but in the absence of modern concepts of property, with only inchoate notions of territory, and before *some* point ignorant of the male role in procreation, the precise nature of the social organisation of the prehistoric world population eludes our comprehension, and it is that elusive order which the artefact record dimly reflects.

For the present, tool classification systems must be judged by their pragmatic value for helping answer specific research questions, and not reified into referring to objectively existing social or economic entities. The Bordes–Binford debate is nevertheless a very real issue, for clearly whether one group makes 14 tool-kits or fourteen groups each make 1 makes all the difference in the world to our image of the Neanderthal's life-style and achievements.

(c) Assembly analysis

Decisions as to industry classification and interpretations of a site's significance hinge largely on the information to be gleaned by analysing the entire assemblage of artefacts, in conjunction with wider site-analysis to which we will turn shortly. This involves tabulation of the relative frequencies of artefact types, from which we may be able to tell if it is a manufacturing site or, say, a butchery site. Substantial debitage implies a work-shop, a low level of debitage matching the number of tools found means perhaps a 'one-off' site where tools were made on the spot for some specific immediate purpose, then abandoned. Some kinds of debitage are identifiable as 'sharpening flakes' implying that portable ('curated') tools were being serviced. Presence or absence of certain categories of artefact may be diagnostic of the industry (e.g. 'tortoise-shell' cores indicating a Levalloisian industry). Even if the artefacts taken individually are not very informative, the profile of their type-distribution may be diagnostic, or valuable in comparison with other assemblages. The material of which the tools are made can also be informative, suggesting trade, or at least transportation, between the site and the parent rock source.

The study of stone artefacts is now one of the most technical and specialised areas pertaining to human evolution, and, being both long-established and well supplied with material to study, it has in some respects perhaps advanced our understanding of it furthest. Increasingly too it is an area where experimental replication is shedding ever more light on manufacturing techniques and usage (see Plate 10), David Crabtree of Idaho State University being one of the major gurus in this, along with Mark Newcomer of the Institute of Archaeology and L. H. Keeley. K. Oakley's *Man the Toolmaker* (1972), published by the Natural History Museum remains the easiest introduction to the topic.

2 *Site analysis*

Use of stone extends beyond the worked implements. The earliest known human occupation site, FLK Bed 1 at Olduvai, excavated by Mary Leakey, and aged 1.7 myr, was identified partly from the arrangement of rocks in some sort of wall to form, it was surmised, a wind-break (though see Potts (1984) for doubts on this). Additionally, rocks are used to weigh down organic material being used as tenting or bedding, as handy ammunition to hurl at prowling predators, and to construct hearths. Much of the advance of archaeology in the present century has involved developing techniques for incorporating such features in the analysis of entire archaeological sites, transcending simple treasure hunting or limited conceptions as to what the discipline could achieve. Such archaeologists as Movius, Garrod, J. D. Clark, Binford and Isaac must be mentioned as leading figures in this as far as prehistory is concerned. The goal of archaeology has now crystallised, after earlier theoretical flux, around the notion of recreating the life-style of the people who occupied the site, utilising the total evidence the site can yield. This has meant an ever-more-cautious approach to excavation and recording with the careful preservation and study of an ever-widening range of material.

In prehistoric archaeology it is now customary to record in obsessive detail the positions of all objects found on a living-floor once it has been uncovered. Artefacts, post-holes, natural objects, bones, and hollows must all be mapped, while soil samples will be taken, e.g. for pollen analysis. What can this reveal? Basically it is aimed at unravelling the organisation of the site, and through that, the kinds of activity which went on there.

Can we identify different activity zones such as a tool-making area with an anvil stone and debitage, or a butchery area with many bones? Do the traces of dwellings as left by post-holes, hearth-sites or rows of stones tell us anything about the size of the group or its social organisation? Analysis of fossilised faeces (coprolites) and food waste may tell us about diet and also about prevailing flora and fauna.

As far as stone tools are concerned it is sometimes possible to reconstruct a considerable portion of a nodule like a three-dimensional jig-saw from its scattered fragments, tools and debitage alike. On occasion this has provided a remarkable insight into just how tools were first made and then distributed around the site. This proved possible at a site near Antwerp in Belgium (Cahen et al., 1979). Of 16,000 artefacts found, 3,000 were refitted, and 6 of the refitted blocks are discussed in detail in this report. Utilising micro-wear techniques as well (Keeley was a co-author), the team was able to produce a penetrating picture of life at the site, as Roe wrote in his Peer Commentary response: 'we can almost feel ourselves standing unseen at the edge of the settlement site, understanding its organisation and actually watching early man at work'.

Analysis of faunal remains has yielded extensive information about hunting and meat-related behaviour, incorporating theoretical ecological perspectives with archaeological methods. A good recent example of this type of work is Gamble (1979) on the interaction between the environment and hunter-gatherer strategies in the central European Palaeolithic.

3 *Prehistoric art*

> There is really no such thing as Art. There are only artists. Once these were men (*sic*) who took coloured earth and roughed out the forms of a bison on the wall of a cave; today they buy their paints, and design posters for the Underground; they did many things in between. There is no harm in calling all these activities art as long as we keep in mind that such a word may mean very different things in different times and places, and as long as we realise that Art with a capital A has no existence. For Art with a capital A has come to be something of a bogey and a fetish. (Gombrich, 1950, p. 5)

The term 'art' here is to be taken as meaning all forms of painting, sculpture, decoration of artefacts and ornamentation. The function of much of this for its creators is a matter of speculation, but certainly extends beyond, perhaps does not even include, the purely decorative or expressive. Its appearance in the archaeological record is almost inevitably interpreted as signalling to us the advent of persons like ourselves. The discovery of this art (see Chapter 2) was treated with scepticism for several decades, but once accepted rapidly came to overshadow, in the public mind, almost all other data. The cave art of the Franco-Hispanic Upper Palaeolithic continues to exert a perennial fascination, and much of it has become firmly entrenched in our own cultural iconography – the Lascaux and Altamira murals being rivalled only by the Mona Lisa, Van Gogh's chair and Constable's *Haywain*. Among the sculptures, the Willendorf Venus has probably been depicted in at least one new publication a year since her discovery (hence I can safely give her a miss!). Inspite of this over-exposure of a few famous examples, we remain largely in ignorance as to the original meaning and significance of Upper Palaeolithic art; it hovers just beyond our grasp, familiar yet alien. Of its artistic merits there can be no doubt, but we need to bear Gombrich's words in mind, that the artists may have had no concept of 'Art' at all. But progress is being made, and the research at Lascaux has clarified much about hitherto enigmatic technical aspects of the mural painting (Arl. Leroi-Gourhan and J. Allain, 1979).

I will first discuss cave and rock painting then turn to other forms.

(a) Cave and rock painting

Cave and rock paintings occur widely in Europe, Africa, Australia and part of America. In Africa the tradition seems to have finally ceased in the nineteenth century with the virtual extermination of its last practitioners, the 'Bushmen', only remnants of whom now survive in the Kalahari Desert. Earliest, and aesthetically unsurpassed to modern western eyes, are the Upper Palaeolithic cave paintings of Europe. This achievement represents a cultural rather than an evolutionary event in the strict sense. It is unique in human cultural development; many societies never engaged in making cave art even under similar environmental conditions, and other cave and rock art traditions which followed are stylistically fundamentally different. The Upper Palaeolithic cave painting is confined to a limited area of France, Spain and Italy. Notwithstanding this regional restriction, its longevity and relative homogeneity are astonishing; it originates around 35,000 bp and finally peters out after 10,000 bp following the last Ice Age. This 25,000-year span crosses dramatic climatic fluctuations and dwarfs any other cultural tradition; at a stretch ancient Egypt lasted 4,000 years and China perhaps 3,000. The continuity may, though, be illusory (Conkey, 1983).

Two names dominate the study of this work – l'Abbé Breuil and André Leroi-Gourhan, although there are numerous other eminent scholars in the field, which is naturally dominated by the French. Breuil devoted his long career to the study of cave art and to recording and publishing as much of it as he could. He developed a two-cycle model of the stylistic evolution of the genre which provided the initial conceptual framework for its interpretation, a model which has since weakened. Breuil's observations and analyses constitute the basis for all later work and the attitude towards him of cave art researchers is only a little this side of idolatry. Leroi-Gourhan, a follower of Breuil, has been concerned with analysing the distribution and spatial arrangements of the superficially haphazard arrays of images. The theories of these two researchers are discussed in Chapter 6.

The rock art of European Mesolithic and Neolithic times (both painting and carving) is wider spread throughout Europe and its forms often resemble those used by extant or recent lithic cultures such as the Australian aborigines (Gould, 1968). Stylistically it is a complete contrast, being highly schematised and abstract. This was long seen as proof of a decline or degeneration from the 25,000-year-long Golden Age, but such a view is misconceived and stems from over-reliance on purely aesthetic criteria of evaluation. Much of it is almost certainly notational in function, the markings usually being referred to as 'petroglyphs' (see Fig. 3.9).

African rock painting extends from the famous Tassili Frescoes of the Sahara (Lhote, 1959) to South Africa. Styles vary widely both across place and time, although the almost academic naturalism of the European Upper Palaeolithic is never repeated and human figures, virtually non-existent in the Upper Palaeolithic, are common. This tradition certainly originates not too long after the demise of European cave art culture, but dating is difficult. The works themselves are fast disappearing through vandalism and an irresponsible tendency of photographers to slosh water over them to brighten up the colours. The Leakeys have long been active in researching and recording this material and Mary Leakey's *The Vanishing Art of Africa* (1983) is the outcome of decades of labour by both herself and Louis. Ritchie (1979) provides a useful summary of the history of the study of African cave and rock paintings, as well as presenting his own theory of its meaning and the link with the Bushmen. Fascinating though it is, the subject is largely incidental for our present purposes, though the African paintings can occasionally provide data on population movement, climatic conditions and vanished life-styles.

The native Australian rock art has some affinities to elements in Upper Palaeolithic work, notably the 'meanders', while their mobiliary art also appears to bear similarities to that on European ivory artefacts. An important work in this area is P. J. Ucko (ed.) (1977), a collection of papers on aspects of schematisation and symbolism in both traditions.

Our principal concern has to be with the Franco-Hispanic (or 'Franco-Calabrian') art, as signifying the culmination of the 'hominisation' process. Even though it is in no way directly ancestral to any known later cultural tradition, its sheer antiquity and proximity to the final stage in our physical evolution render it uniquely important.

The cave painting includes imagery and symbols of five broad kinds: firstly, the realistic depiction of animals, or parts of animals, for which it

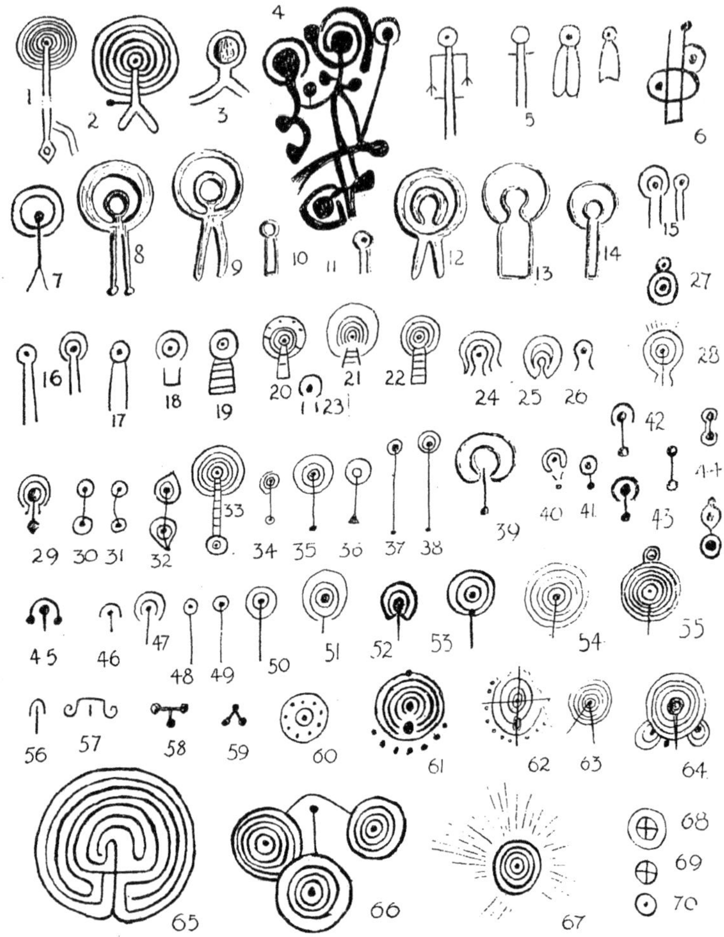

Figure 3.9 Mesolithic and Neolithic petroglyphs; variations on the 'cup, ring and stroke' motif from various sites in the British Isles and Ireland. From Breuil, 1934, p. 316. *Reproduced by courtesy of the Prehistoric Society.*

Selection of cup, ring and stroke signs, many of which (at least until 58) are derived from human figurations. – 1, 16, High Auchinlary (Wigtonshire). – 2, 3, 15, 20, 60, Cardoness. – 4, Kirkclaugh (Kirkcudbright): group of people. – 5, Four similar figures from Bohuslän (Sweden). – 6, Gates of Glory (Dingh, Ireland): type akimbo. – 7, 47. Pitscandly (Forfarshire). – 8, 9, 10, 12, 13, 14, 39, Stronach Ridge, Brodick, Arran. – 11, 58, 59, Arbilot (Forfarshire). – 17, 18, 19, 20, 21, 22, 33: Panorama stone, Ilkley (Yorkshire). – 23, 53, Milton Park (Kirkcudbright). – 24, 26, 40, Robin Hood Bay (Yorkshire). – 25, Tynees. – 27, Ballynasare (Kerry). – 28, Rowton Lynn. 29, near Lough Crew (Meath, Ireland). – 30, Ruddie, Knowes. – 31, 50, 51, 52, 53, Old Bewick (Northumberland). – 32, Letham Grange (Forfarshire). – 34, 54, 63, Carnban (Argyllshire). – 35, Walltown (Forfarshire). – 36, Balbym (Argyllshire). – 37, 38, Ruthven (Forfarshire). – 41, 42, Largy (Argyllshire). – 43, 44b, Bakershill (Ross-shire). – 44a, 45, Kirk Braddoth (Man). – 44c, Letham (Forfarshire). – 46, 48, 49, many rocks. – 55, Chatton Low (Northumberland). – 61, Clauchandolly (Borgus, Kirkcudbright). – 62, Jedburgh (Roxburgh). – 64, near Lough Crew (Meath, Ireland). – 65, Holywood (Limerick). – 66, Torrs (Kirkcudbright). – 67, Nether Linkens (Kirkcudbright). – 68, 69, Mevagh, Donegal. – 70, Lough Crew.

is best known, and these range from mammoth to cattle, cave bears to horses, and a few non-mammalian species also. In a second category, commoner in early sites, are the so-called 'spaghetti' meanders and squiggles, done by drawing the fingers through wet clay, though recognisable animal forms sometimes emerge from these. Thirdly, there is a variety of simple abstract forms – dots, triangles, squares, etc. – which may in some cases be representational. These are referred to as 'tectiforms'. Fourthly, there are a very few, highly schematised, human figures, though more realistically depicted 'sorcerers' occur apparently representing humans wearing animal masks and skins (the best known being at Trois Frères). Finally, there are the hand-prints, either silhouettes probably created by spraying pigment around the hand from the mouth, or positive prints done by laying the pigmented hand itself on the rock surface. In some cases these hands appear to be mutilated, missing fingers or finger-joints. Whilst the first category has been the most researched, the others are receiving increasing attention as being even more puzzling.

We do not know the fate of those who accomplished these masterpieces. The nature and extent of population movements at the end of the Palaeolithic is still obscure, although the conventional picture is of influxes from the east and south-east, a north and westward drift as the ice-cap retreated and possibly southward movement from the Iberian peninsular into north-west Africa. Cultural and ethnic diffusion, though very distinct phenomena, cannot be easily disentangled on the basis of archaeological evidence alone. What seems certain is that the cave-art peoples enjoyed a relatively stable and secure existence, for no culture could last so long otherwise. Their population was small by modern standards, it being estimated that at the end of the period the inhabitants of France numbered only a few tens of thousands. Of 229 cave art sites known, 200 are in France and Spain, 18 in Italy, 10 in Sicily and 1 in the USSR (which is otherwise rich in artefacts). The locations in France and Spain appear to be associated with mineral spas in the southern group from northern Spain across the Pyrenees, and fresh water springs in the Dordogne group (Bahn, 1978).

(b) Mobiliary art

Nomadic and hunting peoples do not like having to carry too much, thus most portable art, '*art mobilier*', 'curated' or 'mobiliary' art, is small in scale and often occurs as 'ornamentation' on functional objects such as spear-throwers. The materials which tend to have survived from the Ice Age cultures are ivory (usually mammoth tusk or tooth) and horn, pierced shells and painted pebbles. Wood, though rare in the record, was undoubtedly widely used. Objects in these materials are found right across Europe and the Russian steppes. The most famous are the 'Venus' figurines and the defecating fawn-headed throwing-stick (or spear-thrower) from Le Mas d'Azil. The former are found fairly widely and usually described as 'fertility goddesses', which is hardly enlightening. After the Willendorf Venus, the best known is probably the more abstract version from Les Rideaux, Lespugue (France).

The most tantalising objects are the numerous plaques and 'batons' studied by Alexander Marshack (Marshack, 1972a,b, 1976, 1977, 1979b). Some of these are puzzling in their own right, the so-called '*batons de commandement*' for example; we simply do not know what they were for. Others are puzzling because of the 'ornamentation' on them. Using intensive microscopic examination and mathematical analyses of grouping within the overall patterning, Marshack has concluded that much of this is actually notational, probably calendrical in function. Again, this is discussed further in Chapter 6. The earliest known engraved artefact is an ox-rib, about 0.3 myr, found at Peche de l'Aze by Bordes in 1969 and described by Marshack (1977). After this comes a huge hiatus until the Upper Palaeolithic tradition. Marshack has successfully demonstrated the interpretational complexities of this 'art', and no longer can such objects be blithely dismissed as 'pendants' or 'cult objects'.

At 32 sites, mostly in France, but also in Spain, Italy and Switzerland, numerous painted pebbles have been found. Of nearly 2,000, 1,400 are from Le Mas d'Azil and many were collected in the 1880s by Piette. No one is sure of their significance, whether they were decorative, notational, religious or playthings, but current views tend towards calendrical notation. Recent research by Couraud has discovered that a disturbing number of these in museum collections are forgeries (Bahn, 1984). Obviously decorative are pierced shells and beads which have sometimes been discovered still lying in the lines of an original necklace or bracelet, the string long perished. Both in Russia and France musical instruments have been found including a mammoth-hip xylophone, with antler hammer, and cave-bear flutes with 3 to 7 holes tuned to a pentatonic scale.

Artefacts do not really fall into the categories of art and non-art, decorative and functional. The Upper Palaeolithic sees a great expansion in the range of non-lithic tools: ivory and horn spear-throwers, throwing sticks and spear-straighteners, harpoons and 'batons' for example, but it also sees less clearly functional figurines, plaques and 'ornaments' appearing for the first time. Their apparent lack of function is due to our own ignorance of the contexts in which they were produced, it is not self-evident. (Is a wedding ring functional?)

It is not surprising that Sollas saw in these peoples the ancestors of modern Eskimos, for their conditions of life were often similar and the adaptive strategies required in the ways of technology frequently converge.

(c) Other 'art' and artefacts

The distinction between painting and sculpture is not always clear in cave art; natural rock-formations are often incorporated heuristically into the figures. Incised outlines of figures shift into relief carving, and full three-dimensional statues are known at a few sites, the best known being the two clay bison in the heart of Tuc d'Audoubert and the headless bear at Montespan. Mention might also be made here of the curious monument unearthed by Freeman and Echegary at El Juyo in Spain, where a stalagmite slab proved to be the topping of a strange structure incorporating pigmented clay pillars and at the bottom of which was a plaque with a face carved on it. (This is described in Pfeiffer (1982), which provides a full review of the art objects of this whole period. In all honesty though his description of the El Juyo find is extremely complex.)

By and large the other artistic activities still identifiable seem to be varieties of 'stage-setting', heightening of the effects of natural rock forms which happen to resemble animals, for instance. Occasionally, archaeologists encounter other artefacts than those already referred to; in Lascaux, for example, l'Abbé Glory found a fragment of twisted string, along with its impression in clay. Such finds have a quite disproportionate importance. Post-holes, too, can prove not only that there was once a structure on the site, but the pattern of post-holes at a site can tell us much even about its social structure, while at Terra Amata in the South of France, a site dating from around 0.3 myr, such data enabled researchers to reconstruct the form of dwellings used (de Lumley, 1969). At this site there were pieces of ochre which must have been used for colouring. Along with the Peche de l'Aze ox-rib, this pushed back the earliest known practice of 'decoration' far into the Middle Palaeolithic.

In Russia, at Moldova near Odessa for example, whole structures made of mammoth tusks and bones have been found dating from the Upper Palaeolithic (Ivanova and Chernysh, 1965; Klein, 1973; Phillips, 1980). The presumed use of wood is occasionally confirmed by a lucky find such as the amazing 272,000 BP wooden spear-tip found at Clacton-on-Sea in 1911 by S. Hazzledine Warren (Oakley et al., 1977), and the much more recent (c. 7,600 b.c.) felled wood uncovered at Star Carr in Yorkshire (Morrison, 1980). By the Mesolithic wood, albeit fragmentary, becomes relatively common, although some argue that these finds have not been fully exploited or wholly appreciated (Coles et al., 1978).

It cannot be over-stressed that surviving artefacts are not a representative sample of those our predecessors made and used. The preponderance of stone tools grossly distorts the picture, especially in later periods. Such tools were in any case frequently only one component of a complete implement the organic parts of which such as a wooden handle have vanished. Other artefacts entirely of organic material would have been used alongside them. The decoration of the human body by pigment, with flowers and feathers, by tattooing and scarification, piercing and deformation, is universally practised and we simply have no way of knowing when this began. Even chimpanzees drape themselves with garlands in their more ecstatic pongid moods, and the known use of ochre goes back, as just mentioned, to 0.3 myr. The advent of clothing is likewise a mystery, as too is its first function. Of course we can assume that our Ice Age ancestors were well clad in sewn animal pelts, since bone needles have survived. The invention of means of carrying – nets, slings, strung gourds – was, in the opinion of some writers (Linton, 1971), a vital factor in hominid adaptation to savannah-dwelling bipedalism, but the material evidence would have entirely gone (see Plate 11).

Trying to reconstruct the social lives of prehistoric peoples from their extant artefacts thus involves an immense amount of detective work and deductive skill. Pfeiffer (1982) reports attempts at reconstructing an Upper Palaeolithic dance from an analysis of the footprints left on the floor of a since undisturbed chamber. A fossilised dance must surely represent the most curious form of artefact yet recovered.

E Molecular biology

Although I shall only be dealing briefly with molecular biology here, since as yet its relevance to behavioural evolution is limited, some of the work in this field has provided palaeontology with its first really direct challenge, and must be noted. In a now classic paper, Sarich and Wilson (1967) attempted to measure the degree of difference between human and pongid DNA. They arrived at the sensational conclusion that they differed by only 1 per cent, a figure indicating a far more recent timing of the divergence between the two lines than previously thought. Instead of around 20 myr Sarich brought the date forward to 5–6 myr. Gribbin and Cherfas (1982) have published an accessible popular account of this from an openly partisan position. The level of current establishment resistance to Sarich's perspective as portrayed here is widely felt to be rather exaggerated. At any rate papers by all parties in Ciochon and Corruccini (1983) indicate that the requisite hatchet burying has taken place and that molecular biologists have now established an important and recognised niche for themselves.

Two issues appear to be involved: firstly, there is the question of the 'molecular clock' itself, and secondly, the question of how to interpret the differences which DNA analysis reveals.

Sarich, Wilson, Cronin and others assume that DNA changes at a constant rate, thereby providing a basis for timing species divergences, since the degrees of difference will be a direct function of time elapsed since the split. Their opponents have argued that this assumption is unfounded. Leading these is M. Goodman of Wayne State University, another molecular biologist, who claims

> the accelerations and decelerations in the evolutionary history of several sequenced proteins tend to invalidate certain of the dates obtained by molecular clock calculations, such as those determining the time course of hominid phylogeny. (M. Goodman, et al., 1983, p. 85)

The Sarich school will have none of this, arguing that the overall constancy of the clock is directly required by the DNA data and not arbitrary, and also the absence of cases where conclusions arrived at via their methods have been proved false. Cronin (1983) notes that 'It is as if molecular interpretations have imperceptibly pervaded and influenced thought processes without some explicit cognizance being taken of their existence' (p. 116). It is claimed that further research on mitochondrial DNA and electrophoretic evidence places the main radiation of modern human populations 'in the order of 100,000–500,000 years ago' (*ibid.*, p. 126). This Cronin paper and the accompanying one by Sarich himself (Sarich, 1983) are the best recent accounts of the topic from the advocates' point of view.

Granted that the 1 per cent difference finding is correct, the second question which arises is how to account for the gross differences between us and *Pan troglodytes*. This degree of difference is, it must be emphasised, about the very minimum required for species differentiation – and in some species of frogs greater levels of difference are found between sub-species. One of the answers offered is that the changes which have occurred since our lineage split have primarily affected the 'codons', those parts of the DNA chain which determine timing of developmental events, rather than their content, facilitating a continuous neotenous drift in hominid morphology (Bolk, 1926). We may be seen as chimps in a state of permanently arrested adolescence. There is nevertheless a lurking uncertainty now concerning the exact role of DNA in ontogenetic development, for example the paper by Lewontin (Lewontin, 1982) discussed below (p. 127).

Although Sarich's molecular biological approach has undoubtedly provided a radical new approach to interspecific comparison, and one not dependent on the chance finds of fossils, its precise long-term role in the study of evolution has yet to be firmly established.

The reader will come across a number of other biochemical and biological kinds of research such as palaeoserology, immunology, chromosome bonding techniques and the like which cannot be dealt with here. Findings from these levels of research naturally enter into cladistic analyses of the kind mentioned earlier. Much of these areas of research are in a pioneer phase and even the keenest advocates would as yet counsel caution in drawing too many firm conclusions.

F Cross-cultural evidence

The study of pre-literate and 'stone-age' cultures, such as they now are, has been an important source of information for human evolution theorists, though no longer because, as a century ago, they are seen as intrinsically less evolved in any biological sense. Their importance lies in the fact that they may reasonably be assumed to have preserved features of prehistoric life-styles which have vanished elsewhere, and by studying them we might obtain a better insight into the making and use of prehistoric artefacts, the economics of the hunter-gathering life-style (Sahlins, 1972), its consequences for social organisation, and more tenuously, the kinds of belief system by which our ancestors construed the world. The information is thus largely heuristic. The areas where these peoples live now do not represent the same environmental conditions as those of the European Ice Age, being for the most part marginal desert, tropical jungles or ice-wastes far more severe than the epiglacial regions of European prehistory.

Caution is required: these cultures are as old as ours and their current forms probably differ considerably from the original ones, innovative adaptation being replaced by ritualised folkways, for example, once the adaptations have been achieved and need only to be passed on between generations, with perhaps a gradual loss of insight into their original rationale. (The attitude that ancestors knew *less* than the present generation, which we have today, is actually extremely unusual and only emerged at the time of the Renaissance in Europe, and later elsewhere.) Some features such as the pre-occupation with kinship among the Australian aborigines may also be quite distinct cultural forms with no similarity to anything in Europe's Upper Palaeolithic past. These cultures cannot then directly illuminate the origins of *Homo sapiens* culture, from which we are all equally distant, only some aspects of its early forms.

Pfeiffer (1978) lists 16 extant hunter-gatherer communities, to which one or two South American Indian tribes might be added. The two most intensively studied of these are the native Australians and the Bush people, the !Kung and San tribes of the Kalahari Desert. ('!K' is the orthographic transcription of the 'click' sound used in their language family.) The remainder are mostly known from more or less one-off in-depth field studies by single anthropologists or anthropological teams, or by smaller-scale research.

Although long the subjects of travellers' writings, the major Australian aborigine research for our purposes was conducted by Richard Gould in the Gibson Desert of Western Australia in the late 1960s (R. Gould, 1969, 1980), but Berndt and Berndt (1965) and a number of papers in Ucko (ed.) (1977) are also extremely important. Of all extant groups the native Australians are perhaps the closest to the Upper Palaeolithic in their life-style. Gould's work has undoubtedly clarified the nature and roles of rock art, stone-alignments and tool-use as well as the overall manner in which hunter-gatherers relate to their natural environment.

Since Laurens van der Post's best-selling *The Lost World of the Kalahari* (1958) the Kalahari tribes, remnants of a once-prolific people jointly exterminated by European colonisers coming from the south and Bantu invaders from the north in the early nineteenth century, have received considerable anthropological attention. Their habitat is one of the least hospitable, apparently, that can be imagined and yet they have achieved a perfectly adapted and not-too-strenuous mode of living. Major authorities are E. Marshall Thomas (1959), G. Silberbauer (1981) and R. Lee (1979). The influential collection of papers edited by Bicchieri, *Hunters and Gatherers Today* (1972) must also be mentioned here.

The pygmy tribes of the Ituri rain forest in central Africa, although highly atypical in many respects (not least physically) are also often cited, their main chronicler being Colin Turnbull (Turnbull, 1968). Eskimos are better documented and fit uneasily into the 'primitive hunter-gatherer' category due to the sheer technological sophistication of their traditional culture. L. Binford has recently worked with them in research on hunting and meat-eating practices (1981).

The most newly discovered tribe, the Tasaday of the Philippines, are described in *The Gentle Tasaday* (Nance, 1975), while the rather different Yanomamo are known from *Yanomamo: the Fierce People* (Chagnon, 1968). This Amazon tribe is cited regularly in discussions of 'primitive' social structure, and has also provided unique data for genetics on the range of human variation on some biochemical features. Another South American tribe now receiving attention are the Ache of Paraguay (Hill and Hawkes, 1983; Hill et al., 1984). Hill and his co-workers have written primarily on dietary and nutritional aspects of the Ache hunter-gatherer life-style.

There is a tendency in books on human evolution (e.g. Pfeiffer, 1978) to side-track completely and devote entire chapters to summarising the daily life of the !Kung or Gibson Desert Aborigines. Since it is widely surmised that our African ancestors occupied not dissimilar niches throughout most of our evolution this is understandable, but nothing is actually proved by substituting the known life-style of a contemporary *H.s.sapiens* culture for the unknown one of *H.erectus*. Much is implied but little established, and where faint evidence of *H.erectus*'s life-style is coming to light (Potts, 1984), it does *not* seem to correspond to that of modern hunter-gatherers. Use of cross-cultural data is best exemplified in the investigation of specific hypotheses on clearly delimitable research topics, for example Hill's use of data on calories obtained per hour spent foraging in a study of the likelihood of our early ancestors engaging in hunting (Hill, 1982), and R. A. Gould's studies of stone-tool-use among native Australians.

Such cultures can illuminate features of hunter-gatherer life which speculation could never disclose, but the fact is that they now number around 10,000 people in marginal conditions and are not an entirely representative sample of the host of hunter-gatherer societies which populated the pre-Mesolithic world. The role of this data can only ever be an interesting auxiliary one as far as the study of human evolution is concerned.

G Primate studies

More than cross-cultural studies, the tenor of contemporary human evolution research has been determined by studies of primate behaviour. The last few decades have seen the publication of a number of ethological and experimental studies of the higher non-human primates which have radically altered our views of their abilities and natural behaviour. The rationale for applying this work to human evolution needs briefly examining. In the first place, if they are as closely related to us as molecular biologists claim we might expect their behaviour to bear some similarity to that of our immediate ancestors, in any case they are indisputably the closest related to us of all mammals, so if ancestral behaviour patterns survive anywhere it will be among them. This argument ought not to be accepted uncritically, reasonable though it is. They too have been evolving since our ways parted, and there is no guarantee that their current behaviour is any more similar to that of our common ancestor than ours is. Indeed both chimpanzees and gorillas are basically jungle dwellers while our earliest known hominid ancestors were already partly at least savannah creatures. The chief value of primate data, as of cross-cultural data, is its importance as a source of hypothesis-generation and data regarding *possible* forms of early hominid behaviour and social organisation. It cannot directly prove that any given behaviour did occur.

Secondly, the detailed ecological observations of recent years have identified the presence among chimpanzees, gorillas and orang-utans of rudimentary forms of types of behaviour hitherto held to be uniquely human: tool-use, front-to-front copulation, warfare, and grief, for example. The challenge to identify wherein human uniqueness does in fact lie, if anywhere, has thus been heightened by our increased knowledge in this area.

Thirdly, in the experimental field, it has been possible to investigate more directly the basic forms of primate behaviour and the mechanisms governing them. This dates back to Köhler's apes (see Chapter 2). In experimental Psychology the work on rhesus monkeys by the Harlows (1959, 1965, 1970) has long been routinely invoked as shedding light on infant attachment mechanisms. Somewhat more savoury experimental research by Gallup and Suarez (e.g. Suarez and Gallup, 1981) on self-recognition in the primates, plus the much-publicised attempts at language teaching, are mentioned below. The cumulative effect of this experimental research has been considerable, but again it sets parameters for hypothesising, or suggests fruitful lines of speculation, rather than unambiguously resolving issues. The major service all these comparative studies have performed is to de-mythologise the topic of 'human nature', a role they began to play in Darwin's work. Neat formulae about our special capacity for self-awareness, our consciousness of death, or our ability to use language have all had to be abandoned or seriously qualified.

The major studies regularly cited in human evolution literature can be considered by species.

1 *Chimpanzees and pygmy chimpanzees (bonobos) (Pan troglodytes and P.paniscus)*

Jane Goodall's *In the Shadow of Man* (1971) is now the dominant text as far as chimpanzee behaviour is concerned. It describes in great detail the findings of an extended programme of field-research in the Gombe (Tanzania) reserve during the 1960s, undertaken with encouragement from the Leakeys. It covers social structure, tool-use and foraging behaviour, mother-child relationships, sexual behaviour, individual differences, indeed the whole range of chimp life. It is likely to remain the major chimpanzee reference for human evolution writers for some time, and the disastrous later history of the troop has been chronicled by Bygott and Goodall herself (J. D. Bygott, 1972; J. Goodall, 1977).

The second body of chimpanzee research which has pervaded current thinking is that on chimpanzee capacity for language learning. The two most famous programmes were those of the Gardners with Washoe and Pettito with Nim. In both cases the chimp was 'taught' American sign language (Amslan). A similar programme with the gorilla Koko belongs in this group of studies (Patterson and Linden, 1982). Mary Morgan's chimpanzee Sarah was taught to 'read' symbols and Lana of the Yerkes Primate Research Center was taught to use a keyboard (Rumbaugh, 1977). Following the initial enthusiasm evoked by this flurry of research, especially over Washoe (who confirmed his immortality by biting a finger off eminent neuropsychologist Karl Pribram in a subsequent encounter), there has been considerable back-tracking on the part of Pettito and

Seidenberg (Pettito and Seidenberg, 1979; Seidenberg and Pettito, 1979) and Terrace (1980) on the issue of whether truly grammatical structuring of utterances has ever been achieved. Sign language and other non-verbal channels were chosen because physiologically primates lack the vocal equipment for speech, but do have versatile gesturing capabilities.

Although long strings of signs have been produced by these pongid subjects, it is unclear whether they are anything more than 'clumped' together, whether they reveal genuine grammatical structure. Lieberman has roundly attacked Terrace, Pettito and Seidenberg on this issue (Lieberman, 1984) and clearly the controversy is far from being settled. Field studies of chimpanzee communication in the wild, both vocal and gestural, suggest a quite high level of sophistication, but not one requiring grammar (P. Marler, 1976). (See Chapter 5 for further discussion of evolution of language.) A sensitive and fairly comprehensive account of the human–ape relationship is *The Ape's Reflexion* (A. Desmond, 1979).

Suarez and Gallup (1981) undertook a research programme involving all three pongid species to investigate 'self-recognition', self-consciousness having long been held to be uniquely human. Using mirrors and a clever technique of dyeing their subject's foreheads while they were unconscious, it was possible to prove that both chimps and orang-utans were able to recognise themselves in mirrors but gorillas were not.

A further oft-cited study is Kortlandt (1972), in which chimpanzee responses to a stuffed leopard were observed. The fact that they used sticks and branches to threaten and attack the leopard is seen as suggestive of the kind of behaviour adopted by early tool-using hominids.

Finally we might mention a number of studies by A. Zihlman (Zihlman, 1979; Zihlman et al., 1978; Zihlman and Lowenstein, 1983) proposing that the pygmy chimpanzee, or bonobo, which inhabits a fairly restricted range of territory in Zaire, can be taken as the nearest living model of our ancestral proto-hominid. She and her supporters view it as similar morphologically to *Australopithecus* and consider its life-style to be akin to that which our first hominid progenitor would have been likely to have adopted, prior to abandoning arboreal life. Johnson (1981) has severely disputed this.

2 *Gorillas*

The major fieldwork on gorillas has been done by Schaller (1963, 1964) and more recently by Dianna Fossey and Alan Goodall (Goodall, 1979; Fossey, 1984). Few animals have had a worse press than the gorilla, and with less justification. Schaller's year-long field study and Fossey's more extended research reveal an essentially placid and easy-going herbivore with a clear social structure. The ferocious displays and charges of dominant males are generally only sustained long enough for the rest of the group (7–20 in number) to make their getaways, and are rarely followed up by actual attack. They are a seriously endangered species occupying a rather narrow environmental niche, and, although in many respects highly intelligent, are not very adaptable. Their role in directly stimulating hypotheses about human behavioural evolution has been far less than that of chimpanzees.

3 *Orang-utans (Pongo pygmeus)*

This most exotic and least researched of higher primates lives in South East Asia and is seriously endangered. The major field study is MacKinnon (1974) and the principle current researcher is Galdikas, set on her way, like D. Fossey and Jane Goodall, by Louis Leakey. Orang-utans have been shown to be capable of using a flint flake to cut string (Wright, 1972), and they sometimes adopt face-to-face copulation, once thought a uniquely human position. They are the most arboreal of the pongids and their locomotion has received particular attention as a possible index of the function of various anatomical features of the wrist, ankle and pelvic structures. Its knuckle walking has been closely studied (e.g. Tuttle and Beck, 1972). Interest in orang-utans is growing with the increased likelihood of their being descended from *Sivapithecus* and Schwartz's claims of close affinities with the hominids (Schwartz, 1984).

4 *Baboons (Papio)*

Although phylogenetically more distant from us, belonging to the *Cercopithecidae* rather than the *Hominidae*, baboons, as savannah-dwelling primates, have attracted considerable attention. The major study by Washburn and DeVore (Washburn and DeVore, 1961; DeVore and Washburn, 1964) was among the first to emerge in the post-World War II ethological tradition, though the principal authorities are now Kummer (Kummer, 1968) and Altmann (Altmann and Altmann, 1970). Among the interesting features they show is a fair measure of flexibility in social organisation according to habitat. South African baboons, living on flat plains, live in large packs or tribes, while those in Ethiopia, occupants of rocky valley terrain, are organised into smaller family units. Due to the relatively greater accessibility of baboons to observation it was among them that behavioural patterns of grooming, dominance hierarchy maintenance and other features of social structure were first studied in primates, and it was to baboons, rather than chimpanzees, that speculative palaeoanthropologists of the 1960s looked for images of proto-hominid behaviour.

Primatology is an extensive area with a now enormous literature. Researchers on topics relevant to hominid evolution include, in addition to those mentioned, Chevalier-Skolnikoff, Suzuki, Fouts, Vauclair, Beck and Gibson. A recent, well-received survey of all primates is Kavanagh (1984). Passingham (1982) is a compendious integration of the work bearing on human behaviour and evolution, of particular value to psychologists.

Several points should be made about the role of primate behaviour in the study of our evolution. Firstly, that finding similarities between them and us cannot in itself shed any light on the differences between us. Secondly, that humans contrast with all other higher primates, especially the pongids, in being highly environmentally adaptable. These species by contrast are isolated, tied to narrow ecological niches by diet and life-style, vulnerable and sensitive to disturbance. It is one of our most notable characteristics that we are the very opposite of all these. Their present highly specialised life-styles have presumably evolved since their divergence from us, rendering them unrepresentative of the Miocene species from which we both descended. Many writers do, as we will see, assume a more-or-less chimpanzee-like life-style for our last arboreal ancestor, but this can hardly be taken as firmly established. Thirdly, in relation to the learning powers of chimpanzees and other primates, a distinction must be drawn between ability to learn a behaviour and ability to generate it. Behavioural innovation is the province of a small minority even in humans, but our communication system and social structure ensure that such innovations spread very rapidly indeed. The fact that a species can be taught a particular kind of behaviour in no way entails that they would ever spontaneously come to exhibit it. Even if chimps could be taught grammatical language by humans, the fact remains that they did not invent it themselves, while we presumably did. This entire topic will concern us again in a later chapter.

What is characteristically *human* in our behavioural repertoire will be the very features which comparative studies will *not* reveal (Wynn, 1982). This is not to counsel despair, for the reverse side of the coin is that the more of our supposedly unique behaviour *is* revealed by comparative studies to be shared by other species, the smaller the uniquely human component becomes.

H Evolutionary theory

The broad consensus among biologists established by the 'Modern Synthesis' in the 1940s came under increasing pressure during the 1970s.

Although the fact of evolution was not challenged, there were growing doubts about the adequacy of the 'gradualist' model proposed by the Modern Synthesis to account for all the data. In particular the process of speciation has come under intense scrutiny. According to the gradualist model, new species arise as the result of a steady accumulation of minor changes which at some point reach sufficient magnitude for it to be necessary to say that a new species has emerged. This new species would not be interfertile with surviving members of the earlier one, or such fertility would be very low and natural hybrids unlikely to appear. The transformation from one species into another can be brought about in a number of ways: if a section of the population becomes isolated then it will tend to diverge as it evolves in adaptation to its new environment; if the environment in which a species lives undergoes change the species will, as a whole, undergo corresponding evolutionary adaptation. In any case no species is entirely static, any more than any environment is, except perhaps parts of the sea, being subject to constant selective pressures of covert kinds which slowly modify its behaviour and morphology. Species are not fixed entities, but temporally transient clusterings around adaptively efficient means of variation. The gradualist school based its model on the numerous breakthroughs in genetics, ecology and mathematical demography achieved during the earlier part of the century. The major problem for this account is that the fossil record still does not seem to bear it out. While fossil-record gaps could at one time be reasonably ascribed to the patchiness of the record itself, many of the expected transitional forms still have not appeared after a century or more of accumulated fossil data. We still find species entering and exiting from the palaeontological stage with a degree of abruptness apparently at odds with the gradualist story. Examples of smooth transition of forms from one species to another are fairly scarce. This is not to deny that evolutionary sequences can be traced, but that they more often take a step-wise succession of speciation 'events' rather a seamless continuum.

It is now customary to differentiate between 'macro-evolution', the rise of new species, phyla and genera, and 'micro-evolution', adaptive change within a species. Micro-evolutionary changes can indeed be tracked over time – brains tend to be slightly larger after a million years or so, teeth more efficient and so on. The problem is speciation. In 1972 Eldredge and Gould published a heretical paper advocating a 'punctuationist' model for evolution, in which speciation events were rapid and different in character to micro-evolutionary changes which occurred within species. This was further elaborated in Gould's major work *Ontogeny and Phylogeny* (1977). On this model species arise rapidly within small fringe populations isolated from the main population and occupying an environment which differs from that to which the species as a whole has become stably adapted. The main population is static, more or less, having achieved an optimum level of adjustment. Mutations within a large population could not in any case spread, however favourable, due to rapid 'dilution' in the large gene-pool. Although isolation of somewhat this type has long figured in gradualist scenarios of speciation, the mechanism involved has been a simple gradual progressive divergence of the two stocks, and it has been seen as only one among a variety of speciation routes. In Gould's theory it is the principal mechanism, and the rate of divergence in the small isolated group is rapid. Selection pressures are intense and favourable (i.e. reproduction-enhancing) mutations can quickly become established. If successful this new species will eventually oust the parent species when their habitats reconnect. Species are seen in this model as possessing a holistic coherence, and arising from rapid internally generated changes at the genetic level, 'to present fundamentally new features as *faits accomplis* to forces of selection (that may then accept or reject) or drift (that may then fix or eliminate)' (Gould, 1983b, p. 363). The traditional mechanisms of selection and genetic drift are thus relegated to a micro-evolutionary role, important, but inadequate to explain speciation itself.

The literature on this topic is too vast to consider here, but a recent important critique of Gould's position is Ayala (1983). Ayala claims that micro-evolutionary processes as now understood are compatible with either gradualism or punctuationism, and that there is no intrinsic conflict between macro- and micro-evolutionary theory in Modern Synthesis terms, nor any need therefore to invoke Gould's third kind of evolutionary process to explain speciation. The gradualist v. punctuationist matter is empirically open. Perhaps Ayala's most telling point is that the suddenness of speciation in the punctuationist model is still long enough in time for orthodox micro-evolution to bring about the observed changes. The geologically 'sudden' still represents thousands or perhaps millions of generations. In some ways stasis is more puzzling than change, for why do certain species persist apparently unchanged for millions of years (e.g. sharks and many shell-fish) even though their genetic material changes constantly? The usual answer is 'stabilising selection' – any appreciable degree of deviance from present form is too detrimental to the individual's reproductive fitness for it to have any offspring. This is not, though, our species' problem!

The punctuationist model of speciation is itself problematical at a number of points: the fossil record cannót show reproductive isolation, nor can it differentiate between morphologically similar species. For speciation to be observable in the fossil record it must, almost by definition, involve major morphological change. Even where this is seen, there is no proof that it does not involve the entire phyletic lineage *without* splitting having occurred. The status of the Eldredge and Gould model is still a matter for eager debate among evolutionary theorists, therefore, with neither camp clearly in the ascendant.

For us the issue becomes important as it affects the background orientations of palaeontologists when considering the four or five major transition points: the *Australopithecus* emergence and hominid-pongid split; the *Australopithecus* to *H.habilis* transition; and the subsequent *habilis-erectus*, *erectus-sapiens* and Neanderthal-Modern transitions. If Ayala is right that there is no way of deducing from micro-evolutionary processes alone which model is correct, we will probably have to await a solution from elsewhere than within the direct study of human evolution to the problem of hominid speciation.

Gould's are not the only radically new ideas to have been proposed. Dover (1982) proposes a 'molecular drive' model, explaining changes primarily by processes at the molecular level. Of more impact on human evolution thinking, though, has been the 'sociobiology' school. This emerged from ethology in the 1970s and was launched by E. O. Wilson in a massive and ambitious work: *Sociobiology: The New Synthesis* (1975), followed by another leading exponent's *The Selfish Gene* (Dawkins, 1976). Although their models of issues in human evolution will be evaluated in due course, we need to say a little about their general stance.

Central to sociobiology is the doctrine that behaviour patterns, as well as physiology, are determined at the genetic level in a fairly inflexible way. The presence of a given form of behaviour in a species' repertoire is to be accounted for by identifying how it enhances either individual 'reproductive fitness' or 'inclusive fitness' – the enhanced transmission of the genes possessed by the individual so behaving (this latter may involve a given individual *not* reproducing in order to enhance reproduction of genetically similar kin, in which case the phenomenon is termed 'kin-selection' rather than individual selection). Most controversial is their belief that behaviour is genetically determined on a far wider scale than previously accepted. Whereas insect behaviour, stereotyped behaviour such as nest-building by birds and reflexes such as blinking have long been accepted as genetically 'wired-in', sociobiologists extend this to cover social structure, psychological differences between the sexes and phenomena such as altruism and aggression.

This strong 'hereditarian' commitment has roused tremendous controversy, as it seems to imply a pessimistic or fatalistic view of the human condition and the possibility of improving it. (This is counter, in fact, to the classic Darwinian model of evolution in which constant variation implied the very opposite of fatalism – a species could steer away from the legacy of previous adaptations if they threatened its survival.) The 'Selfish Gene' model of Dawkins views animals as virtually no more than puppets of these hidden entities and has been criticised from numerous directions, from philosophy (Midgley, 1979a, b) to technical genetic grounds (Fuller, 1978), leaving aside the ideological furore. It is a more speculative model than at first appears, even ignoring the fallacious personifying of hypothetical constructs, such as genes are, by Dawkins. Although the genetic basis for behaviour in a very fine-grained and hard fashion is basic to the theory, the way the relationship between the two levels is mediated is relatively unexamined. Sociobiology has nonetheless offered some intriguing analyses of certain forms of behaviour, such as altruism in particular, which will require closer attention in Chapter 5 below.

J. M. Smith (1978, 1983), a leading theorist in the field, has explored another tack which may be considered as sociobiological: the application of Games Theory to social behaviour (some of this may sound very old hat to social psychologists familiar with the work of people like Thibaut and Kelley (1959)). He has analysed the effects, in terms of pay-offs for participants, of adopting 'hawk' or 'dove' (or 'sheep' or 'wolf') strategies in conflict situations. This has shown how populations may comprise an optimal combination of timid and aggressive individuals, or optimum rates of adopting Hawk and Dove policies, which ensure minimisation of escalation and injury. He terms these 'Evolutionarily Stable Strategies' (ESS's). (Bennett and Dando (1983) have already seen a use for this in studying power bloc confrontation in international politics.) Smith, though confessing sympathy for Dawkins, is open as to the extent to which this is explicitly coded for genetically. There would, however, seem to me to be a lot of work yet to be done on all the other faunal strategies (what of limpets and snakes?)!

A recent development among some evolutionary theorists, notably Lewontin (1982) Plotkin and Odling-Smee (Plotkin and Odling-Smee, 1982; Odling-Smee, 1983) challenges the very assumption that genes and environment (or nature and nurture to use traditional terminology) can be sensibly differentiated anyway. If this move succeeds it will make the sociobiology account look rather simplistic, and also, harking back to the previous issue, will perhaps provide a route by which macro-evolutionary theorising can be developed. Lewontin is an associate of Gould's at Harvard, while the other two are at University College, London.

This new development, in Lewontin's 'strong version', as it may be called, sees older concepts of evolution as handicapped by commitment to two underlying metaphors for development – 'unfolding' and 'trial and error'. The first, transformational, metaphor is rooted in embryology; the second, variational metaphor comes from 'natural selection'. Both are inadequate as models of either ontogenetic or phylogenetic development. The way a genotype develops is far from being a totally specified 'unfolding'; rather it is susceptible from the start to variations in the 'environment' which can sometimes switch it from one pathway to another.

> The problem with trying to sum up the embryological development of an individual in any simple scheme is that the relation between genes, environment, and organism is extraordinarily diverse from one species to another, from one organ or tissue or enzyme to another, from one genotype within a species to another. Some organisms show some regulation of development some of the time. Some species develop virtually the same morphology in any non-lethal environment, while others are remarkably plastic. (Lewontin 1982, p. 156)

But neither does simple trial and error, or natural selection, operate in a straightforward adaptive fashion; traits may appear for a host of reasons other than their 'adaptive' value, from fortuitous chromosomal linkage to a trait that *is* being selected ('hitch-hiking') to allometry. While also 'it is the organism itself that determines the actual object of natural selection' – whether cold-resistance, for example, is met by increasing the fur or subcutaneous fat, or increased wing-size (in *Drosophila*) is achieved by more cells or bigger cells. In fact, Lewontin argues, the organism is always actively engaged with its environment, constructing, interpreting and changing it continuously, not merely passively adapting *to* it. The adaptive value of a trait is not fixed, and will even change according to the nature of the gene-pool of the species as a whole. (To be three feet high will have quite different consequences depending on whether the rest of the population is on average two feet or six feet!) For Lewontin both the notions of a clear organism-environment distinction and 'adaptation' as previously conceived are inadequate. Genetic and environmental levels interpenetrate in the organism; adaptation is as much active as passive.

Plotkin and Odling-Smee also consider the traditional nature-nurture divide to be misconceived and obsolete. Instead they identify a hierarchy of levels at which 'life-units' (from the DNA molecules through to societies) can store information, with varying degrees of feedback between them. This encompasses both learning and instinct as hitherto understood (see Figure 3.10). It will be seen that this incorporates Lewontin's notions by including 'variable epigenesis' (flexible developmental processes) and feedback from individual learning to gene-pool level (but *not* to DNA molecules!). It does not so clearly represent the active changing of the environment to which Lewontin refers.

In the light of these new developments in evolutionary theory particularly 'evolutionary epistemology' (see Chapter 5), we may expect to see considerable modification in the sociobiology camp, and the general way in which macro-evolutionary processes are conceptualised will be subject to much revision in the near future. One consequence of this, hopefully, will be to open up the issue of human and hominid behavioural evolution at the theoretical level, permitting a broader range of theoretical moves

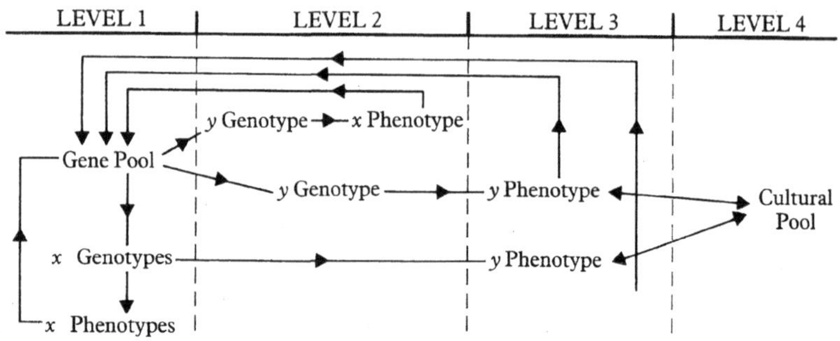

Figure 3.10 The nested hierarchy of knowledge-gaining processes. The prefix *x* indicates non-information-gaining units. The prefix *y* indicates flexible, autonomous information-gaining units. From Plotkin and Odling-Smee, 1982. *Reproduced by permission of H. C. Plotkin and Academic Press.*

and hypotheses than were possible under the original Modern Synthesis consensus. Plotkin (1979) provides a useful survey of the connections between evolutionary and behavioural studies prior to the late 1970s, advocating research into brain-behaviour relationships as an important level in clarifying the way in which genetic factors are translated into actual behaviour.

This section might usefully be ended by stressing that no serious biologist at present disputes the evolutionary perspective in broad terms. The creationist back-lash in the USA has made a spurious distinction between theory and fact, equating them with speculation and certainty. This betrays such a fundamental failure to grasp the nature of scientific inquiry, such a misunderstanding too of the very meanings of all four terms, and their use in western thought, that one is inclined to leave the entire issue to one side. Although some philosophers of science would not be happy in every detail, the following quote from a group of US scientists countering the creationist stance as exemplified in a decision by the Texas State Board of Education may I think stand:

> There is a slow progression toward increasing certainty . . . but at no point do 'theories' become 'facts'. Evolution theory has been subjected to vigorous verification over the last 100 years and in its broad outlines is as certain as anything we have in science. We are as sure that biological evolution has occurred, and continues to occur, as we are that matter is composed of atoms. (quoted in *Scientific American* 250 (3), 1984, p. 54)

Conclusion

Having followed this survey of the material and methods on which the study of human evolution is based, the reader will appreciate what a massive job of synthesis is in store if the evolution of our behaviour is ever going to be comprehensively understood. Such a synthesis is not at present possible. What we do have are a number of more or less broadly based accounts of the process by workers in a variety of the major contributing disciplines, plus some outside of them. It is with these that the rest of this book is concerned.

CHAPTER 4

Brains and sex, meat and reason

Introduction

We now turn from the physical evidence to its behavioural interpretation, concentrating on the initial phase of hominid evolution, from the advent of bipedalism to the appearance of *Homo erectus* and the associated increases in cultural complexity and brain size. Regarding this 3–4 million-year-long period (on current assessments) agreement is more restricted than popular accounts might lead one to imagine. There are though a number of points on which there is a fair degree of unanimity.

1 Hominids probably owed their origin to the circumstances attending environmental changes at the Miocene-Pliocene transition period, notably decline in dense forest cover, mountain-building, and a temporary separation of Africa from the Eurasian land-mass. There is increasing reluctance to accept a simple 'trees-to-savannah' model for this early period. Although a savannah environment is accepted from about 4 myr onwards, it is felt to be inadequate to account for the characteristic hominid traits such as bipedal walking.

2 The immediate ancestor is widely, if not universally, assumed to have been an arboreal primate, ancestral to the pongids, and resembling the chimpanzee or the pygmy chimpanzee in life-style, though being smaller than the former. *Ramapithecus* has fallen from favour as candidate for first hominid since about 1980 (see previous chapter), although was still frontrunner when several of the position-statements discussed below were being written.

3 Dietary and food-acquisition factors were central determinants of the basic characteristically hominid adaptations, notably bipedalism and tool-use, although the respective importance of these and social organisation or reproductive strategy factors with which they were interwoven is disputed.

4 It is generally, though not quite universally, held that significant *expansion* of brain size did not occur prior to the advent of *H.habilis*. Significant changes in brain *organisation* on the other hand *are* assumed to have taken place by the time Lucy appears on the scene.
5 Similarly it is generally, though not quite universally, held that hunting as such appeared with *H.habilis*, although meat-eating, to an unknown degree, is assumed to have developed previously via scavenging.
6 The life-style of the gracile australopithecines and *H.habilis*'s immediate precursors is normally depicted as a gathering-scavenging one, with male provisioning of females and young, especially with meat, but beyond this there are a variety of detailed models on offer, and it is a highly contentious area.

When we try to bring the picture into clearer focus there is disagreement on nearly every substantive issue. To give some idea of the range of this discord the following are among the matters in dispute; the circumstances in which bipedalism first arose and the adaptive advantage it originally conferred; whether early hominids were monogamous or polygynous, and if the latter, when and why the shift to monogamy eventually occurred; whether males or females or both initiated such developments as bipedalism and tool-use; how important meat was in the proto-hominid diet and how it was obtained; the degree of male parenting 'investment' and its explanation; all aspects of the evolution of language; whether social relations or tool-use were the primary factor in the evolution of 'intelligence'; the ecological context in which hominids made their advent; levels of sexual dimorphism; levels of female dependency on males (and vice versa); adequacy of 'neoteny' as an explanation of physiological behaviour-related changes; various aspects of brain evolution; the role of recapitulationist models in theorising on behavioural evolution; relevance of comparative studies of primate behaviour; and what the first stone artefacts were for.

The possible solutions to such issues are permutated in numerous ways by different authorities, but there are few neatly identifiable theoretical 'schools'. Rather we find a continuum of variations on the central themes of food and diet, sex and social organisation, tool-use and language, and models of how they interacted to produce positive feedback loops capable of eventually generating the continuous technical, intellectual and cultural advances associated with *Homo*. Table 4.1 summarises the positions adopted on a number of key issues by some of the leading theorists. I have disregarded most of the proposals made prior to 1978 as these have either been updated by their authors or become obsolete in the light of later work.

TABLE 4.1 *Models of hominid behavioural evolution up to* H. habilis

Issue	*Author* Hill[1]	Holloway[2]	Isaac[3]	Lovejoy[4]	Morgan[5]	Parker and Gibson[6]	Potts[7]	Tanner[8]	*Notes*
Monogamy (M) or Polygyny (P)	P	M	?	M	–	–	–	P[a]	[a] though restricted
Importance of male provisioning High (H), Medium (M), Low (L)	H	M	M	H	–	–	–	L	
Female dependency on males (H/M/L)	H	M	M	H/M	L[a]	L	–	L	[a] Morgan (1972)
Tools Primary (P) or Secondary (S) in hominisation	P	S	S	S	–	P	–	P	
Use of home base – Yes (Y)/No (N)	Y	–	Y[a]	Y	–	–	N	Y[b]	[a] uncertain [b] fem. domestic group
Primary factor in bipedalism	male prov.[a] + tool use	–	unsure	male prov.	aquatic pre-ad.[b]	extract. foraging	–	female carrying	[a] prov. = provisioning [b] pre-ad. = pre-adaptation
Food-sharing within group	Y	Y	Y	Y	–	Y	Y	Y	

Continuous female receptivity	Y	Y[a]	?[b]	Y	N	Y[a]	–	N	[a] qualified [b] uncertain
Cause of hominid sexual dimorphism	epigamic	epigamic[a]	–	epigamic	aquatic[b] pre-ad. epigamic/ m & c	–	–	m & c[b]	[a] to limited extent [b] m & c = mother and child interaction
Importance of meat-eating (H/M/L)	H	H	H	M	–	H	H	L	
Language present early? (Y/N)	–	Y	?[a]	N	Y	N	–	?[a]	[a] uncertain
Hunting present, and important, early? (Y/N)	Y	–	?[a]	N	N	N	N	N	[a] uncertain
Brain enlarged early? (Y/N)	N	N	N	N	Y[a]	N	–	N	[a] Martin (1981) also gives a qualified Y. Morgan (1984) modifies to limited Y only
Scavenging of meat (Y/N)	Y	Y	Y	Y	–	Y	Y	Y	
Social (S) or Tool (T) origin of intelligence	T	S	S[a]	S	S	T	–	T/S[b]	[a] by implication [b] interwoven

[1] Hill (1982); [2] Holloway (1981a); [3] Isaac (and Crader) (1981, 1983); [4] Lovejoy (1981); [5] Morgan (1982); [6] Parker and Gibson (1979); [7] Potts (1984) discussed in Chapter 6; [8] Tanner (1981).

In this chapter I propose to deal firstly with some of the problems and controversies arising in connection with the interpretation of hominid brain evolution, this being at the heart of any consideration of behavioural evolution. Secondly, the specific issue of the origin of bipedalism will be discussed, as this constitutes one of the central riddles of the entire topic. Thirdly, the major position-statements referred to in Table 4.1 will be reviewed. Finally, I will discuss Elaine Morgan's *Aquatic Ape* (1982) since, although it has been treated dismissively by most academic palaeoanthropologists, in the present author's opinion it deserves a further airing and warrants serious consideration as offering a parsimonious solution to a number of otherwise anomalous hominid traits. Fuller treatment of (a) the genes-culture relationship and (b) the evolution of language is reserved for Chapter 5.

A Brain evolution

Although a 'brain-first' model of human evolution is no longer on the cards, it still constitutes the obvious starting point for any account of the evolution of human behaviour, the brain being the organ primarily responsible for controlling and determining behaviour. Current research in this area is fast-moving and plentiful, with numerous controversies and a diversity of methodologies. Major figures include Harry Jerison whose *Evolution of the Brain and Intelligence* (1973) is the principal reference point for modern research, albeit much of it has since been revised by both the author and others. Jerison's ideas can be tracked forward from this text through numerous subsequent papers and peer commentary letters. Ralph Holloway, whose endocast research was mentioned in the previous chapter, is a major force in the area and his wider view of behavioural evolution is discussed later. As far as brain research is concerned, he is a firm advocate of studies on brain organisation and form as well as simple size. Though confining herself somewhat more narrowly to brain evolution research, Dean Falk has contributed in similar areas to Holloway and is at present in dispute with him over australopithecine cerebral anatomy. Workers long eminent in the field include Paul D. MacLean, Leonard Radinsky, and palaeontologists such as P. V. Tobias. Recent major contributors include R. D. Martin, Richard Passingham, Este Armstrong and Bennett Blumenberg. For more extended accounts of the issues readers are referred to Passingham (1982), Blumenberg (1983), Armstrong and Falk (1982) and Ciochon and Corruccini (1983).

For many years evolutionary theorists saw the human brain as not only unique with respect to size, but the very key to our self-evident supremacy. A. R. Wallace felt unable to explain it by natural selection since its abilities exceeded its actual performance. 'Brain-first' models

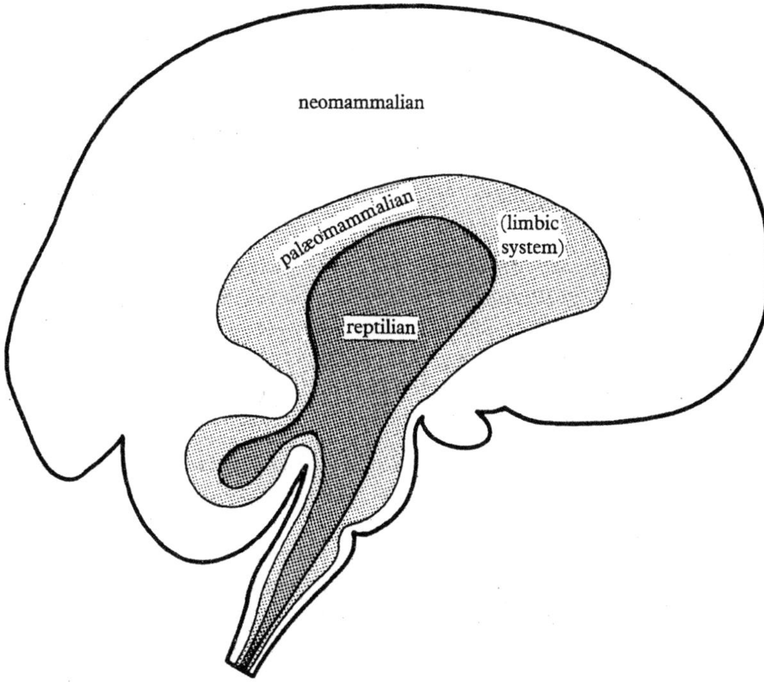

Figure 4.1 The triune brain.

dominated early twentieth-century thinking as we saw in Chapter 2. The current picture is more complicated and in flux.

But first, let us remind ourselves of the nature of the organ in question. The modern human brain is on average 1330 cc (and 1.3 kg) but ranges from 870 to 2150 cc. Structurally it may be described as 'triune' (MacLean, 1982) (see Fig. 4.1). The central 'protoreptilian' brain is concerned primarily with basic social communication, recognition and survival routines such as foraging, homing, etc. MacLean includes here mating and courtship rituals, hierarchy establishment and territorial defence displays. Encasing this is the 'palaeomammalian' brain, usually referred to as the 'limbic system'. Closely bound up with sexual, feeding and fighting functions, he considers it to be the seat of the emotions and maternal behaviour as well as, interestingly, feelings of conviction, truth and falsity. Neither of these levels possesses speech. Thus so far we have a core of basic behaviours governed by the reptilian brain, overlain by a limbic system which seems to add a dimension of emotion, feeling and evaluation. At the outer level is the 'neomammalian' brain or neocortex, a phenomenon which 'mushrooms progressively in the higher mammals'.

The role of this can be summed up as 'information processing', including learning, language, integration of motor-behaviour and handedness. MacLean notes a difference between *H.s.sapiens* and *H.s.neanderthalensis* regarding the apparent extent of development of the 'prefrontal cortex', which gives us our high foreheads, though this has not gone unquestioned; Holloway (per. comm.) suggests it is a misleading visual impression due to Neanderthal brow-ridges, and that no evidence of actual brain deficit in this region exists.

> Clinically, there are indications that the expansion of the prefrontal cortex affords an increased capacity to relate internal and external experience and thus to identify one's inner feelings with those of other beings. (it) . . . is also recognized clinically to play a fundamental role in relating past, present, and future in regard to 'looking ahead' and making possible both anticipation and choice. (p. 311)

This 'triune' differentiation is marked, MacLean claims, by clear physiological differences in cellular structure, though the limbic system is widely interconnected with the other two.

Over the last century, and especially since the pioneering work of Penfield (Penfield and Rasmussen, 1950; Penfield and Roberts, 1959), the functions of various parts of the brain have yielded somewhat to investigation, as MacLean's model indicates. Even so, it is a topic still having much uncertainty, and even such well-advertised and long-established connections as those of Wernicke's and Broca's areas with speech production and comprehension respectively are not exactly cast-iron in detail though research is rapidly clarifying this particular issue. (Fig. 4.2 and Table 4.2 give the connections as presently understood.) Large tracts of the brain are fairly unspecialised. As a result of technological innovations in methodology, advances in this area are proceeding quite rapidly, though at a high price in terms of primate lives.[1]

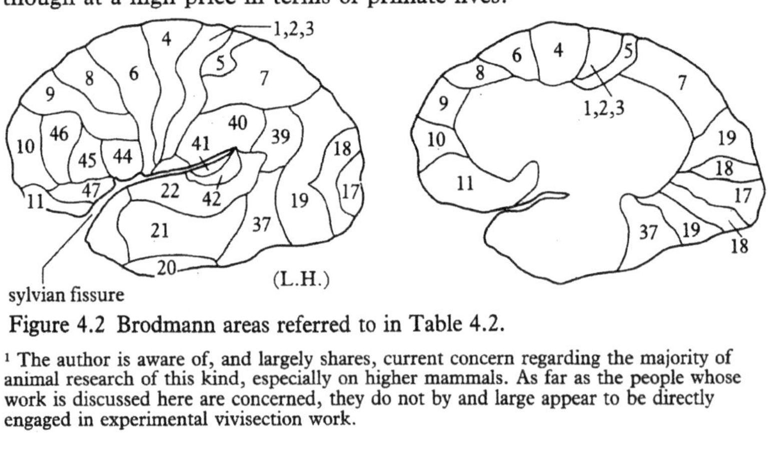

Figure 4.2 Brodmann areas referred to in Table 4.2.

1 The author is aware of, and largely shares, current concern regarding the majority of animal research of this kind, especially on higher mammals. As far as the people whose work is discussed here are concerned, they do not by and large appear to be directly engaged in experimental vivisection work.

TABLE 4.2 *Functions of the human cortex*

Function	*Brodmann area*[1]	*Lateral differences*
Vision		
Primary[2]	17	
Secondary[2]	18–21, 37	Naming L.H. Visual recognition, etc. R.H.
Audition		
Primary	41	Speech recognition 41, 42 L.H. (Wernick's Area)
Secondary	22, 42	Musical recognition 22, 42, R.H.
Body Senses		
Primary	1–3	Awareness of illness 7 R.H.
Secondary	5,7	Localisation and naming of body parts 7 L.H.
Sensory		
Tertiary (association areas)	7, 22, 37, 39, 40	Short-term memory 37, 40 L.H. Spatial ability 7, 40 R.H. Right-left differentiation 7, 40 L.H.
Motor		
Primary	4	
Secondary	6	
Tertiary	9–11, 45–7	
Eye-movement	8	
Speech	44 (L.H. 'Broca's Area')	
Language-comprehension	22 (L.H.)	

[1] See Fig. 4.2. Named after the German physiologist who identified them in 1909.

[2] Primary projection areas are those receiving direct input from sensory systems or project to the spinal motor systems. Secondary are those to which primary areas send fibres, or from which they receive them (depending on whether they are sensory or motor areas respectively). Beyond those are the 'tertiary' or 'association' areas.

(Adapted from Kolb and Whishaw, 1985)

In studying brain evolution we are, of necessity, compelled to focus on gross external characteristics of size and form, inferring as best we can from these to their functional or structural implications. In general we will be constrained to features of the neomammalian cortex, the 'grey matter' surface layer of the brain which is related most closely to characteristically human functions of language, learning and motor-co-ordination. Passingham (1982) identifies three broad issues: size, organisation and the presence or absence of special features.

1 *Size*

Although it is taken for granted that human brains are unusually large, this apparently common-place observation has proved surprisingly difficult to formalise. Passingham and Ettlinger (1974) concluded that our brain was three times the size to be expected of a primate of our build. The problem in assessing the degree of atypicality in a given species' brain size is establishing a standard of comparison. Absolute brain size cannot be used. Whales and other cetaceans have brains up to five times the size of the human brain, while the elephant's is four times the size of ours. This difference can be accounted for by the gross animal size – such colossal brains are required for controlling their larger bodies, in an allometric fashion. At the other end of the spectrum, smaller animals have *relatively* big brains; adopting a simple brain/body weight ratio we come out as inferior to mice and even to some primates. The truth is that although brain weight increases with body size, it does not increase at the same rate, and relative brain size decreases. The regression slope as recently recalculated by R. D. Martin (1982) is about 75 per cent, a figure replacing the oft-cited 66 per cent figure of Jerison (1973). There are further complications: although this allometric function itself might be constant, there is a degree of phyletic variation in mean brain size for a given body weight, thus predator species tend to have relatively bigger brains than prey and non-predator species (while in bats, fruit-eaters have bigger brains than insectivores). In primates, furthermore, females have relatively larger brains than males. The Passingham and Ettlinger estimate given above rests on data from 39 primates (including *H.sapiens*) (Stephan et al., 1970), which provides a proper primate standard of comparison for humans.

The term now used to refer to a species' relative brain size is 'encephalisation', and this is measured by the 'Encephalisation Quotient' (EQ) which we have met before. This EQ requires a little attention in its own right, but before this there are some other questions to consider. Firstly, sticking to the physiological level, what does brain enlargement actually involve? There are three possibilities; (a) an increase in the number of brain cells (b) an increase in the interconnections between them (neurons), (c) an increase in both. It has now been established (Steele Russell, 1979) that brain size increase is accomplished primarily by (b). The number of brain *cells* remains remarkably constant throughout the *Mammalia*; what increases is their degree of interconnectedness, the cell-density drops and the complexity of the neuron network increases.

A second question is: what parts of the brain have expanded? Have some regions enlarged more than others? Or have they all increased at the same rate? The former would suggest some kind of specialisation in addition to enlargement, the latter that allometric increase is more or less sufficient to account for change in brain form (i.e. allometry considered in relation to brain size itself; we can ask, that is, 'given a primate brain of size *x*, how big should region *y* be?', just as we can ask, 'given a primate of size *x*, how big should its brain be?'). This is a less clearcut matter. Jerison has controversially argued (Jerison, 1982) that the relative sizes of the various parts of the brain are such as one would expect of a primate brain that size, its surface convolutedness being the most economic solution to the problems of increasing neuron-density within the volumetric constraints imposed by human anatomy. The olfactory lobe seems to be an exception, it having suffered a relative decrease in size, i.e. it has expanded more slowly. Armstrong (1982) has presented a more complex picture, suggesting a considerable degree of 'mosaicism' in hominid brain evolution. In particular the data seem to show a preferential increase in neurons in a part of the limbic system called the 'anterior principal nucleus', and less than expected in the thalamus. Clear psychological implications of such findings have not as yet been identified. The message must be, though, that there are interspecific differences among the primates as to the regions which are most enlarged or most conservative, even though these differences can only be detected at the microscopic rather than the gross morphological level.

Thirdly, an important question as yet unasked: does size matter anyway? The equation of brain size with 'intelligence' is an implicit assumption which provided much of the rationale for early research on the subject. In the nineteenth century the size of the brain came to be taken as an index of intellectual virility in a way not unlike the significance popularly ascribed to another organ. Races, sexes, and social classes were compared. Anthropologists established a mania for craniometry which has never entirely died out, though no longer in the original craniolatrous context (craniolatry-idolisation of the cranium, a term I introduce for the first time, as far as I know). Unfortunately, although there is obviously some broad relationship between a species' brain size and its brightness (as viewed anthropocentrically), the correlation within humans between brain size and intelligence is so low as to be virtually insignificant (above pathologically low levels) and no studies appear to have been done in which body size and social class are controlled for. Some very stupid people have had very big brains, and vice versa. Keith (1948) suggested that there was a certain critical brain size which represented a 'Rubicon' above which human levels of intelligence were possible, and he placed it at 750 cc. This notion has fallen into disuse (see Jerison, 1973, p. 388). At the same time it is universally agreed that the complexity of the human cortex is the physiological underpinning of our enhanced learning and memory capacities and our unique linguistic ones. But is this a matter of size or organisation? Even if we cannot move in a simple fashion from size to usage, the fact remains that hominid evolution has been marked by a progressive enlargement of cranial capacity and that measures of this have always figured largely in efforts at tracking hominisation in the fossil record. But the EQ issue must be tackled before we can turn to brain size evolution as such.

To measure relative brain size you need at least three items of data: a brain weight, the body weight of the individual it belonged to, and a standard of comparison. As Jerison defined it 'The encephalization quotient EQ is the ratio of actual brain size to expected brain size. . . . The expected brain size is a kind of "average" for living mammals that takes body size into account' (1973, p. 61). (In fact his 60 per cent regression mentioned above.) More recently Holloway and Post (Holloway and Post, 1982) have reviewed the EQ literature and the various equations offered by Jerison and, earlier, Bauchot and Stephan (Bauchot and Stephan, 1969) for obtaining an EQ measure. Comparing ten different EQ formulae, they show how outcomes are affected by the data points used to derive them in the first place. Although 'fossil hominids are always intermediate between the extant pongids and modern *Homo* . . . *the degree of intermediacy depends on the data base selected*' (p. 59, their italics). And later on,

> . . . placements of hominids with regard to modern *Homo*, or chimpanzee, can vary by almost 20%. These kind of shifts can effect (*sic*) interpretations of time-related shifts in the nature of hominid mosaic evolution, leading to a polarisation of conceptual viewpoints regarding the role of the brain in human evolution. (p. 73)

They suggest in this paper a new formula in which *H.sapiens* EQ = 2.87, as compared to 6.93, using Jerison's formula. They then propose that the most useful measure is to be obtained by dividing the EQ for a specimen by this human EQ to arrive at a percentage figure describing its relative brain size in comparison to ours. (*H. sapiens* relative EQ thus = 1.00 by definition.) It is this procedure which Walker et al. (1983) used to arrive at 48.8 per cent for *Proconsul africanus*. The fundamental difficulty with obtaining reliable EQ measures for hominid fossils is a simple one – lack of post-cranial data on which to calculate the body-weight figure. Palaeontologists are really stuck here, relying on inferences from post-cranial material assumed to come from the same species, though not the same individual, and guesses based on more tenuous assumptions of comparability with modern species.

In looking at the course of the evolution of brain size then we are largely reliant on gross brain sizes rather than the relative ones a reliable EQ measure could provide. But estimates of EQ are needed to answer some central questions, in particular the timing of *relative* enlargement in ancestral brain size. To answer this thoroughly we would need to have complete sample skeletons of the succession of ancestral species, plus the same for non-hominid/non-hominoid primates contemporary with them to establish a valid comparison. Such samples are simply not there, to our knowledge, and we have to rely on hints and informed guesswork. Temporal tracking ideally requires a regular chronological spacing of samples, instead of which they are scattered quite erratically between the Miocene and the Neanderthal demise, with serious gaps in the late Miocene and at 3–2 myr in the late Pliocene/Pleistocene transition. In the face of what must in all honesty be considered pretty hopeless odds, the researchers have done what they can to track brain-size evolution. Their conclusions, tentative of course, can be summarised as follows: some degree of relative brain enlargement was possibly present among the hominoids even prior to bipedalism (Martin, quoted in Lewin, 1982). But the australopithecine brain size remains virtually constant from Lucy to *H.habilis* aside from a small increment from *A.afarensis* to *A.africanus*. Such slight increase as there is only reflects a general tendency for all mammals slowly to evolve larger brains over long periods. With *H.erectus*, the rate of increase accelerates in a roughly curvilinear fashion down to Moderns, but the data is so patchy as to be consistent with gradualist, punctuationist, or intermediate, scenarios (Godfrey and Jacobs, 1981). The largest human brains seem to have been those of the Neanderthals at around 1600 cc, although their shape was somewhat different to ours. There is a possibility that this is a false impression resulting from sampling factors in some way. The 'gracilisation' which accompanied the emergence of *H.s.sapiens* did not result in an appreciable fall in brain size.

Notwithstanding current scepticism over the 'Rubicon' concept, it surely remains a possibility in principle that the increase in brain size associated first with *H.habilis* did take the species through what mathematicians call 'a catastrophe'; a quantitative change which, though in itself very marginal, flips the system into a new state altogether (e.g. the freezing point of water). To repeat a point made elsewhere, innovation requires a very small proportion of the population, often only one, to be capable of the innovation, provided the rest can learn subsequently. It is the upper end of the ability distribution curve which determines the upper limit of a group's cultural and technological capacity, not the mean (this is not to endorse ideological elitism, there is a great plurality of such curves, not just one.) Size alone is nevertheless clearly not enough, we need to look instead to organisation. The size issue itself is something of a paradox. Humans *are* remarkably encephalised, and the process of 'hominisation' has been accompanied by their becoming so. And yet size as such has a relatively low correlation with intelligence among humans at the *intra*-specific level. The important difference between us and our nearest relatives might on balance have less to do with our brain's size, and more to do with how we use it.

2 *Organisation*

Adult brain size alone is, says Holloway (1983), 'the most distal expression' or 'sum' of the real developmental complexities which are far more interesting

> . . . and that *sum*, I would argue, is really far less interesting than the parts that went into making the whole . . . by all means, we must and will continue to study brain size, but let us not lose our perspective as to what *that* phenotypic expression implies. For one thing, it implies a large number of supportive or complementary social behavioral adaptations and biological sequelae to accommodate the final product. The blood supply, pelvic diameters and flexibility during parturition, social care and nourishment, 'learning', behavioral patterns emergent with extended growth and developmental periods, both in terms of extracting adequate energy resources from the environment (gathering and hunting), and a socially responsive network of caretakers, are only some of the most obvious supportive adaptations one can imagine. (p. 218)

This view of brain evolution as embedded in a total sociobehavioural field typifies Holloway's model of evolution, to which we return later in the chapter, and provides a bridge between the anatomical and physiological concerns of specialist palaeontologists and the broader psychological issues. It leads Holloway to concentrate, as far as brain evolution is concerned, on the more interesting 'parts', on such phenomena as asymmetry, presence or absence of 'speech' areas and the like. In pursuit of this he has developed a complex technique of 'stereotaxic' mapping of brains, or more usually endocasts, to provide a route to the comparative study of their morphology. The number of hominid fossil specimens which he has been able to study so far is 37 (1983), and the number of endocasts available for study altogether he gives as 40 to 50 'and most of these are incomplete'. Before reviewing his provisional conclusions, we need to backtrack and consider ways in which brain organisation can express itself at the gross morphological levels amenable to palaeoanthropological study.

It is now well known that the human brain is somewhat asymmetrical, an asymmetry widely held to be related to a 'lateralisation of function' which, if not unique to our species, we have taken to a further degree than any other species. The extent of hemispheric specialisation is, though, a matter of controversy. There are wide individual differences among us as to the extent of our lateralisation, but as a very broad generalisation the left hemisphere is typically ascribed the functions of language, and 'convergent' cognition, while the right is concerned with 'visuo-spatial' thought, imagination and 'divergent' thinking. This idea was so widely romanticised and inflated during the 1970s that the entire woes of the race were being attributed to a western suppression of, or failure to develop, the right hemisphere. Women were more 'right hemisphere' than men, orientals than occidentals, artists than scientists and so forth. Jaynes (1976b) published a book in which the origins of consciousness and language were explained in terms of the relationship between the hemispheres (of which more in Chapter 5). Recently psychologists and neuropsychiatrists have tended to adopt a more downbeat attitude, but there is agreement that lateralisation of function occurs, and that it typically takes somewhat the form just mentioned – which we may sum up as reason to the left and imagination to the right. The consequences of brain damage, differential electrical stimulation of hemispheres and differential presentation of stimuli to right and left halves of the visual field are often complex, enigmatic and bizarre, seeming to imply multiple consciousness or separate identities within the same individual. This is especially so if the commissural fibres connecting the two hemispheres are damaged or severed. Another aspect of the topic receiving increased attention is the extent of sex-differences in lateralisation. For a recent review of the localisation of function area see Bradshaw and Nettleton (1983) or Kolb and Whishaw (1985).

For us the interesting question arises as to why such a differentiation should have evolved in the first place. In considering this question Passingham (1982), though not the first to do so, points out the advantages of having complex behaviour directed from a single hemisphere:

> . . . such an arrangement (i.e. control by both hemispheres) would not be optimal for the execution of complex sequences of movement as in the production of speech. In such a case we would surely expect the highest skill to be achieved if the sequence was directed by one central programme, located in a single hemisphere, rather than by two separate programmes which must use the long commissural pathways between the hemispheres to co-ordinate their instructions. (p. 103)

Reorganisation could of course take forms other than lateralisation, for example the process of becoming bipedal, must, as Holloway insists (1983), have involved 'some restructuring of the nervous system . . . to permit the operation of a new locomotory system' (p. 226). It is interesting to speculate regarding the form this could have taken: if the newly bipedal animal was immediately preceded, as widely assumed, by a tree-dwelling primate similar to a chimpanzee which used all four extremities as 'hands' in a quadramanous fashion, the shift could have had dramatic consequences. Areas of the cortex previously required to integrate the posterior 'hands' with the anterior ones would be largely released from this function in the light of the gross simplification of the former's use once specialised

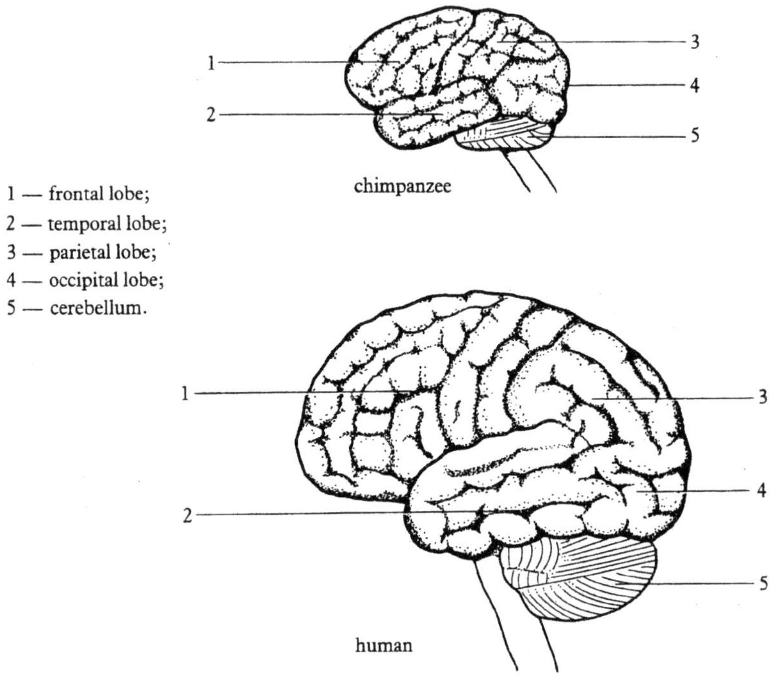

Figure 4.3 Chimpanzee and *Homo* brains, illustrating the apparent relative enlargement of the frontal and temporal lobes, the rearward shift of the parietal lobe and relative decrease in size of the occipital lobe.

1 – frontal lobe 2 – temporal lobe 3 – parietal lobe 4 – occipital lobe 5 – cerebellum

for walking. A surfeit of processing capacity would become available for the co-ordination of the remaining two hands. Holloway sees brain evolution as involving, in addition to simple allometric enlargement, both internal reorganisation and changes in ordering and structuring of its ontogenetic development, providing numerous points at which natural selection would be acting.

The most obvious difference between modern human brains and other primate brains, other than size, is the extent of both the neocortex and the cerebellum (see Fig. 4.3). Whether this difference is, at the gross level, greater than we would expect, on allometric grounds, for a primate brain our size has been questioned, as noted earlier (Jerison, 1982). There does, though, seem to have been a *relative* reduction in the occipital lobe, which governs visual association, by contrast with the remainder of the neocortex, particularly the temporal and frontal lobes. This has had the consequence of pushing progressively rearward the fold known as the *lunate sulcus*. A forward location of this fold has been taken as the hallmark of the pongid brain by comparison with the human one. Regardless of brain size, therefore, we could take a posterior lunate sulcus as evidence that the brain had begun to evolve towards its modern form. It would imply that the parietal lobe in particular had begun its expansion (Falk 1980a, b; 1982), an area long known to be related to learning and memory functions.

Is there any evidence for such a posterior migration of the lunate sulcus among our earliest hominid ancestors? We do have the Taungs skull endocast, Dart's *Australopithecus*. Unfortunately the majority of the relevant area is missing, but there are clues in what remains, notably the last few millimetres of a fold which would have extended back from the break. Dart himself believed this to be the lunate sulcus, as have most later investigators, including Holloway. Falk, though, has taken issue with this interpretation and believes the fold in question to be another landmark, the lambdoid suture.

> Contrary to current consensus, my study of all seven of the currently available South African australopithecine natural endocasts indicates that the sulcal pattern of the australopithecine brains appeared to be ape-like rather than human-like. . . . I believe there is little or no evidence with respect to the lunate or other sulci (i.e. superficial evidence) that elucidates the nature of brain evolution in early hominids. (1982, p. 221)

Holloway has vigorously disputed this. The outsider can only look on as evidence and counter-evidence are presented in what is becoming a classic example of a scientific controversy. Falk does not though believe that australopithecine brains were the same as pongid ones, but that the differences were not manifested at 'the level of cortical sulcal patterns'. She believes *Homo*-like sulcal patterns appear earliest in KNM-ER 1470 (*Homo habilis*) at 1.8 myr (Falk, 1983), a point with which Holloway in fact agrees (Holloway, 1976).

The existence of asymmetry is more amenable to investigation than the minutiae of sulcal patterning. Other primates do in fact show slight degrees of asymmetry, but none to the extent of the human brain, nor in the same directions. One of Holloway's most successful studies in this area (Holloway and La Coste-Lareymondie, 1982) found that fossil hominids did seem to possess a human-like pattern of asymmetry. This research involved studying the 41 available hominid endocasts and 135 pongid ones. All hominids, regardless of brain size, showed evidence (somewhat weak in the case of the australopithecines due to the poor condition of the endocasts) of the same pattern of asymmetry, technically described as 'left-occipital, right-frontal petalias' (see Table 4.3). Such pongid asymmetries as are found are different in pattern to that shown throughout the hominid sample. This finding, tentative though it is for the australopithecines, though not for later hominids, can be plausibly interpreted as supporting an early hominid adoption of handedness (which is correlated with brain asymmetry, though not perfectly, in humans), and of gestural communication. The literature on lateralisation and asymmetry is now considerable and readers are referred to Dimond and Blizard (1977) for further papers on their evolutionary implications and roots. A more recent account of asymmetry in pongid brains is LeMay et al. (1982).

TABLE 4.3 *Principal anatomical asymmetries of the human brain*[1]

Favouring left hemisphere	*Favouring right hemisphere*
Greater specific gravity	Heavier
Longer sylvian fissure[2]	Longer internal skull size[3]
More grey matter (neocortex)	Wider frontal lobe[3]
Larger planum temporale[4]	
Larger inferior parietal lobe[2]	
Wider occipital lobe[3]	

[1] Drawn from Kolb and Whishaw (1985), whose list is rather longer. Features referred to here are those which might feasibly be indicated by endocasts or are non-technical in nature.

[2] See Fig. 4.2.

[3] Features identified by Holloway and La Coste-Lareymondie (1982).

[4] This is an internal feature determined by asymmetry in length (and angle) of sylvian fissure, it is however a particularly clear asymmetry (see Passingham, 1982, pp. 93–4).

There are wide individual differences in the extent of these asymmetries.

So far we have been looking at changes in the neocortex, the expansion of which contributed so considerably to the great increase in hominid brain size after *H.habilis*, and the re-patterning of which leaves at least some traces in endocasts. But it is not only the neocortex which is so much greater in size than is found in pongids; as mentioned earlier, the cerebellum is enlarged too. This is the site of the limbic system, the 'palaeomammalian' brain related, in MacLean's model to emotional functions and to the sex-feeding-fighting complex. Other researchers have noted how in humans this system is also related to memory, not as a storehouse but 'as a logic system which is indispensable for the proper functioning of the memory machine' (Barbizet et al., 1978, quoted in Vilensky et al., 1982). The connection between memory and emotion is also of course extremely intimate; not only do our memories invariably have an emotional dimension but in some instances, such as those frustratingly intangible evocations sometimes triggered by smells, sensation, memory and emotion seem to fuse at some non-verbal level. Furthermore, as the size-data showed, the advent of bipedality and the socio-behavioural changes associated with it come *before* neocortical expansion. Perhaps we have been misled once again into seeing purely intellectual functions as the key to hominisation, concentrating on the neocortical enhancement of learning, information processing and behavioural control to the exclusion of something more fundamental; the more basic motivational and social organisation level. Since a 'neocortex first' model is as erroneous as the older 'brain first' one, the possible role of the limbic system in initiating the 'hominisation' process must be brought into consideration. This view has been endorsed by Holloway and is expounded more fully in the Vilensky et al. paper just cited.

That the difference between humans and pongids concerns more than the neocortex is shown by the fact that those with pathologically reduced neocortices, or with brain damage affecting neocortical operation, do not all of a sudden start behaving like chimpanzees, let alone like reptiles.

Anticipating briefly the content of later sections of this chapter, although divergent in the scenarios envisaged, most models of the early stages of hominid evolution include radical changes in hominid sexual behaviour, social organisation and feeding. *All these are areas in which the limbic system seems to play a more important role than the neocortex.* These changes could have preceded neocortical expansion by perhaps 3 or 4 million years though this is probably unknowable. On the anatomical side, Stephan (1969), Stephan et al. (1970), Andy and Stephan (1976) and Armstrong (1981, 1982) have shown that 'although limbic structures comprise a relatively small percentage of the human brain, they are relatively larger in humans than in almost all other primates' (Vilensky et al., 1982, p. 453). And more strongly (p. 454):

> It is worth emphasising, as does Armstrong . . . that in terms of number of neurons, humans diverged further from other hominoids in the limbic nuclei that in the association nuclei which are traditionally related to 'higher' cortical functions. Armstrong concluded ' . . . human cognitive functions may not have been superimposed on a conservative limbic substrate, but the two substrates may have been differentiated and expanded together in an integrated manner.' (1981, p. 381)

There is some evidence (Armstrong and Onge, 1981) that type of social organisation in primates is correlated with number of neurons in the 'anterior thalamic complex' region of the limbic system.

Physiological research is revealing far more complex interactions between the limbic system and other brain regions than the fairly schematic 'triune brain' model suggests; indeed it is so heterogeneous a region that some have doubted the utility of retaining the term. For the topic of brain evolution there now seem to be firm grounds then for giving primacy to enlargement and re-organisation of the limbic system and its connections with other parts of the brain rather than stressing the much later mushrooming of the cortex. Such changes in behaviour as would accompany limbic system changes relate to basic social behavioural and survival routines. They may be seen as laying a motivational basis for the forms of social life which, it later turned out, constituted fertile conditions for the elaboration of higher cognitive functions to evolve, but the evolution of such higher functions need not have substantially figured at all to start with. Such changes in cognitive functioning as did, undoubtedly, accompany the new bipedal life-style could have been achieved by changes in the limbic system's connections to other brain regions, and internal re-organisation generally, such as lateralisation, re-allocation of processing capacity released by the 'enslavement of the foot' and the like. Only when the limits of what could be achieved by re-organisation had been reached would the energetically costly (Martin, 1981) expansion of the neocortex become advantageous.

The third of Passingham's issues, presence or absence of special features, pertains primarily to the speech areas of the brain and will be discussed in the next chapter, when we come to the evolution of language issue.

3 *Blumenberg's model*

The post-*H.habilis* enlargement phase is the focus of attention for a far-ranging paper by Blumenberg (1983). Much of this paper need not concern us here, since it is concerned with reviewing material discussed elsewhere. What is distinctive about his treatment of the appearance of 'the advanced hominid brain' is his incorporation of molecular biological approaches and ideas from current evolutionary theory. Although he presents the paper as a radical new integration of data from all levels, the novel element only really lies in his account of the initiating process. He does not see the advanced hominid brain as owing its origin to any adaptational event, since he finds no grounds for positing environmental change at the time and place in question. (He does not, by contrast to Holloway, consider the sociocultural unit as providing an environment for its members in its own right that might possess a dynamic independant of events in the natural world.) Rather it owes its origin to 'stochastic molecular genetic mechanisms' at the nucleotide level. Relatively simple changes at this level affecting brain biochemistry (particularly neurotransmittters) would be sufficient to initiate changes in information-processing and nerve cell growth:

> The mutations that led to the advanced hominid brain likely mediated cytological changes of considerable complexity that affected the overall pattern of nerve cell differentiation, details of dendritic differentiation and synaptic interconnections, number of axons per unit of cortical volume, number of glial cells, and the rate of myelination during the first year of development. (p. 596)

He then outlines the way in which such a random mutation or set of mutations, could spread throughout a population and how behaviour 'could have had effects that reinforced, accelerated, and canalized physiological changes that proceeded from mutational events' (p. 598). An example would be the effects of increased glucose metabolism enhancing functioning of the visual cortex which would in turn enhance scavenging/hunting competence, bringing about the dietary changes required by increased glucose metabolism in a positive feedback fashion.

His broader evolutionary model strives for a compromise between gradualism and punctuationism, sharing with punctuated equilibria models an episodic nature, and stochastic origins of mutations, but with gradualism, undirectionality and lack of environmental change. Reception of his paper was mixed, Holloway seeming somewhat unfairly grumpy, while Jerison found very little to disagree with. However, only one contributor to the peer commentary, Kurt Fristrup, tackled Blumenberg's molecular biology and he was sceptical:

> At issue here is the magnitude of the role assigned to chance: it is one thing to assert that stochastic processes can alter the direction of selection by moving a system into a different evolutionary domain – the slopes of a new adaptive peak. It is quite another to assert that the summit itself is reached by chance . . . Adopting a stochastic explanation for the doubling of brain volume is betting against long odds. (1983, p. 603)

It is also perhaps interesting that he fails to discuss the limbic system role dealt with just now. Nevertheless, Blumenberg's paper clearly represents an important attempt at synthesising information across a dauntingly wide range, from biochemistry to demography.

4 *Summary*

What emerges from all this might be described as a two-stage evolution of the human brain. Firstly, it seems probable, came a phase of reorganisation involving the limbic system and hemispherical lateralisation with some degree of gross relative enlargement, but only modest in scale. Much later, with *H.erectus*, comes a rapid increase in the neocortex associated, archaeologically, with the advent of stone tools and anatomically with an initial increase in body size (though this remains stable henceforth, reversing in the final stage of 'sapienisation'). The resultant brain is several times the size which would be predicted for a primate of our stature, though the extent to which its actual morphology is different from what we would expect of that size of primate *brain* is debatable, with Jerison (1983) still sceptical, but other evidence pointing to a degree of mosaicism in the rates of enlargement of some structures which contribute little to gross weight.

Given that the sample of hominid endocasts is only about 40-odd, and many of them fragmentary, one can only be impressed that so much has been gleaned. Yet most authorities remain properly tentative in their conclusions, and many issues are still quite obscure. There are few clear 'schools' in the broad sense in this area, though a size v. organisation division is apparent, with Jerison and Holloway representative of the two camps (Jerison has also, at least in 1973, been highly critical of the value of endocast research). An 'early' v. 'late' division regarding appearance of hominid brain morphology also exists, basically in the Falk v. Holloway dispute over the lunate sulcus. Most writers hasten to distance themselves from the notions of a critical brain size, Keith's 'Rubicon', separating human from earlier hominid.

We can now move from the brain to its output, behaviour. Though that is not to imply that the brain alone determines behaviour, behaviour is limited and channelled by the gross physiology of the animal. It is not for want of cortex alone that a horse cannot knit. The chronological relationship between brain and behavioural changes is, in the final analysis, obscure in the evolutionary context, though most writers now tend to give behavioural changes the priority, as providing the circumstances in which such changes in brain physiology as facilitate the behaviour further acquire selective significance. At the same time every behavioural innovation is possible only because existing physiology fails to rule it out.

B Origins and consequences of bipedalism

Both the morphology of the Hadar fossils and the Laetoli footprints confirm that hominids were more or less bipedal by just after 4 myr. Whether the footprints were made by *A.afarensis* is a matter of some dispute, but they were certainly made by a protohominid around 3.7 myr. There agreement ends. The factual question is not now whether bipedality was present this early but the extent to which gait had acquired a modern form; was protohominid walking the same as ours or was it significantly different? On this issue the field is at present wide open, with two major schools of thought. The 'modernists' believe that Lucy and all later hominids were walking in a basically modern human way, any differences being due simply to their smaller size. This position is advocated by Johanson and Lovejoy in particular, supported by their respective academic departments at Berkeley and Kent State University in Ohio. Opposed to them are a group based primarily at Stony Brook, part of the State University of New York, represented principally by Stern, Susman and Jungers. This group believe the evidence shows *A.afarensis*, and the maker of the Laetoli footprints, to have been less efficiently bipedal than moderns, and that some degree of arboreal adaptation was still present. Stern and Susman (1983) even suggest that there was some sexual dimorphism in behaviour, with males more terrestrial than females. This, they argue, is implied by the morphological differences found between the robust and gracile specimens from Hadar, which are assumed to be members of the same species. The anatomical points at issue are the continued presence of curvature of the metacarpals (typical of arboreal primates) and the biomechanics of hip and knee joints. The controversy in fact overlaps, though is not identical with, the 'one' versus 'two' species debate discussed in the previous chapter, with the sceptics being somewhat more inclined to a two-species position if sexual dimorphism of behaviour is unacceptable. Tuttle has expressed doubt as to whether Lucy's species was responsible for the footprints.

The gait which Stern and Susman believe *A.afarensis* to have adopted was still somewhat bent-kneed, and biomechanically less efficient than modern walking. Modern humans, in walking, place their feet virtually one directly in front of the other, but *A.afarensis* seems to have had to adopt a 'side-to-side' style akin to, if less extreme than, walking in flippers. Contrary to both positions is a view that female walking at this time was actually more efficient than it later became, when pelvic modifications to cope with giving birth to large-headed babies accompanied significant increase in brain size. This is unlikely in view of the fact that human neonate brain size is normal for a primate our size at 12 per cent neonate body-weight (Sacher, 1982).

A resolution to this must probably await further post-cranial hominid fossil remains, though as noted before, inference from morphology to behaviour is far from easy when dealing with 'generalised' species. Anatomically the upright posture has required major reorganisation, and as Krogman (Krogman, 1951) long ago described, this has rendered us especially vulnerable to lumbar and circulatory complaints. It is perhaps underemphasised in most texts that the *range* of human locomotion modes is very much broader than in other species, and while upright posture is indisputably of paramount importance it has been achieved at the same time as retaining reasonable degrees of facility with earlier modes (e.g. climbing) and adding new ones (swimming and jumping). The image of human puniness has in the past been badly overdone; although humans are not supreme in any single mode other than sustained bipedal walking, they can operate with a fair level of efficiency across the whole spectrum (bar flying), with adequate training. It would be interesting to know the evolutionary course of this amplification of the locomotor repertoire – could Lucy dive? How good a mountaineer was *H.habilis*? As yet such questions barely figure in the literature, central though they surely are. M. D. Rose (1984) proposes that full modern gait occurs only with *H.erectus*.

If the precise nature of *A.afarensis* gait, and that of protohominids in general, is still a little obscure, why walking was adopted in the first place is positively shrouded in confusion. The bounteous benefits of the bipedal mode once adopted are abundantly clear: our hands are freed to carry, make tools, and gesture, our forward facing heads atop a vertical spinal column maximise our range of vision, hands can add another sound channel to the vocal one by clapping, drumming, etc., while the whole posture facilitates incorporation of a range of others (an upright animal may stoop where a bowed one cannot straighten). But these many advantages provide an embarrassment of choices when it comes down to trying to identify which, if any, of them *originally* enhanced the fitness of bipedal walkers.

Given the early date at which bipedalism appears, it now seems unlikely that the more sophisticated advantages of 'freed' hands figured much in its original adoption, and since the brain's enlargement only really gets under way after two million years of walking have elapsed the two developments, contrary to what was once believed, must be considered as separate. There is a broad consensus that reduction in forest cover and need to adapt to an environment with substantial tracts of open space was one vital factor in the situation, but though it might have provided a necessary condition, it did not provide a sufficient one. Why did the protohominids not, like baboons, who too made a transition from forest to savannah, revert to some form of quadrupedalism? Why, like the great apes, did they not adopt knuckle-walking? No other savannah species of mammal is bipedal, although the giraffe of course has made a remarkable vertical extension! (The best current evidence is that the proto-hominid environment was not in fact savannah-like, but the variegated mosaic terrain described earlier in Chapter 3.) There seems to be no reason for believing that upright walking evolved at any stage from knuckle-walking – the great apes are not simply behind us on the same evolutionary track; they took a different one.

Isaac (1983) identifies eight candidates for the role of prime causative factor. In the first group the causes of bipedalism are more or less identical with its consequences and these he lists as: carrying of weapons and tools, carrying of food, carrying of provisions in and between trees, hunting and, finally, improved vision. In the second group the causes are different from the consequences: eating of grass-seed, feeding from bushes (involving upward stretching, etc.) and aquatic adaptation. Isaac's own preference is, we will see later, for provisioning of food. We do not know then whether right from the start bipedalism was being selected for because it was associated with those behaviours it now so obviously facilitates or whether it first served a fairly humble function, probably to do with food-acquisition, the rest only gradually following in its wake, as consequences not causes.

Isaac's alternatives are not exhaustive. Two recent papers have suggested yet further factors involved in the origins of bipedalism. Merker (1984) has drawn attention to its efficiency in connection with stalking, citing work by Geist (1978) in support

> . . . who points out that the human foot, leg, and associated sensory, postural and locomotor mechanisms are superbly designed for silent, slow stalking of game to a distance where the prey comes virtually within arms reach, that is to a 'stalking hunt' *without* a chase. The 'stalking hunt' comprises such features as frozen, immobile balancing on one foot; tactile exploration and guidance with the foot along the ground for its secure and silent planting while the eyes maintain steady unbroken fixation on the prey; and slow as well as rapid shifts of the centre of gravity during advances. (p. 113)

Merker suggests that stalking, as opposed to the more energetically costly chasing, may have been the principle protohominid hunting strategy, and he sees this in the 'male provisioning' context, and as related to the sexual dimorphism in bipedalism identified by Stern and Sussman (*op. cit*) as one possible interpretation of the Hadar hominids foot morphology.

Wheeler (1984), considers bipedality and loss of functional body hair to be interrelated. A central problem for mammals is the risk of hyperthermia, overheating, and this becomes especially acute for those with relatively large brain sizes. Primates lack the cavernous sinus which cools the brain's blood supply in most other mammals. So in hot climates our ancestors would have faced considerable problems of temperature control and regulation. Bipedality and hair-loss solve this in the following way; bipedality reduces the amount of exposure to the sun, 'when the sun is overhead, an upright hominid presents only about 40% of the area it would if it was in a quadrupedal position' (p. 94). The combination of naked skin and subcutaneous fat meanwhile facilitates the rapid dissipation of heat and cooling by sweating during the day, while providing, in the fat, insulation during the much cooler nights. The establishment of this very efficient system of temperature regulation, ensuring 'extreme body temperature stability' is also, he considers, likely to have been a necessary condition for the later expansion of brain size.

Whatever the cause, having adopted this mode of moving our ancestors were faced with new problems, such as carriage of infants and various sequelae for social organisation which ensued from the carrying and manufacturing life-style it engendered. Models of these are discussed later. Although we can only speculate about the details of protohominid life, the bipedalism move does deserve further analysis, especially in the light of many writers tending to glide over some of the behavioural aspects of the transition to terrestrial life with catch-all phrases like 'freeing the hand' and ready analogies with existing species of hunter-gatherer tribes.

It is reasonable to assume that trees figured largely in the protohominid life-style long after they ceased to be the primary occupation zone. After all they still do. They would have continued in use as sources of food, wood, refuge from non-climbing predators and look-out posts. Many of the most important basic manual 'schemata', to use Piagetian terminology, would have already been evolved prior to the abandoning of the arboreal life-style, as they have been by chimpanzees and orang-utans. Gripping and releasing, throwing, picking, pulling, waving, hitting, dropping, knocking together, scratching, poking and so on would all have been within the protohominid's capacities. The integration of these into the higher order secondary and tertiary reactions would of course be far more limited (Parker and Gibson, 1979; Vauclair and Bard, 1983). A recent study of the hand morphology of *A.afarensis* (Marzke, 1983) shows that a high proportion of modern grips were possible, only certain 'power grips' and full cylinder grip (as we use when holding a straight-sided glass) were beyond Lucy. I have already noted the possible re-allocation of neocortical processing resources following 'enslavement' of the foot, but two other aspects of the situation need to be noted, although they receive little attention in most texts. Firstly, the muscular energy previously required for suspension and climbing now became available for manipulating other objects (e.g. detached branches); forelimb musculature, in other words, was preadapted to strenuous swinging, hitting and lifting, etc. Secondly, the new terrestrial environment would, to a psychologist, seem to offer new categories of objects, notably bones and stones, for assimilation into *existing* behavioural schemata, rather than immediately requiring dramatic behavioural innovation. The behavioural changes accompanying the shift to ground-dwelling were thus likely to have been rather smoother than often depicted and it took many generations for the physiological changes required for the perfection of such genuinely new behaviours as accurate throwing to take place. Taken together, these points suggest that some increase in the complexity of manual behaviour and change in overall behaviour patterning could be achieved by reallocation and reorganisation of existing resources, neocortical, muscular and cognitive, without any immediate increase in brain capacity being necessary.

None of this explains how it was that our protohominid ancestors opted for bipedalism in the first place. On balance we must assume that it was because they had already diverged behaviourally from the great apes' ancestors in some important respect that they were able to take the ground-dwelling option rather than remaining in the trees. Lucy's anatomy and the Laetoli footprints prove to the satisfaction of most that *A.afarensis* was already firmly bipedal, notwithstanding disagreement over its precise gait and continued arboreality. It is hard to see how the origin of bipedalism can be explored further at present, barring dramatic new fossil finds, and even knowing 'when', the question of 'why' will remain unanswered. The Latin poet Ovid ascribed our upright posture to Prometheus who 'made man stand erect, bidding him look up to heaven and lift his head to the stars.' As an insight into the profound psychological consequences of the event, this has not been bettered.

C Models of early hominid behavioural evolution

Before looking at the specific theories and models of our early behavioural evolution it is worth spending a page or so considering why these attempts at reconstructing early hominid life-styles so often come to focus on dietary issues. Food is essential to life. For any species the manner in which it obtains its nutrition is the central factor in determining its morphology and ecological niche, indeed largely defines these. Beyond this level it is bound up also with the animal's social organisation, if any. Changes in diet and in food-related behaviour are crucial in exposing a species to new selection pressures. Such changes result primarily from changes in the environment in which the species, or a sub-population of the species, lives; climatic, geological and ecological factors may all be involved in this. The old food source, or acquisition strategy, no longer supplies enough nutrition; they must innovate or perish. Or a new food source supplies a surfeit, facilitating higher energy-use adaptive strategies. Evolutionary processes are governed by differential rates of successful reproduction, ability to reproduce successfully is in turn dependent on success at keeping well-fed and healthy, and many mating rituals reflect this in one way or another. Sexual attractiveness and ability to provide nourishment are often closely interlinked, as evidenced by the remarkable parallelisms between feeding of females by males in many bird courtship rituals and in those of humans. The areas of the human brain governing the two are also closely linked (MacLean, 1982). Although other evolutionary mechanisms, such as genetic drift and Dover's 'molecular drive', supplement natural selection of the classic kind, it is still to natural selection that we look first in trying to understand a given evolutionary development. Food is an absolutely central factor in determining species life-style, and thus the dietary habits of our ancestors are of special importance. On this hinges the whole nature of protohominid social structure and the construal of archaeological and palaeontological evidence.

Nevertheless, placing food in this pivotal role is not universally accepted as the key to hominid evolution. Some, for example, see social relations as the logically more fundamental issue (Fishbein, 1979) or fit food into a broader context of demographic forces (Lovejoy, 1981).

How, though, can we make progress in researching this topic? There are a number of lines of investigation which can be followed. Firstly, assuming that our immediate ancestor was somewhat chimpanzee-like in life-style – arboreal, primarily fruitivorous and vegetarian but opportunistically carnivorous – we can track the likely dietary adaptation of such an animal under the novel conditions which are assumed to be associated with the shift to more ground-dwelling life-styles in the late Miocene: shrinkage of forest and environmental diversification. Secondly, we can consult the morphological evidence directly by studying the dental and mandibular remains in the fossil record. Thirdly, there are the archaeological sites themselves, a handful of which in the Olduvai Gorge and Koobi Fora seem to represent hominid living or occupation floors at which food remains can be identified. Fourthly, more theoretically, optimal food-extraction strategies can be calculated for the environments in which protohominids are assumed to have dwelt, using, for example, calculations of nutrition yield per unit of time and energy invested, basing this on comparative data from modern hunter-gatherer tribes and non-human species which resemble hominids in some respect. It is nevertheless an elusive topic, requiring a careful balancing of plausibilities in evaluating the minimal direct evidence of protohominid feeding behaviour.

The specific question as to the presence, and importance, of meat in the early ancestral diet is lent additional force by the central role given to it in some influential theories (Dart, 1957; Ardrey, 1976). As Isaac and Crader (1981) say, in a paper to which we will be returning shortly,

> The view has often been expressed that the gradual development of many important human abilities, including technological skill, insight, ability to co-operate, mobility, and the ability to plan, has been advanced by a pattern of natural selection which over many generations favoured the most successful hunters. This view in its simplest form would be valid only if it were true that hunting success was an important prop for protohuman subsistence during the long span of evolutionary development. (p. 92)

Unfortunately direct evidence for hunting as such, as opposed to scavenging, is virtually unobtainable for the pre-*H.sapiens* period, while direct evidence of meat-eating of any kind prior to the appearance of tool-use around 2 myr is as yet barely possible. Only since about 100,000 years ago is the evidence for regular meat-eating unambiguous, though evidence for *some* meat-eating is now solid as far back as 1.75 myr.

The situation is further complicated by the fact that the general direction of hominid evolution has been to greater environmental adaptability and behavioural flexibility, thus it is unlikely in the extreme that early hominid diets were any less complex than the already fairly varied diet of their immediate primate ancestor, and by the Plio-Pleistocene transition period of 2½–2 myr there were possibly differences both between hominid populations in different areas and within them between different seasons. The question is not simply a choice between 'vegetarian', 'carnivorous' and 'omnivorous', but granting the greatest likelihood of their being 'omnivorous' in a broad sense, how significant meat was in the diet, how it was obtained, how its distribution was organised, and what meat it was. It is these which have crucial consequences for social organisation. We now turn to the first of our 'position-statements'.

1 *Isaac and Crader*

Glynn Ll. Isaac and Diana Crader (1981) 'To What Extent Were Early Hominids Carnivorous? An Archaeological Perspective'
Glynn Ll. Isaac (1983), 'Aspects of Human Evolution'

The late Glynn Isaac was among the most eminent of modern palaeoarchaeologists. Having first worked under Louis Leakey in 1961, he became responsible for interpreting several of the most crucial East African sites where hominid activities and animal remains are apparently associated. He eventually developed a 'food-sharing' model of early hominid social organisation in which the practice of food-sharing was seen as the 'initial kick' for hominisation. The 1981 paper can be taken as representative of his position and a convenient starting point for our review of current models. The 1983 paper offers a useful résumé of his perspective on the whole topic of human evolution and contains helpful synopses of his model, insofar as he was able to develop it prior to his premature death in 1985.

First, the factual question of whether or not early hominids ate meat at all. After discussing the whole range of evidence, anatomy and physiology, environment, fossils and archaeology, Isaac and Crader feel that most are 'inherently inconclusive' about whether early hominids were more carnivorous than other primates. The evidence from tooth morphology suggests that it is

> unlikely that our phylogeny includes any prolonged period of exclusively carnivorous diet . . . . Toothwear patterns can show that meat was *not* the exclusive food of a given fossil hominid, but they are unlikely to help us determine whether meat comprised 0%, 5% or 20% of its food. (pp. 40–1)

The only hope of extending our knowledge further seems to rest on detailed analysis of a handful of sites in East Africa, notably at Olduvai Gorge and Koobi Fora. Even so the conclusions obtainable from these must be limited.

Olduvai Gorge contains 'by far the largest body of evidence bearing on both the technological capabilities and the behaviour of early proto-humans, and is thus the crux of any review of evidence for meat-eating in the Plio-Pleistocene' (p. 48). There are 3 sites in particular (from 26) which are of special interest; these are the FLK, FLK NN Level 6, and FLK N 11 sites, but can be referred to less technically as the *Zinjanthropus* floor, the Elephant site and the *Deinotherium* site respectively.

(a) *Zinjanthropus* floor

This is the site at which O.H.5 and O.H.6 were found. There is a distinct patterning within the scatter of debris, including the arrangement of larger stones interpreted by Mary Leakey as remains of a windbreak or wall of some kind – the earliest known hominid structure. The bones are mostly broken, but unweathered, and in smaller pieces than common at other fossil sites in this bed (Bed 1), 16 taxa are represented. There are 2,275 artefacts but only a small proportion show signs of wear (73 flakes) and a mere 60 are shaped tools. There are also 62 nodules, cobbles and blocks which show signs of heavy use.

On the face of it this would appear to indicate a site where hominids used stone artefacts to break bones of a variety of species, but of course there are alternative interpretations.

(b) Elephant site

A somewhat disarranged elephant skeleton was found mingled with 130 introduced stones, mostly unutilised flakes, but including 14% utilised heavy stones, 4% utilised flakes, 4% core tools and 5% 'manuports'. 'On the face of it, this site would seem to provide good evidence of "butchery" activities by tool-making hominids, but other possibilities cannot be entirely eliminated' (p. 61). Note that this find in no way implies hunting rather than scavenging.

(c) *Deinotherium* site

39 introduced stones were found among remains of a *Deinotherium* (an extinct relative of the elephants) which appeared to have died upright while stuck in a swamp. Regarding this, it is concluded:

> Although this site probably represents an example of hominid meat-eating, it is not impossible that the co-occurrence of the skeleton and the stones could be fortuitous. We regard this site as less satisfactory evidence than that of the FLK N 6 elephant. (p. 63)

Since the paper was written, the evidence from the *Zinjanthropus* floor has hardened considerably, with three papers published in 1981 (Bunn, 1981; Potts and Shipman, 1981; Keeley and Toth, 1981). These reported on a number of investigations on the surfaces and broken edges of bones from this site, plus a site at Koobi Fora. Bunn describes his findings:

> A series of very fine linear grooves on bone surfaces constitutes the most unequivocal evidence of hominid involvement with the bones. These grooves are cut-marks made by hominids using knife-like stone flakes to remove skin from carcases, separate articulated bones and detach meat adhering to bones. The grooves occur singly or in multiple sets and vary in length from several millimetres to several centimetres. Typically they are straight sided and V-shaped in cross section. . . . (they are) indistinguishable from experimentally induced cut-marks. (pp. 574–5)

8.6 per cent of the bones at the 'Zinj' site (about 300) have such cut marks. Keeley and Toth provided additional evidence by demonstrating the presence of meat-polish micro-wear on stone artefacts from Koobi Fora. The original Isaac and Crader conclusion that 'there is good *prima facie* evidence for the involvement of early tool-making hominids in meat eating' (p. 83) thus seems to have been vindicated, although Mary Leakey has confessed herself somewhat sceptical about the interpretation of 'cut-mark' data.

But was the meat hunted or scavenged? Present-day non-human primates, though sometimes predatory, do not actually scavenge, while large predators do both. Were our ancestors more like modern social carnivores or modern primates in this respect? A further extraordinary find by Shipman (Lewin, 1984) is of a bone from the 'Zinj' floor on which a cut mark can be seen to overlay a carnivore tooth mark (Plate 12). This is the first positive evidence for meat *scavenging* – though of course it tells us nothing of the extent of scavenging nor does it disprove hunting. In this case however there is no doubt that the carnivore had the bone before the hominid did. That *H.habilis* (presumably) practised scavenging is virtually certain in the light of this discovery. In South Africa, Elizabeth Vrba has tried to unravel the situation by examining differences between the kinds of bone assembly left behind by hunters and scavengers. Hunters would leave a disproportionately high number of juvenile bones and their kills would be in a fairly narrow weight range, scavengers, by contrast, would leave remains randomly distributed throughout the weight range and disproportionately lower in juveniles (these being taken by hunters). On this basis at least one South African site (Sterkfontein Extension Site Member 5) appears to provide evidence of hominid scavenging later followed by hunting, but these are tentative conclusions.

Very provisionally the authors ' . . . favour the view that the early tool-making hominids were opportunistic scavenger/hunters and that, given the simplicity of the technology of the time, the flesh of medium and large prey was probably obtained more by scavenging than by hunting' (p. 86).

Such a conclusion is consistent with the 'food-sharing' socio-economic model advocated by Isaac. Similar, but certainly not identical to the life-style of modern hunter-gatherer societies, this envisages small social groups occupying temporary base-camps from which individuals or sub-groups travel over a home-range each day foraging. Food surplus to their immediate needs is brought back and shared. One innovation implied by this is the invention of containers – baskets, slings, bags or whatever – for easy transportation of what has been gathered. This has actually been considered by one writer as 'the basic human invention' (Linton, 1971), and Lovejoy (1981) also considers it crucial (Plate 11). The broader picture of the evolutionary implications of this active food-sharing system are described in the following passages:

> The system cannot operate without some simple *equipment and tools*, namely containers for carrying food and knives to cut up carcases. The whole complex is interconnected with an evolutionary change in anatomical-locomotor arrangements, namely *bipedalism*, which facilitates *carrying* things about. A social system involving exchange of energy in the form of transported food puts a premium on the ability to exchange information and to make arrangements regarding future movements of group members. It also increases the importance of regulating social relations among individuals. All these influences might be expected to favor the evolutionary development of an effective communication system, such as a protolanguage, and of sharpened sociointellectual capabilities. (p. 91)

> . . . certain of the fundamental innovations differentiating human from non-human primate behavior had begun to be established by Plio-Pleistocene times (i.e., 2m.y.B.P.), but it does *not* mean that the tool-making hominids of that time would be recognizable as 'human'. We surmise that language and systems of cultural rules and codes were at a level that would make us react to these hominids as non-human, were we to confront living representatives. In fact, they probably represented a mode of life that has no living counterpart. (p. 92)

In this system the role of meat is as a part of a complex behavioural system:

> We thus favor a model in which the active delivery of some meat to fellow members of a social group developed in a reciprocal relationship with the practice of transporting and sharing some surplus plant foods. We see the model as representing a functionally integrated behavioral complex, in which any attempt to isolate one or another component as an initial or prime mover is probably misleading. (p. 93)

This would involve some division of labour between the sexes and the establishment of long-term mating bonds, 'combined with the establishment of joint responsibility for sustaining offspring' (p. 95). Hunting as such does not seriously enter the Isaacs and Crader model and they think that it could not alone account for the behavioural innovations which mark hominid evolution. Meat itself, though, being portable and intensively nutritious, 'helped establish an adaptive complex that involves the transport and sharing of food obtained by the complementary endeavours of different members of the same society' (p. 95).

A later summary of his 'food-sharing' hypothesis appears in Isaac's 1983 paper. This suggests the importance of a home-base as a focus for distribution activities. He is aware of the risk of projecting too-human a picture of this life-style, stressing that this food-sharing sets the preconditions favouring future hominisation, stimulating enhanced communications and complexifying social relationships. He is wary of committing himself to saying that sharing was 'deliberate' or consciously motivated.

> The food-sharing model has been widely misunderstood as implying that by two million years ago there existed friendly, cuddly, cooperative human-like hominids. This need not be so. The attractiveness of this model is that it seems entirely feasible for such a behavioural system to come into existence among *non human* hominids that had brains no larger than those of living apes, and it is my strong suspicion that if we had these hominids alive today, we would have to put them in zoos, not in academies.
>
> Clearly, this initial configuration can very readily be plugged into models involving kin-selection, and/or tit-for-tat selection patterns that would provide plausible, if hard-to-test models of the subsequent elaboration of brain-speech-culture-society systems. Amongst other things, the provisioning and division of labour implied by the system would make bonded male-female reproductive modules highly adaptive, if they did not already exist at the outset. (p. 536)

Lewis Binford (1981) has cast a cold eye over the whole question of the interpretation of supposed kill-sites and damage to fossil bones generally. It is a masterpiece of taphonomic research, and while, in the light of the findings of Keeley, Bunn, Potts and Shipman, his queries about Olduvai can probably be fended off, he is keeping a healthy pessimistic presence in the area.

2 *Tanner*

Nancy Makepeace Tanner (1981), *On Becoming Human*

An alternative perception of protohominid social life, with affinities to Isaac's model nonetheless, is offered by Nancy Tanner of the University of California at Santa Cruz.

A number of women writers have, over the last decade and a half, attempted to redress what they see as the excessive male-centredness of most accounts of human evolution (Morgan, 1972; Reed, 1975, for example). This is the most comprehensive of the recent works of this type, focusing on the earlier phase down to the emergence of *H.erectus*. Her thesis is in two major parts; firstly a systematic review of the chimpanzee studies to date (taking in work by Goodall, Nishida, Teleki, McGrew, Kortlandt and Zihlmann – the last being a close associate), and a rationale for taking these as a reasonably accurate model of the ancestor from which hominids diverged. That is, she assumes this ancestor to have had, by and large, a similar life-style and behavioural repertoire to modern chimpanzees. The problem then is to envisage the circumstances under which such a species would evolve in the hominid direction. The second part of the book consists of an attempt at relating this 'chimpanzee' model to the palaeontological and archaeological evidence.

Again the ecologically varied late Miocene is the setting for the first phase, providing opportunities for group isolation facilitating speciation on the Gould model. Extrapolating from primate tool usage, Tanner sees the primary, or 'basal', savannah adaptation as a use of tools for plant foraging by females, on whom nutrition pressures were greatest as they were responsible for bearing and nurturing offspring. The pressure on males was less intense and they continued to forage for themselves in a more traditional way. She envisages a matrifocal group of a few females with offspring, including older juveniles, as the central social unit. The adult males are more peripherally involved. She does not see males as especially necessary for predator protection, as female chimpanzees are equally as competent as males at frightening off or avoiding predators. Hunting does not figure seriously at all at this stage, the species being physically too small in her opinion to engage in serious planned hunting, although smaller animals would obviously be part of the diet, plus, probably, scavenging of larger carcasses.

This situation has a number of outcomes: it provides a constant selection pressure for the more intelligent and resourceful females, and also for stamina and carrying ability. Upright posture affects sexual signalling, while the new form of social organisation places females in an enhanced position from which to exercise sexual selection. Tanner suggests that this will result in females selecting the more sociable, least aggressive, more friendly and co-operative males to mate with. Perhaps the profoundest consequence of this apparently loosely knit savannah and woodland life-style is that it progressively increases the complexity of social bonding and communication. Sexual dimorphism decreases as a result of female sexual selection and relative independence (male canine tooth size, for example, steadily drops). Tanner stresses though that 'It was the mothers who had reason to collect, carry, and share plant food; at this time males were likely still foragers, eating available food as they went' (p. 141). Males, though, would learn to share 'with those with whom they were travelling or resting' to strengthen or enhance social relationships.

There would of course be a positive feedback loop between improved provisioning and the possibility of extended socialisation which, in turn, enhanced cognitive capacities and provisioning ability. Following the *A.afarensis* basic hominid type, she envisages the gracile australopiths as beginning to use tools for food *processing* as well as foraging, while the robust species adapted by enlarging their tooth size in order to process larger quantities of plant-foods of relatively low nutritional value. Food processing by pounding and so forth enabled the gracile species to dodge this cul-de-sac, as it eventually became, though there is some evidence of later robust australopiths at Chesowanja (KNM-CH 1 cranial fragment in particular) adopting this strategy somewhat belatedly, since their teeth are remarkably wear-free. Since the earliest *Homo* specimens appear in some respects to share both gracile and robust features, it is not impossible in her opinion that the two varieties were still capable of interbreeding, and indeed did so. She in fact views the entire australopithecus family as a single polytypic species with constant gene-drift and gene-flow between different groups, a somewhat unusual position and at variance with her earlier preference for a Gould-like model of speciation to explain the emergence of the first hominids. Two brief critical notes ought to be made at this point: firstly that the *A.afarensis* material suggests, on one interpretation, that males, not females, were the most bipedally efficient; and secondly, the current favoured line of descent is direct from *A.afarensis* to *Homo*, not via *A.africanus* either gracile or robust (see discussion in Chapter 3 of McHenry and Corruccini (1980)). These of course only weaken, they do not disprove, Tanner's basic picture.

Only on the eve of the emergence of *Homo*, she believes, does hunting appear. Prior to the special throwing adaptations and more developed technical and social skills of this period, hunting would have played a minor part in hominid food-acquisition.

Tanner's focus of concern is thus with the mother-child unit and its ramifications. Males follow the bipedal plant-foraging female into the savannah, they do not lead her there, or even escort her. There is a curious contrast between this image and the next.

3 *Hill*

Kim Hill (1982), 'Hunting and Human Evolution'

Once again the key question is the nature of the pongid-hominid divergence. Why did the hominids split from the ancestral pongid stock, which Hill too sees as chimpanzee-like? The critical factor for Hill is male provisioning of meat. Arguing from a mass of empirical data concerning the contribution of meat to the diets of various animal species and hunter-gatherers, Hill draws very different conclusions to Tanner.

What is curious is the difference in the way they view sexual relations, for Tanner males are helpful occasional visitors to the matrifocal group; on entering this busy domestic world they had better make themselves useful and behave, for the females exercise a high ethical standard in evaluating them and choose to mate accordingly. From this angle the male's food-gifts are gestures of good-will to 'enhance' social relationships. Hill sees it all quite differently. The females are far more dependent on the meat-carrying males, and eager to acquire their share of protein, they trade sexual favours – a sort of 'prostitution' model. The females thus learn to be sexually available all the time. A quote from Tanner is apposite on this point 'The old argument is that with loss of estrus females became receptive at any and all times. This would appear to be more in the realm of wishful thinking than a description of actual sexual interaction in any known society' (p. 154). Hill is not, though, alone in accepting 'constant female sexual receptivity' as a fact, as Table 4.1 shows. Hill's model of hominid divergence is summarised under eight points:

(a) A sub-population of Miocene apes found itself in an ecozone where easy prey made predatory specialisation advantageous.
(b) Males meet their own nutritional needs easily and use their free time to acquire meat for females, especially estrous females, in exchange for copulation privileges. At this stage monogamy is not implied, provisioning by males is only partial, sexual dimorphism remains high.
(c) Increased use of hands as weapons leads to reduction in large canine size and increase in meat chewing (which is hampered by large canines).

(d) Females become permanently sexually receptive.
(e) Increased reproductive fitness from provisioning selected bipedal locomotion and the ability to carry food items easily to females.
(f) Increased tool manufacture leads to changes in hand morphology.
(g) Tool-use and predation (especially on other primates) increases selection for intelligent organisms and a slightly enhanced brain/body-weight ratio. This results in reorganisation of the nervous system. (In general, as noted before, predator relative brain size is greater than that of non-predators.)
(h) Decline in infant mortality leads to increase in mean longevity, increase in juvenile dependency, establishment of a role for grandmother care-taking and appearance of female menopause. Birth interval is slightly shortened.

For Hill, the dependency of females under the burden of child-bearing and care-taking gives male provisioning a central role in the protohominid food economy. For Tanner, the female's nutrition needs are what drives her towards independence and initiative-taking in food-acquisition. While both see food-acquisition as central in selecting for bipedalism, for Hill it operates on the male, for Tanner on the female. For Hill the sexual deal is sex-for-meat, for Tanner it is sex-perhaps-for-good-behaviour (bringing meat being good behaviour of course). For Hill tools are by implication a primarily male invention for enhancing food-extractive efficiency while foraging, and their key feature is their portability. For Tanner the reason for tools is the same, but in female rather than male hands.

Hill's data-base is rather different from Tanner's, and in some respects more firmly rooted in empirical data, certainly his analysis of the nutritional intake factors would seem to raise genuine problems for Tanner's model in its present form. As far as hunting is concerned, even without weapons human hunting abilities are surprisingly good, more than off-setting deficiencies in sheer speed by cunning. By and large, however, current thinking seems to be moving against pre-*Homo* hunting. Although the Tanner and Hill models are at loggerheads over several crucial aspects of the hominisation process, some of their viewpoints seem at least to the present author to be seriously affected by their own sexually determined perspectives, and the contradictions more apparent than real. Hill is, it may be noted, reluctant to incorporate group and kin-selection mechanisms into his model, preferring to stick to the individual as the unit of evolution, a position which Isaac, for example, would not hold.

The Tanner and Hill models are derived from fundamentally different data-bases, and raise the problem, which will surely become more acute in the future, of the relative weighting to be given to such different orders of evidence as primate ethology and comparative nutritional studies (though there is a modest overlap). It ought perhaps to be made clear here that Hill's research is very different to Ardrey's *The Hunting Hypothesis* (1976), which is now generally dismissed.

4 *Lovejoy*

Owen Lovejoy (1981), 'The Origin of Man'

Lovejoy, a physical anthropologist of high standing and Professor of Anthropology at Kent State University, Ohio, put forward his model of human evolution in this influential *Science* paper. It has been taken as a reference point, and point of departure for many later writers. While it has many points in common with the models already discussed, its influence is in part due to his having accepted the apparent inadequacies of many earlier models which hinged on hunting, a savannah-dwelling transition and brain-expansion or tool-use as critical factors or initial kicks in hominisation. Discoveries of the last decade and a half have undermined the credibility of most of these as candidates for that role, and Lovejoy turns instead to social and sexual, or reproductive, factors. Following a critical review of these earlier theories, his own argument starts with a consideration of the demographic situation facing primates, and mammals in general: the interactions between such factors as duration of infant dependency, gestation period, life-span and birth-spacing. In particular he notes the increasing adoption by the primates of a 'K-type' demographic strategy, i.e. the bearing of relatively few young, with long dependency periods, associated with a long life-span and adapted to maintain a more or less constant population level. This type of strategy evolves under stable environmental conditions, provided high individual survivorship can be ensured. (The opposite of this, 'r-type strategy', is characterised by large numbers of rapidly maturing offspring and is typical of environmentally unstable conditions where population levels are subject to dramatic environmentally induced collapse, with the complementary need, therefore, of very rapid population recovery.) This attempt at incorporating the theoretical demographic perspective gives Lovejoy's paper as a whole an impressive theoretical depth.

In bringing this viewpoint to bear on the mosaic late Miocene environment of north-east Africa, Lovejoy envisages an ancestral primate population in which life-span has already increased, along with strong social bonds, high intelligence, intense parenting and the long period of learning jointly required to reduce environmentally induced mortality – a necessity for slowly reproducing, long-lived species if stable population levels are to be maintained. Under the K-selection conditions presented by the Late Miocene/Pliocene environment these ancestors have already, he imagines, achieved the limit of what is achievable in reproductive success without dramatic behaviour change. Increased parenting and social bonding can go no further, while 'all other features are direct linear functions of mammalian developmental physiology and could not be altered' (p. 344). For the protohominids only two means of increasing reproductive rate are possible: a yet further increase in survivorship (i.e. reduction in environmentally induced mortality) and decrease in birth-interval (i.e. increase in birth-rate). The challenge then is to figure out how these were achieved:

> . . . the existence of successful hominid clades in Pliocene mosaics suggests that both birth space reduction and elevated survivorship has probably been accomplished. This is without explanation unless a major change in reproductive strategy accompanied occupation of novel environments by these hominids. Yet neither brain expansion nor significant material culture appear at this time level. . . We are in search of a novel behavioral pattern in Miocene hominoids that could evolve from typical primate survival strategies, but that might also include important elements of other mammalian strategies, that is, a behavioral pattern that arose by recombination of common mammalian behavioral elements, and that increased survivorship and birth-rate. (p. 344)

Lovejoy finds the answer in a novel form of food-foraging and social patterning hinging on sexual separation in feeding. Such a strategy is known to be advantageous if three conditions are met: (a) feeding is done outwards from a fixed base, (b) sex ratio is on a par, (c) feeding is limited by search-time rather than handling time (i.e. it takes a long time to find the food, but not to eat it, as opposed to circumstances where the food is abundant but needs long preparation or consumption time). The ecology of the period in question would seem to meet, or enable to be met, each of these and also provided a setting which encouraged generalised feeding. The life-style described is similar to Isaac's, though with some differences in emphasis, notably the importance of the fixed base and monogamous mating. Females would remain together as a group, foraging near the fixed base, thereby reducing exposure of themselves and offspring to predators and enhancing their parenting, while males forage further afield and engage in provisioning of the females. It can be shown, however, that for this to work as an evolutionarily viable system, monogamous pairing is necessary, since polygyny leads to disruption of the sex-ratio and would favour males who did *not* forage far afield. Monogamously paired males are seen as provisioning their own mates and offspring, not the female group as a whole. The better he was at this, the healthier his mate and the greater therefore her reproductive rate, she being able to manage a larger number of offspring. This would confer an advantage on such males, by selecting for more male offspring of high-provisioning monogamous males – and more daughters for females capable of establishing pair-bonds with such males. Thus enhanced survivorship is achieved by the 'home-base' set up, while improved nutrition as male provisioning augments female foraging enhances the birth-rate, all this being possible by virtue of monogamous mating.

Once established, such a pattern has major long-term effects. Male provisioning will require regular, prolonged and extensive carrying, thereby creating a strong selection for bipedality. Only such a sustained pressure could select for the skeletal alterations required. 'A prolonged and extensive period of regular and habitual use of simple (primitive) carrying devices could eventually allow the co-ordination and pattern recognition necessary for a more advanced reliance on material culture' (p. 346). Thus both bipedalism and advanced tool-use can, Lovejoy believes, be explained on this model. With monogamous mating, greater male investment in parenting will ensue, eventually creating the 'bifocal nuclear family'. In other primates monogamy is only found in gibbons and siamangs where male parenting is also high.

From this Lovejoy leads us on to sexual behaviour and anatomy. Humans are, he observes, 'the most epigamically adorned primate', but our sexual dimorphism is not of the kind typically associated with polygyny. This has often been misunderstood, since sexual dimorphism is commonly assumed to be associated with polygyny, and thus primitive polygyny has been inferred for hominids (but see Clutton-Brock (1983) for a more critical consideration of this original assumption). Usually, though, sexual dimorphism takes the form of gross size-differences and differences in dental morphology such as enlarged male canines, yet these are just those areas of dimorphism which are actually reduced in hominids. Hominid differences pertain to visual traits like hair-pattern, breast and buttock forms, and skin texture. Like many others he stresses the 'continual sexual receptivity' of hominid females, though modifies it on one occasion to 'relatively continual sexual receptivity'. This is linked with a need for 'copulatory vigilance in both sexes in order to ensure fertilization' (p. 346) in the absence of overt signs of ovulation. This higher sexual activity would also serve to strengthen 'pair-bond adhesion and serve as a display of that bond' (p. 346). Our unique 'epigamic elaboration' is also interpreted in this context, leading him to suggest that '. . . these epigamic characters are highly variable and can thus be viewed as mechanisms for establishing and displaying individual sexual uniqueness, and . . . such a uniqueness would play a major role in the maintenance of pair bonds' (p. 347).

Thus even the origin of personality differences might be considered as coming within the embrace of this theory. Against the argument that sexual behaviour is in fact culturally variable and hence environmentally determined, he counters cryptically 'the more that culture can be shown to dominate the mating structure and processes of recent man, the more ancient must be the anatomical-physiological mechanisms involved in the formation and maintainence (*sic*) of pair bonds' (p. 347). It is unclear whether this means they are so ancient they are now weakening under cultural impacts, or so deeply embedded as to be strong enough to withstand them.

Turke (1984) puts forward a comprehensive model of how ovulatory concealment, group synchronisation of menstrual cycle and continual receptivity interact to maintain a pair-bonded monogamous system within small social groups, such as early hominids are assumed to have lived in. This can be read as an amplification of the Lovejoy model.

The intense social interaction and regulation of social relationships now required is, in Lovejoy's view (citing V. Reynolds, 1976), a more likely single cause for the origin of human intelligence than any other. Others have also noted that demands on intelligence are greater in dealing with the social world than in tackling the external environment. As in the primates, intelligence is most likely to result from social and reproductive behavioural adaptation. Tool-making and hunting are to be considered as vehicles for intelligence, not its cause. The result of the monogamous bonding structure is the rise of the nuclear or 'bifocal' family (i.e. focused on both parents, not mother alone), a 'prodigious adaptation central to the success of early hominids' (p. 347).

Summarising Lovejoy's case, the key factor in hominid adaptation was the demographically necessary emergence of a novel social-reproductive strategy suited to the diversified world of the Miocene-Pliocene transition in north-east Africa. In this, males foraged from a central home base, while females remained close to it, the males provisioning their monogamously selected mates. This had the combined effects of increasing safety for females and young, while enhancing the reproductive rate. It led directly to bipedalism as an adaptation for sustained male carrying, and to the bifocal family unit, along with the intelligence necessary for management of these more complex social relationships. The central factor in hominisation for Lovejoy is, then, clearly in the area of reproduction and social relationships rather than in any single behavioural or physiological innovation.

This 1981 paper provided an impressive and authoritative integration of then current knowledge within the framework of demographic and evolutionary theory. Some weaknesses are evident nonetheless: (a) Notwithstanding Stern and Susman's (1983) suggestion of *A.afarensis* dimorphism in the nature of bipedal adaptation, it is surely not entirely convincing to ascribe this to the exclusive pioneering efforts of either sex – we surely need a theory to account for the wholesale adoption of bipedalism by the entire species. But there is no reason why female carrying could not be incorporated into Lovejoy's scheme, since they are envisaged as undertaking local foraging, would need to carry infants and might also wish to bring back food to base for sharing either with their mate or with other females. A broad diet would enhance the likelihood of reciprocal sharing, especially if individual food preferences had already appeared. (b) The explanation of all sexually dimorphic characteristics as 'epigamic' (i.e. secondary sexual display features) needs to be reconsidered in the light of numerous feminist critiques of such interpretations (which have been doing the rounds since Morris, 1967). Female hair-length, breast softness and prominent buttocks might also be interpreted as providing the neonate with something to grasp, providing a soothing and comforting neonate and infant resting place and nutrition storage respectively. Holloway (1981) is also sceptical about breasts as epigamic features. And he's a male. (From a Freudian perspective it comes to the same thing anyway, I suppose.) (c) As I will be describing later, the way in which the concept of 'intelligence' has been used here, and in most writings on human evolution, has major shortcomings from a psychological point of view.

Which leads us on to our next model. Contacts between Psychology and those disciplines more centrally concerned with human evolution, have, in recent decades, mainly taken the form of those in the latter drawing on research on animal behaviour. A major step towards directly integrating Psychology and evolutionary theorising came in the 1979 paper by Sue Parker and Kathleen Gibson to which we now turn. In addition I will describe subsequent work with affinities to their approach by Thomas Wynn. A further paper by Sue Parker on the evolution of higher intelligence in *Homo sapiens* is discussed in a later chapter.

5 *Parker and Gibson; Wynn*

Sue Taylor Parker and Kathleen Gibson (1979), 'A developmental model for the evolution of language and intelligence in early hominids'

Both authors had for several years been researching the cognitive development of primates and, in Gibson's case, comparative anatomy, particularly of the brain. This paper was the culmination of various trends in their earlier work and presented a model of hominid behavioural evolution combining a Piagetian theoretical framework with a version of recapitulationist evolutionary theory. Their opening seems to echo the quote from Köhler given in Chapter 2:

> We can hardly determine the stages of the evolution of intelligence before we know which achievements are more advanced than others,

> which are prerequisites for others, which are correlated with others, and most basically, which imply intelligence. (p. 367)

They claim to have identified the key behavioural innovation from which the evolution of intelligence and, eventually, language stemmed. As in other theories, food-related behaviour is at the heart of the matter;

> Feeding strategies are primary determinants of mating and parental care and, hence, of social structure.
>
> We suggest that feeding strategies are . . . primary determinants of intelligence and that intellectual adaptations for social life are probably secondary to those for feeding. (p. 370)

This last view, as we have seen, would not be universally endorsed. The behaviour in question is 'extractive foraging', that is, foraging using 'tools' to extract food from some kind of matrix in which it is embedded – a nut from a shell, an edible root from the ground or marrow from a bone. They hold that this became a dominant hominid foraging strategy rather than, as in chimpanzees, remaining in a minor role. Such a shift could have occurred as protohominids moved from seasonal to year-round reliance on foods acquired in this way. Their small physical size prevented them using their hands and teeth in this task to the extent that modern larger pongids do, and so, they speculate:

> The first hominids had a basic tool kit consisting of perishable wooden and other organic tools and unmodified stones; pounding stones . . . digging sticks . . . hitting sticks . . . probes . . . leaves for cleaning and wiping grit from food; natural shell dippers and perhaps natural containers for collecting and transporting. (p. 371)

Other strategies of gaining food would not have been challenging enough to account for the outcome:

> The seed-eating and vegetable gathering models . . . do not provide an adequate challenge for the selection of intelligence and language nor do they provide pre-adaptations for hunting and construction. (p. 371)

Typically for late 1970s writers, they cast *Ramapithecus* as likeliest protohominid and hence first extractive forager of this type, but this is not central to their theory.

The unique feature of the model, however, is that they analyse in considerable depth, in Piagetian terms, the logically necessary course that subsequent evolution of intelligence should take, and relate it to the archaeological and fossil record. They believe that the ontogenetic sequence offers additional evidence in its own right since a broadly recapitulationist model is valid in the context of *intellectual* development. This commitment has been one of the most criticised aspects of their model and will be examined further in due course.

The nub of their case is a comparative analysis of primate and human cognitive behaviour. Piaget's basic sequence of stages of cognitive development runs through a sensori-motor period (up to about 2 years), a 'pre-operations' period with two sub-phases, symbolic and intuitive, for 2–4 and 4–7 years respectively, a 'concrete' operations period from 7 to 12 and a 'formal' operations period from 12 years on. (Those unacquainted with Piaget's theory will find a handy synopsis in Beard, 1969.) This has been heavily criticised over the years from many directions, some (Brainerd, 1978) have taken issue with the whole notion of 'stages', while others have queried the universality of the model and the reality of the stage distinctions. The fact remains that Piaget's terminology and classification system still provide the most comprehensive scheme (to use one of his own terms) for tackling this area, and Parker and Gibson do not uncritically accept the entire Piagetian theory. The stages focused on are the first two, sub-divided in finer-grained fashion to track the increasing complexity of cognitive behaviour, each step logically emerging from the previous one. Their survey of primate abilities confirms, they argue, that the evolutionary hierarchy from prosimians to *Homo* in terms of ability matches the ontogenetic course. Their Table (Table 4.4) presents this most effectively. More recent work by Boesch and Boesch (1981, 1983) is cited by Blumenberg (1983) as evidence of 'intuitive' sub-stage abilities, though Parker disputes this, since the research does not, as Blumenberg claims, show aimed throwing in the full sense (Parker, per. comm.). The work of Vauclair and Bard (1983) would seem to confirm the Parker and Gibson picture.

We can now look, selectively, at their treatment of a few of the various specific issues in behavioural evolution their paper covers.

(a) Food-sharing

'We suggest that food sharing first arose as a secondary adaptation for extractive foraging with tools, rather than as an adaptation for hunting and gathering' (pp. 372–3). This of course is quite the reverse of Lovejoy's position on the matter. They picture the situation as one in which the young require a prolonged learning period in order to master extractive skills, during which they would, like chimpanzees, 'depend on their mothers to nurse them, share food with them, and act as models for observational learning of local subsistence technology' (p. 373). The greater reliance of protohominids on hard-to-process foods required in their model, would intensify this dependency. They observe further that

> The importance of food-sharing in hominid adaptation is indicated by its apparently universal occurrence among human children as

> young as one year of age. In their second year children begin feeding their parents with real and imaginary food and greeting strangers with food and other objects. (p. 373)

The appearance of food-sharing in the sensori-motor period implies, they claim, that it began with extractive foraging, although the reasoning behind this particular conclusion is not actually very clear. It does though suggest an early origin for food-sharing.

(b) Throwing and tool-making

The advent of stone tool-making associated with *H.habilis* around 2½ myr is again related to food-acquisition, this time with the adoption of hunting 'as an additional subsistence strategy'. The larger *H.habilis*, a lakeside dweller, learned to prey on larger animals, driving them into traps or bogs, but more than chasing was required, since most of these fauna could outrun them. This extra was achieved by aimed missile-throwing. Here we see the authors applying the Piagetian framework directly to a specific behavioural item, and the passage in question may be quoted *in extenso* as representative of their overall approach:

> Aimed missile throwing was a significant innovation, and a much more complex and difficult one than it appears. It requires the construction of a straight line between the thrower and the target through line-of-sight aiming. This ability does not emerge until four to six years of age in human children – before that time the child is unable to draw or construct with sticks a straight line between two points unless he does so along the edge of a table or another guide – and it does not seem to occur at all in great apes. The emergence of aimed-throwing games at four to six years of age is consistent with the notion that aimed throwing is dependent on line-of-sight straight line construction emerging in the intuitive period.
>
> The adaptive importance of aimed throwing in hominids is also suggested by the ubiquity of aimed-throwing games among human males. (p. 375)

After noting the rarity of accurate aimed-throwing among primates they continue,

> . . . there is a report of unaimed missile-throwing at adult animals to isolate immature prey for seizure (Plooij, 1978). This is apparently an intelligent application of social tool-use. It is very likely that the first hominids regularly engaged in this practice, and that this behaviour was a pre-adaptation for the practice of aimed missile-throwing to drive and stun game. . . .
>
> On the basis of the importance and complexity of aimed throwing for hunting game, and its association with the development of line-

TABLE 4.4 *Primary adaptive function of primate intelligence, by grade levels*

Kind of intelligence	*Prosimian*	*Old-world monkey*	*Great ape*	*Early hominid*
Sensorimotor intelligence				
Stages 1 & 2				
Simple prehension, hand-mouth co-ordination	manual prey catching, branch-clinging, climbing-by-grasping			
Stage 3				
Hand-eye co-ordination		manual foraging		
Secondary circular reactions*			object play for tool use	same
Stage 4				
Co-ordination and application of manual schemes on single objects	manual food preparation and cleaning, manual grooming			
Stage 5				
Object permanence		food location, memory (?)	same	same
Object–object co-ordinations, trial-and-error investigation of object prop. (tertiary circular reactions), discovery of new means (tool use)			trial-and-error discovery of tool use for extractive foraging on embedded foods	same
Stage 6				
Deferred imitation of novel schemes			imitative learning of tool-use traditions	same
Mental representation of images of actions			search for new embedded foods, insightful tool-use	same

PREOPERATIONAL INTELLIGENCE		
Symbolic subperiod		
Topological preconcepts of enclosure and proximity	search for rare embedded foods	same, plus shelter-construction
Make-believe games		practice of subsistence roles
Intuitive subperiod		
Euclidean and projective preconcepts of straight line and angle		tool manufacture, tool use in butchery, shelter-construction
Classification and seriation		food division
1 to 1 correspondence		
Construction games		practice in shelter-construction and tool manufacture
Aimed-throwing games		practice for aimed-throwing in hunting and defense

*Not present in Macaques – arose as retrospective elaboration in great apes
From Parker and Gibson (1979), reproduced by permission of S. Parker and Cambridge University Press.

> of-sight aiming during the intuitive subperiod . . . in human children, we suggest that the projective and Euclidean preconcept of the straight line constructed through line-of-sight aiming arose in *Homo habilis* as an adaptation for aimed missile-throwing at prey. (p. 375)

In conjunction with this particular skill, the new hunting strategy required the manufacture of missiles themselves, while the associated need to develop butchery techniques also generated special tools. Spheroid tools found at Olduvai might represent the first stone missiles. The production of the new range of stone butchery tools – choppers, scrapers and the like – again implies the entry of novel schemata into the hominid cognitive repertoire.

> Each artifact implies a set of intellectual operations. Creating a sharp-edged stone tool requires the notions of sharpness or angle and of sectioning solids, which emerge only during the intuitive subperiod. These notions of angle and section emerge in conjunction with the construction of the straight line as a part of a complex of emerging projective and Euclidean spatial preconcepts (Piaget and Inhelder, 1967). Using percussion to create a geometric section requires a notion of the transmission of forces through object contact, which begins to emerge at this time (Piaget, 1974). (p. 375)

Thus these concepts entered our cognitive repertoire as adaptations for stone-tool manufacture.

(c) Shelter construction

'The importance of shelter construction in hominid evolution has not been widely appreciated' (p. 376). A need for shelters arises in the context of increased bipedalism and ground-dwelling, where defence against both predators and wind and rain are required. Such sensori-motor schema as stacking of objects on top of one another were the first to be involved in this, but eventually far more abilities were developed and co-ordinated.

> Shelter construction by *Homo habilis* required some degree of planning to collect and transport building materials; a propensity to form collections of simple materials . . . and seriate them by size; a propensity to construct straight lines and simple geometric figures; and the ability to tie knots and intertwine materials. Construction also required hierarchical organization of elements. All these abilities emerge during the intuitive subperiod . . . in human children. (p. 378)

Hominid shelter construction is to be contrasted with both the stereotyped building of nests by birds and insects, which is clearly under genetic control, and the crude nests made by great apes which are generally no more than simple 'body-sized concave structures' made by treading down a number of small branches.

This selection from the many topics tackled by Parker and Gibson should be sufficient to indicate the underlying thrust of their approach. (Their account of language will be discussed elsewhere.) Intelligence is not seen as evolving in a totalistic fashion, but, starting from the needs of extractive foraging, as being tied to a succession of specific behavioural tasks associated with food-acquisition and processing and the life-styles generated by these.

> . . . the variety of functions of intelligence in modern man does not suggest that intelligence arose as a general adaptation for non-specific functions. Mutation and natural selection work very specifically: they generate specific adaptations that may turn out in the future to be pre-adaptations for new functions. If this occurs, these new functions may obscure the original function. (p. 378)

Hominids advanced beyond the primates initially by developing extractive food foraging techniques using tools, which led to food-sharing and 'protolanguage', the extended learning required to perfect the new motor-skills reinforced social learning and gave rise to 'symbolic play'. In *H.habilis* the appearance of intuitive intelligence was due to its facilitating hunting and the attendant constellation of related activities – butchery, tool-manufacture and the like. Each stage nevertheless logically springs from its predecessor, so that we could not have evolved hunting prior to extractive foraging. This means that the recapitulationist assumption is valid for cognitive behaviour.

> Given the fact that in human children the abilities of each stage of intellectual development . . . are logical and structural prerequisites for the emergence of the abilities of the succeeding stage . . . and given the fact that the abilities of each ancestral species were logical and structural prerequisites for the evolution of new abilities in descendent species, we must conclude that intellectual abilities develop in the same sequence in which they evolved. (p. 380)

Criticism of this pioneering paper focused on three principal issues: the acceptability of Piaget, the validity of recapitulation, and the primacy of tool-use over social relations in triggering the evolution of intelligence. The first two of these are though interwoven as far as the evolution issue is concerned, since the major criticism of the Piagetian framework *in this context* is its applicability at all in cross-species studies. The adaptations of each species are unique, and it is disputed whether their intelligence can be ordered in the ascending fashion the Parker and Gibson model requires. Snowdon and French, in the Peer Commentary claim,

> . . . there is no simple relation between phylogenetic level, ontogenetic level, and intelligence or language. Behaviors that we human beings consider 'intelligent' or 'protolinguistic' have appeared at several points in evolution and failed to appear at other points. . . . Since each biological species has adapted to a unique habitat, one would expect not just quantitative differences in behavior or cognition between any pair of species, however closely related, but also qualitative differences reflecting the qualitative differences in habitat. (p. 398)

For psychologists the entire Peer Commentary to this paper provides an invaluable, if confusing, introduction to the range of issues, and viewpoints, involved in reconstructing behavioural evolution, and contributors included Gould, Jerison, Isaac and Marshack.

The recapitulation issue is, in this connection, a complex one. In the first place there is the apparently logical unidirectional unfolding or expansion of intellectual abilities which Piaget's model claims to track; simple schemata and motor operations are prerequisites for more complex ones which integrate them in progressively more sophisticated ways, these in turn generating higher-order logical concepts. Secondly, there is the biographical context in which this happens – this cognitive complexification is occurring in parallel with non-cognitive physical and emotional maturation processes which, so to speak, determine the input or content of whatever cognitive operations are going on. As Hume said, 'Reason is and can only be the slave of the Passions' (though there are philosophical grounds for considering this image obsolete! (Midgley, 1979a)). The passions of a toddler are not those of an adult; the behaviour of mature species members is, from a motivational point of view, quite different from that of dependent infants and juveniles, regardless of their 'cognitive capacities' or 'intelligence'. Stupidity is not immaturity. Thirdly, there is the Snowdon and French query regarding species comparability, while Ettlinger, also in the Peer Commentary, makes a fourth, asking 'in what respects are the factors determining the optimal development from birth to maturity within a species likely to be the same as those determining the evolution of a new species?' (p. 384).

At the heart of all these is a profounder philosophical problem beyond our present scope – how far is 'reason' as we understand it in some sense 'objective' a realm to which *Homo sapiens* has gained access by virtue of having travelled, albeit unwittingly, the one logically possible route, and how far is it a species-specific behavioural adaptation, one of any number of possible 'reasons' which are incommensurable? In brief, there is some tension between the 'species-specific evolutionary adaptation' and the 'logically necessary unfolding' components of the model. Both are compatible with recapitulation; indeed the latter requires it. We might perhaps view the specific environmental contexts identified by Parker and Gibson as providing the necessary conditions for the evolution of intelligence, the actual progressive structural development of which is governed by an autonomous logically necessary dynamic of its own. The very specific adaptational events eliciting each new development initiate a process with a momentum provided by the recursive operation of 'accommodation' and 'assimilation' mechanisms such as Piaget postulates for individual development. The individual's recapitulation of the stages is perhaps more apparent than real, representing their passing through the one logically possible sequence of stages necessary for the development of adult intelligence. Even so, if it *is* the one logically necessary route, it can still be validly used for the kinds of inference Parker and Gibson make, but the need to wrestle with the theoretical problems of recapitulationism as a genetic model disappears.

I have left these last aside in any case; although there is the knotty problem of reconciling the 'additive' nature of recapitulation with the 'neoteny' models of physiological evolution now in favour, conceptualisation in this area has now become sufficiently sophisticated to cope with this using concepts such as 'retrospective elaboration' (S. J. Gould, 1977).

On the question of interspecific comparability, accepting that it is an essential component of research in this area it is worth reiterating a point made previously, that what we share with other species is precisely what is not unique to us and 'Since it is the evolution of precisely those behaviours typical of humans that we wish to trace, the comparative evidence supplies us with no analogues' (Wynn, 1982).

On the issue of the primacy of the social or technological, there are very influential voices against Parker and Gibson, notably Isaac, Lovejoy and Holloway. The present writer's view is that we have here a classic example of the pitfalls which occur when one discipline uncritically deploys concepts from another, a peril exacerbated when, as is the case with psychological concepts, they are terms in general use anyway. In this case 'intelligence' is the difficulty. By espousing the Piagetian framework Parker and Gibson have failed to make fully explicit the alternative psychometric insight that there is more than one sort of intelligence. This is notwithstanding their emphasis on the very particularistic character of the initial appearance of specific cognitive operations in evolutionary contexts. Admittedly the Piagetian scheme integrates everything from mathematics to morals and language into one grand dynamically unfolding structural system, but there is no single thing called 'intelligence' pervading everything, and to some degree the development of each of the major arenas is autonomous, if structurally similar in its course. The 'nature-nurture' debate regarding intelligence is, as I have described elsewhere (Richards, 1984), marked by deep conceptual confusion, but leaving that aside, there would be little disagreement that 'general intelligence' – even if identifiable – does not account for everything.

The longest recognised split, surely antedating modern Psychology, and now with an apparent brain anatomy parallel in hemispheric lateralisation, is that between verbal intelligence and spatial intelligence. Although there is a tendency for those high on one to be high on the other, there are plenty of exceptions. It has often been noted that transmission of mechanical, technical, skills is primarily managed by observation and imitation. The direct verbal component is restricted to short utterances like 'harder', 'other way', and non-verbal gruntings of encouragement, disapproval, caution or praise. The verbal component becomes more important in placing the mechanical activity in its social setting, and in labelling task-related objects. Not to pre-empt the discussion of language in the next chapter, I will note here that the evolutionary origin of verbal intelligence could well lie in the social relations area, while that of mechanical intelligence could lie in the foraging equipment area. These two would never be entirely distinct, since technology enters into social life, and social life determines the demands made on technology, but talking about tools and using or making them are different activities. Language of some sort might have been a necessary precondition for the elaboration of technology, but it need not in itself have been centrally involved in that elaboration. Conversely, tool-use might have been a major factor in creating the social circumstances in which language arose, but language is not in itself technology (at least, the view that it *is* in some sense technology struck a quite revolutionary note when Wittgenstein articulated it). We surely need to clarify how far a unitary model of the origin of 'intelligence' is in fact required, and whether or not the two principal forms of intelligence might have different evolutionary roots, however intertwined they now are. Given the incredible current disparities between our species' performances in the technical and social domains, a degree of 'mosaicism' as far as intelligence is concerned is more plausible than not.

For the most part these points are also relevant to the work of Thomas Wynn, of the University of Colorado, to whom we turn next. He too has addressed the applicability of a Piagetian framework to the evolution of intelligence in three papers (Wynn, 1979; 1981; 1982), but restricting his attention to stone artefacts, our only direct evidence of the evolution of our behaviour. The two earlier papers are concerned with Acheulean tools from Isimila, in Tanzania, and Oldowan tools from Olduvai respectively. Although the Acheulean industry lies later than the period being covered in this chapter, the two papers are complementary and will both be dealt with.

Wynn tries to identify the motor-operations involved in the manufacture of these two Palaeolithic industries in the belief that this gives an indication at least of the cognitive sophistication of their makers.

The timing of the appearance of modern levels of intelligence in hominid evolution is of major importance, bearing upon all attempts at model-building and theorising. In the Acheulean paper a more specific issue lies in the immediate background: whether or not the Upper Palaeolithic transition was an evolutionary or cultural event, whether it was correlated with the attainment of a modern level of intelligence or post-dated this by a significant margin. Wynn adopts the latter view, arguing that Acheulean tools display evidence of the attainment of full operational thinking. The Upper Palaeolithic transition does not concern us at the moment, but Wynn's approach does. The tools he studied are from a Tanzanian site dated to between 330,000 and 170,000 years ago (well prior to the transition). The internal geometry of artefacts reflects, says Wynn, the 'infra-logical regulations respected by the maker'. The simplest relations are purely topological, basic 'inside'/'outside', 'near'/'far' distinctions, for example. Following this come projective geometry, e.g. ability to co-ordinate different viewpoints (Piaget and Inhelder, 1967), and Euclidean geometry – schemata of symmetry, equivalence and three-dimensionality. In artefacts four characteristic schemata, diagnostic of attainment of adult operational thought as they are absent in the pre-operational stage, may be identified; the 'whole-part' relationship enabling one to conceive of wholes as comprised of a number of separate units which can be assembled in various ways; 'qualitative displacement' – the ability to relate elements to each other in specific reversible ways such as being in a series, being opposite one another, arranged in size order, etc.; 'spatio-temporal substitution' – recognition that relations are interchangeable in achieving an equivalent result (e.g. although the scene may look different, we might recognise a landmark coming from a strange direction – we grasp that the systematic alteration in the elements in the scene adds up to the same result); and finally 'symmetry' which implies attainment of 'Euclidean' geometry.

Wynn's analysis revealed the following in the Acheulean artefacts:

(a) Whole-part relations are implied by the evidence of retouch in shaping; the maker anticipated a final shape and knew how to achieve it, and which component operations would construct it.
(b) Qualitative displacement is harder to recognise but Wynn believes that *intentionally* created straight edges suggest this as they are equivalent to a series of points in a line.
(c) Spatio-temporal and symmetry characteristics are clearly present as evidenced by both bilateral symmetry of some hand-axes and regularity of cross-sections: 'During flaking the modification of the surface to regularise the cross-section from one point of view

must not be allowed to ruin other cross-sections, most of which are not directly observable' (p. 380) ('spatio-temporal substitution'). (See Fig. 4.4.)

The 'minimum competence necessary' to create these tools then must involve operational intelligence at the concrete level. But what of formal, or 'propositional', intelligence, which marks the culmination of the Piagetian sequence? Obviously we cannot know whether this was achieved, but Wynn notes that current evidence is far from clearcut that it is universal even among contemporary cultures. It might relate to certain levels and types of education, an arena in which intelligence is applied perhaps rather than a distinct level.

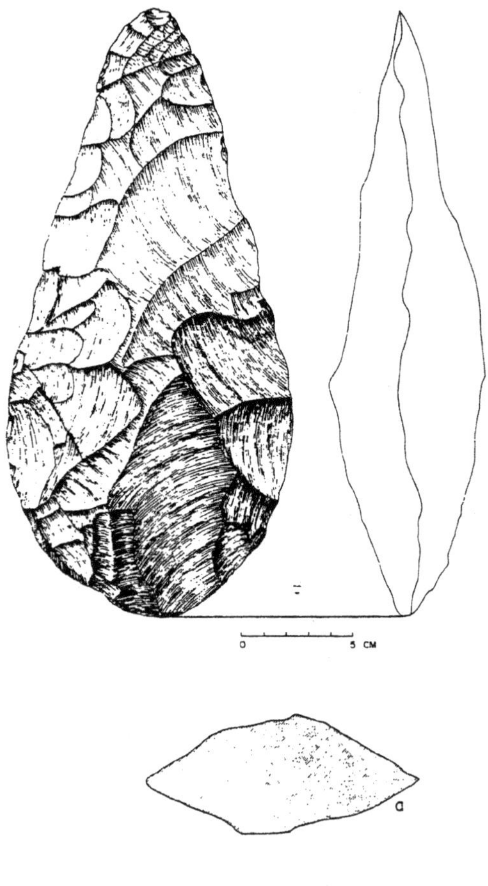

Figure 4.4 Acheulean hand-axe from Isimila, Tanzania showing bilateral symmetry indicative of attainment of full 'concrete operations' competence. From T. Wynn, 1979. *Reproduced by permission of T. Wynn and Academic Press.*

Speculating further on the implications of attainment of operational intelligence levels for other areas of behaviour, Wynn concludes that the makers of Acheulean tools *could* at least have classified kinship relations, mapped and structured their geographical environment in some form around distinct landmarks and had a notion of causality. Thus even the earliest grades of *Homo sapiens* were in possession of a near-modern intellectual *capacity*. The realisation of that potential is probably, in his view, better conceived of as a cultural than a physical evolutionary event.

The Oldowan artefacts, described in his 1981 paper, present a very different situation. Although tool-making itself implies some 'semiotic ability' – an ability internally to represent the task – this is within the capacity of pre-operational intelligence. Using the same analytical approach to that in the Acheulean paper, Wynn shows that Oldowan artefacts from Olduvai Bed 1 and FLK N (the 'Zinj' floor again), nearly 2 myr, required neither Euclidean nor projective geometry for their manufacture, only simply pre-operational topological schema of proximity, and direction. The symbolic and intuitive subperiods of pre-operational thinking (see Table 4.4) would have sufficed. Their crudity is not a result of lower manual dexterity as such, fine control being evident in some of the flaking, but of their being constrained within the elementary logical framework. (See Fig. 4.5.) This illuminates the difficulty archaeologists have found in classifying the early Oldowan artefacts into types at all. These 'hominids were incapable of creating classes in the modern sense' (p. 537). They simply did not have a typology in mind when making the tools, but bashed away at the same general area of the stone until the kind of edge they required appeared, the most complex level of organisation being to knap in two points successively as a pair-sequence. Wynn states baldly that their intellectual level was on a par with modern pongids, 'The absence of a pongid lithic technique does not result from lack of organisational ability . . . it reflects a lack of motivation or a lack of precise enough control over manual dexterity' (p. 538). Nevertheless, constraint within the pre-operational intellectual framework does not imply immaturity in any other sense: 'Oldowan hominids were not human children. They had neither the same experiences nor the same motivations. The parallel lies only in the principles used to organize behavior, not the actual behavior' (p. 537).

He concludes that regarding *H.habilis* (presumed maker of these tools) that there 'is no clear tendency toward a human intelligence and away from a general hominid intelligence evidence in the archaeological record' (p. 539). This conclusion places Wynn in a different camp to those like Jerison (1973, 1982b) who believe that more sophisticated levels of learning, planning and the like may be inferred from plausible models of hominid life-styles at this time, even though absent from their artefacts. The first evidence of a further shift comes around 1.6 myr in some

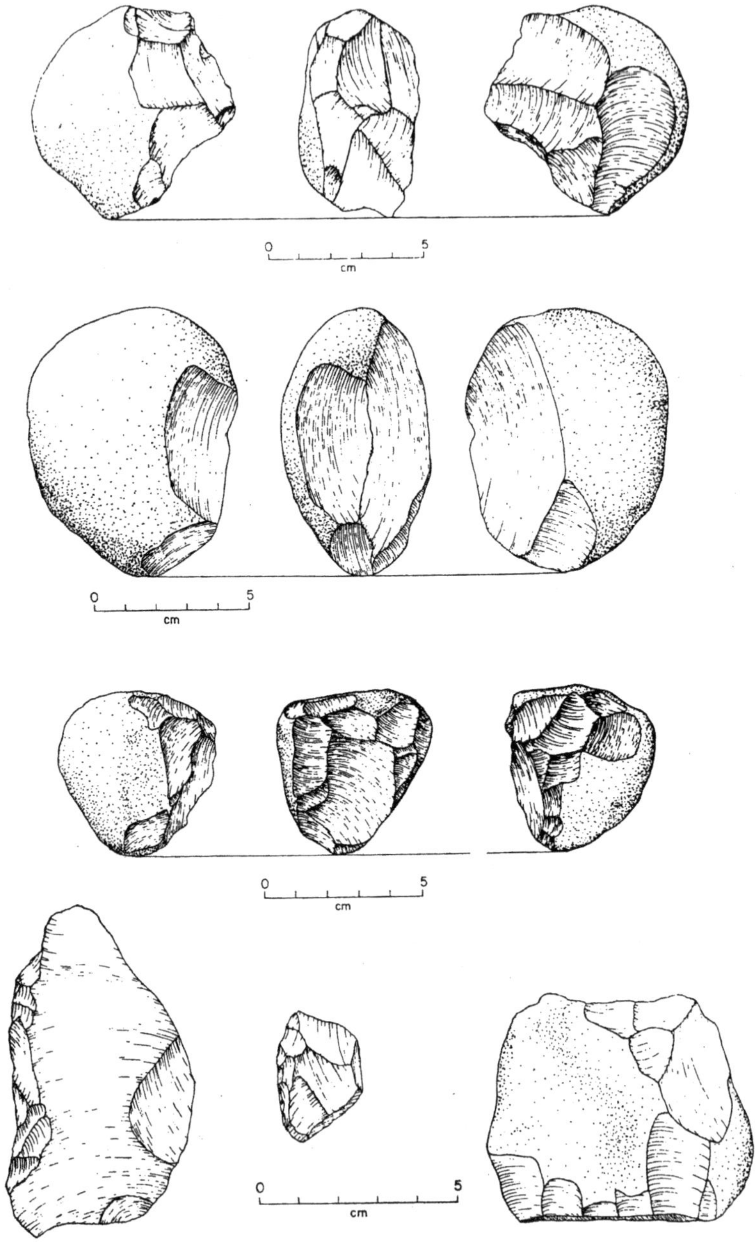

Figure 4.5 Oldowan artefacts requiring only rudimentary motor schemata for their manufacture. From Wynn, 1979. *Reproduced by permission of T. Wynn and Academic Press.*

Developed Oldowan and Early Acheulean artefacts where some indications of limited attempts at achieving symmetry in one dimension can be found.

The 1982 paper, as well as reviewing his previous research, tackles some broader issues. Of relevance here we might mention first that he is sceptical of recapitulationism, 'while a parallel is generally acknowledged, a simple correspondence between ontogeny and phylogeny is far too naive a solution'. He is also doubtful, as an earlier quote indicated, as to the value of comparative studies. For him, the value of Piaget's approach lies in its definition of intelligence as 'organisational ability'. Unlike other definitions, this is sufficiently broad to permit cross-species comparisons, while his developmental sequence does provide a 'rough outline of a sequence of real categories of intelligence'. It is workable for archaeology, though not ideal. More seriously problematical, though, is the fact that in Piaget's model the individual 'constructs' their intelligence using the principles of accommodation, assimilation and equilibration, a process which is neither 'environmental' nor 'genetic', but the result of dynamic structuring principles. This 'constructional' approach is hard to apply to populations. 'How, in any sense, do populations construct increasingly more complex structures of intelligence or, to put it another way, how do assimilation, accomodation and equilibration work to increase the organizational ability of the group?'. (This reiterates Ettlinger's point cited above.) Neither he nor Parker and Gibson have found it possible to carry over the 'constructionist' mechanism into their evolutionary models and Wynn's own views on the matter are at present in flux (per. comm.). (A possible route for tackling this might be provided by ideas currently being developed in the philosophy and history of science fields about the relationship between conceptual change and social structure of the group, notably the applicability of Mary Douglas's notions of grid-group dimensions (Douglas, 1970) as explored by Bloor (1978). These relate the collective response to novelty, or to ideas and phenomena which cannot be fitted into current theory, to the social structure of the group itself. This would determine whether the theory was changed (accommodation) or the novelty somehow incorporated into the existing theory (assimilation). (To pursue these themes further here would, though, take us too far afield.))

6 *Holloway*

Ralph L. Holloway (1981), 'Culture, Symbols and Human Brain Evolution: A Synthesis'

Holloway has periodically supplemented his anatomical research on brain evolution with broader accounts of his understanding of the hominid evolutionary process as a whole. This paper is the most recent of these. His principal concern is to stress the 'mutual-causal' or interactional nature of the relationships between physiological, socio-cultural and natural selection processes, particularly the need to recognise that socio-cultural factors define the meaning of human behaviour. Only in this setting can we interpret the role of the brain in our adaptation adequately. For Holloway, as for Isaac and Lovejoy, the social behavioural changes are more fundamental than technological ones:

> . . . it was not the tools themselves that were the key factor in successful evolutionary coping. Rather the associated social, behavioral and cultural processes, directing such activities as tool-making, hunting and gathering were basic. (p. 288)

> My contention is that the most significant evolutionary changes leading from an ape-like hominid to humankind were in the area of *social behavior* and, moreover, that the fossil record provides us with the clues about the reticulation of the totality of these changes with social behavior. (p. 287. Italics in original)

The importance of the cultural level lies in the unique relationship which humans have with their environment: we *create* our environment by imposing a structure on it, a structure generated by our '*arbitrary* symbol system' (italics R.L.H.). Thus the '*standards* of forms' (of stone tools for example) 'are *social* and *arbitrary*, not instinctive or innate' (p. 288). Elsewhere he refers to 'the sheer anthropocentrism of the human being, the ability (and perhaps need) to order experience and environment as he sees fit' (p. 288).

There was no simple 'initial kick' to the hominisation process as far as Holloway is concerned. The 'kick' was a broad

> selection for social behavior based on increasingly co-operative and less aggressive behavior within social groups. It is speculated that these changes were based on endocrine-target tissue interaction that affected not only the development of co-operation and the diminishment of aggression, but also changes in growth rates affecting brain and body size, sexual dimorphism, and social behavior. (p. 289)

But these endocrinal level changes would only be adaptive in the context of selection for the behavioral changes correlated with them. It is vital to appreciate that the 'cultural' contexts now generated are themselves part of the environment to which subsequent adaptation is geared. Holloway greatly stresses the arbitrary, even fanciful, nature of the 'ideational systems' which we impose on, and by means of which we actually construct, our environments. Once established, our symbol systems, especially language, enhance our facility for learning, elaborating and controlling social relationships and generally transcending the strictly biological level.

Stone tools are important indications that linguistic capacity (if not practice) has been achieved, as the following quotes indicate:

> My contention is that any theoretical model that describes language *also* describes stone tool-making. (p. 290)
> Both processes utilize a limited number of basic units that are combined in a finite number of ways . . . and there is an overlying set of rules, or syntax . . . about how units combine and concatenate. (ibid.)
> The stone tools . . . thus become our most important empirical evidence about the processes (and their evolution) of early humankind's growing perceptions, cognitive processes, and control of the environment. (ibid.)

Wynn, of course, would only accept this as true for the period after about 1.6 myr. His denial that early *H.habilis* could classify tools into types or possessed a 'whole-part' relationship schema is a denial of precisely that capacity to which Holloway is here referring. However, Holloway believes that 'the beginnings of human cultural behaviors preceded the major enlargement of the brain which thereafter seems to parallel the evolution of cultural complexity' (p. 288). On anatomical grounds the process is seen as originating probably just prior to the australopithecines: 'fossil hominid endocasts, from *Australopithecus* on, do show a typical *Homo* pattern of left-occipital, right-frontal petalial asymmetry, which has been strongly correlated with right-handedness in modern humans' (p. 291).

Once the novel hominid social behaviour pattern has emerged, facilitated by appropriate physiological changes, a 'dialectical process' of 'deviation amplification' or positive feedback sets in. Increased brain size and organisational complexity and behavioural complexity continue to reinforce each other, until the limits of brain size increase set by female reproductive physiology or metabolism (Holloway, per. comm.) are reached. This dialectical process is 'non-linear'; there is no simple cause-effect chain but a continual dynamic interplay between social, behavioural and anatomical/neurological levels. As already described, once capable of 'arbitrary symbol systems', hominids generate the *cultural* environment in which physiological evolution henceforth happens. We cannot say that behaviour is 'caused' by physiological adaptations or that physiological adaptations are 'caused' by the environment in any simple fashion, for the reciprocal relationships are also present. Holloway's model is thus 'structuralist' in the sense that Rychlak (1977) and Piaget (1971) use the term to refer to non-linear or synchronous theories such as Piaget's, Lévi-Strauss's, Chomsky's and, to some extent, Freud's.

Human behaviour becomes therefore 'a relatively open system capable of creating new and different constructs at rates well above the replication of genetic instruction' (p. 294).

Environmental complexity is not a straightforward 'objective' fact. The complexity of an animal's environment is really the range of stimulus discriminations which the animal is capable of making within it. As hominids improved at classifying and discriminating, the phenomena in both their cultural and natural worlds became more complex. Environments, says Holloway, 'shrink or expand *depending on the organisms awareness of them*' (p. 294, italics R.L.H). There is thus a clearly phenomenological aspect to Holloway's thought, rare among writers in this area, (and reminiscent in this passage of Koffka's *Principles of Gestalt Psychology* (1935)!). The evidence as to any expansion of the hominid environment is archaeological: stone tools, living sites, and the plant and animal remains associated with tools and shelters. The following passage depicts Holloway's vision of how environmental complexity increased:

> Weather, soil, rock, water sources, game spoors, animal (prey and predator) behavioral habits, seasonal variations, animal anatomy, etc. add to environmental complexity as the awareness and utilization of each contributes to positive selection for their being perceived. The increasing complexity of stone tool types, the proliferation of assemblages and the plurality of animal remains in different archaeological contexts all lead us to assume an expansion of external environmental complexity for the hominids. *Awareness* of relationships expanded; the hominids became better and better at the ecological game. (p. 294, italics R.L.H.)

Stone tools may be conveniently seen as environments themselves, as well as means of coping with the environment. (Similarly computers and cars are part of our environment, not just our ways of coping with it.) 'Stone tools in a sense represent fossilized behavior: they "tell" us something about the perceptual, cognitive and motor actions of their past makers. They suggest certain relationships and attributes that must have been important . . .' (p. 295).

Although this is essentially the same view as Keeley, Newcomer, Wynn and many others would presumably express, Holloway's picture is distinctive because these 'relationships and attributes' are cultural environmental institutions – ideas transmitted and maintained within the group as important, they are not just internal psychological capacities.

Returning to the 'initial kick' phase, Holloway's speculations on the precise social behavioural patterns selected for are similar to several discussed previously, although it must again be stressed that these are associated with the physiological changes which facilitate them, particularly in the brain:

> The foci for natural selection were the glandular (hormonal) and developmental processes that resulted in greater degrees of complexity-management at the level of neuron size, density, dendritic branching, glial-neural interaction, and other possible subcortical and neurochemical changes. *The outward manifestation of these changes was a 1000 ml increase in cranial capacity*. (p. 297, italics R.L.H.)

> The basic shift was probably related to a change in endocrine-target relationships; as aggression and sexual dimorphism were affected so were prolonged periods of gestation, growth and offspring. (p. 298, all italicised in original)

Behavioural and behaviour-related features include sexual division in the food quest, co-operative sharing between and among sexes, social nurturing of offspring, decrease in sexual dimorphism of size, increase in dimorphism of epigamic features, higher aggression threshold within the group, relatively permanent female sexual receptivity and a new way of transferring information about the environment by gesture and language. The timing for the take-off of the 'deviation amplification' situation now created is immediately prior to *Australopithecus*. Holloway is thus in the 'early' camp as far as language capacity is concerned; there is nothing in the 'earliest cadres of *Homo*' or australopithecines to rule it out.

The changed pattern of sexual dimorphism is related perhaps to more permanent bonding, as is the permanent sexual receptivity of females. Prolonged dependency and nurturing is related to post-natal brain growth and learning, and requires use of protein-rich foods. Summarising this complicated initial kick (which by now more resembles a quintuple *entrechât*), he described it as,

> a complex reticulation of anatomical, physiological, and behavioural changes, which also related to bipedalism (securing protein-rich sources by scavenging and/or hunting), the hand, stone-tools and the brain. It is, however, impossible to set down this reticulation in any linear sequence or simple concatenation of events. (p. 300)

> Increased effectiveness in differentiating out from the surrounding environment those features important to survival and social construction, such as suitable rock materials for tools of different kinds and purposes, habits and tracks and spoors of game and possible predators, sites suitable for camping, water and food supplies (including vegetable sources), would have encouraged larger brains, increased communicative skills, which in turn required not only larger

> and better brains, but more effective and affective social nourishment for the transmission of culture in the making. (p. 301)

The qualitative 'initial kick' is followed therefore by a process of continuously enhanced 'complexity-management', as he calls it. Throughout the paper Holloway emphasises

(a) the role of our unique ability to create symbols systems and *construct* our environment in a relatively arbitrary way.
(b) The multi-causal nature of the hominisation process; 'It is impossible to understand the unique evolutionary past of our species without holistically integrating behavioral (cultural) and neural complexity, and the cognitive basis for both.' (p. 302)

Three mutually dependent levels of neurology, behaviour and cognition are thus identified, and his brain evolution studies discussed earlier in the chapter are therefore integrated into a more comprehensive structuralist model of the hominid and human evolutionary process.

A number of observations can be made regarding Holloway's perspective as expressed in this paper. Firstly his stress on the *arbitrariness* of symbol systems (though he accepts the operation of iconic representations too) needs some clarification. At the level of basic classification of the contents and properties of the environment, to be functionally adaptive these would have to bear some correlation with the 'real' contents and properties of that environment, though this does not imply that there is one unique correct classification. In fact though most languages have been shown to classify animals into the same broad classes, and the expansion of this classification is also regular in form (generating, typically, categories roughly identifiable as 'bird', 'fish', 'snake', 'wug' (= worm + bug), mammal (including all large four-legged animals)) (C. H. Brown, 1979). Holloway is not, one must stress, concerned here with arbitrariness in quite this sense. Arbitrariness at the morpheme level – the fact that neither 'chien' nor 'dog' resemble their referent in any way, and are arbitrarily assigned labels – does not imply arbitrariness of reference. This kind of morphemic arbitrariness is that identified by Hockett (1960) as a 'design feature' of human language. Nevertheless, arbitrariness in the more conventional sense does though emerge at the higher cultural levels of religion, ideology and cosmology (though it is not of course an entirely random variability). My impression is that, in relation to the technical basis of human psychological evolution, it is the capacity for arbitrariness in the Hockett sense which Holloway considers crucial (as he spells out in an earlier, 1969, paper), *but* he also accepts a high degree of variability between cultures in the way they construe the world, a variability also arising primarily from differences in their 'symbol systems' – including of course language. There is a long-standing controversy in sociolinguistics over the extent of relativism in linguistic structuring of the world (the 'Whorfian' or 'Sapir-Whorf' hypothesis as it is usually known after its original advocates). Holloway does seem to be in the Whorfian camp; however, 'relativity' rather than 'arbitrariness' is what is entailed by this position.

Secondly, he is far less inclined to indulge in detailed model-building about protohominid social behaviour than some other writers, being fully conscious of the sheer range of factors involved, while his appreciation of the role of the 'relativistic' cultural level would in any case render such model-building more problematic.

Thirdly, it should be borne in mind that as far as timing of developments is concerned he tends towards 'early' rather than 'late' positions, especially in the matters of the advent of language and evidence of cultural complexity in archaeological material. Readers should remain alert to future statements from Holloway, who uniquely combines a structural, culturally oriented, overall theoretical perspective with a detailed anatomical expertise derived from the main body of his empirical research over two decades.

Conclusion to Part C

The material selected in this section has necessarily been at the expense of a number of prominent figures such as P. V. Tobias, Pfeiffer, Wolpoff and indeed Richard Leakey and Johanson. We have been concentrating, however, on current ideas about early hominid *behavioural* evolution. Insofar as they discuss this, the other authorities tend to present simplified versions of the positions described here, often, in the more popular works, glossing over important differences, and they are in any case usually more concerned with the physical anthropology side.

What we are seeing in these recent papers (in fact 1981 seems to be the crucial year!) is a coalescence of archaeological, physiological and behavioural science expertise being brought to bear on the data and exposing much of its fundamental ambiguity. We now move to a theory which might be called the 'Joker' in the pack.

D The aquatic-phase hypothesis

The hypothesis that our early ancestors passed through an aquatic phase first received publicity in 1960 when it was proposed by Sir Alister Hardy, who was impressed by the pattern of hair direction on the human foetus, as depicted by an earlier biologist, Frederic Wood Jones: 'such an arrangement of hair, offering less resistance, may have been a first step in aquatic adaptation before its loss.' Various other similarities with marine animals such as streamlined shape and subcutaneous fat also seemed to him to render the notion plausible, in addition to our fairly high swimming abilities and general widespread attraction to water. Hardy has continued to lend support to the idea, but its principal exponent has been Elaine Morgan. Having first taken it up in a book entitled *The Descent of Woman* (1972), one of the influential feminist accounts of human evolution which appeared during the 1970s, she has more systematically restated it in *The Aquatic Ape* (1982), and it is with her theory as presented in that text that I will end this chapter. Elaine Morgan is a freelance dramatist and writer rather than a professional physical anthropologist, and the mainstream academics have treated her position very warily. Glynn Isaac (1983) says of it 'Palaeoanthropologists in general judge the aquatic hypothesis to be highly implausible or impossible, but it cannot be formally eliminated yet, and it is fun to keep it on the list in the meanwhile' (p. 530). Gowlett (1984) is frankly dismissive, considering it only slightly less far-fetched than Erik von Danniken's extra-terrestrials ' . . . it has no basis in fact. If . . . we had evolved successfully in water . . . why should we abandon it? Why do we not have flipper-like adaptations, like other aquatic mammals such as seals and walruses?' (p. 17) It is hard to believe he has actually read the work in question, since both these rhetorical questions are fully dealt with, and to say of Morgan's position that it has no basis in fact is to beg the entire question of how the evidence she offers is to be interpreted. It must be stressed nevertheless that Hardy's original mooting of the idea was, according to at least one authority who was present, meant as a joke and that the present author is taking considerable risks with his academic credibility by devoting any space to it (Day, per. comm.).

In outline, Morgan's argument is that (a) a number of the most characteristic human behaviours, notably bipedalism and speech and (b) a number of our most apparently idiosyncratic physiological features such as hairlessness, subcutaneous fat, sweating, weeping, nose-form, breast-form, ventral orientation of the vagina and our highly developed sense of balance can be economically explained by an aquatic or semi-aquatic phase in hominid evolution immediately prior to the appearance of *Australopithecus* and lasting perhaps for a million or two million years. Given our genetic similarity to the chimpanzee, the gross differences between our species are extremely puzzling. The orthodox models which try to explain our whole suite of unique characteristics as adaptations to savannah life leave many of these features enigmatic, and in some cases imply quite opposite directions to those which we actually took. The evolution of the human mind 'can hardly be accounted for by an evolutionary history which differs in no essential particular from that of the savannah baboon' (1982, p. 114). And although neoteny is clearly evident in some of the physiological developments, it is in itself a mechanism rather than an explanation. It does not say why particular neotenous shifts came to have survival value. Nor does neoteny as a mechanism account for all the hominid features in question. In short, the unique combination of behavioural and physiological traits possessed by the hominids is a result of an equally unique sequence of selection pressures, an idiosyncratic evolutionary history.

Morgan envisages a situation in the late Miocene when rising sea levels around north-east Africa isolated a number of primate populations on islands where dwindling resources forced them into either extinction or novel adaptations. Small isolated populations like these are known to be ideal 'forcing houses' for evolutionary change (especially on Gould's model). Among these was our Aquatic Ape ancestor which shifted to eating marine foods such as shellfish and crustaceans, and gradually became more and more aquatic in life-style. At a later stage the sea level dropped again and the population in question returned to a more terrestrial mode, though continued to prefer river and lakeside habitats which enabled it to preserve some of its aquatic adaptations. All this is plausible from a geological point of view, for sea levels did rise and fall in this manner at the period in question and one writer, La Lumière (1981) has proposed a quite specific site, a 'Danakil Island' where the present Danakil Alps are, south of the Red Sea. This, though, is purely speculative. *H.habilis* is regularly depicted as a lakeside dweller and *A.africanus* as a river bank dweller. There is nothing intrinsically far fetched about Morgan's scenario here.

The aquatic phase left a legacy of unique physiological traits and a number of features pre-adapted to the later developments of bipedalism and vocal language. The supposedly odd human traits which are so difficult to account for in a terrestrial evolutionary account fall neatly into place when viewed as aquatic adaptations, since they are normal and widespread among aquatic mammals. What we have, then, is a double transition, from trees, to water and then back from water to land. Morgan's major points may be taken in turn.

1 *Hairlessness, subcutaneous fat and sweat*

We are the only primates who have evolved a temperature maintenance system which integrates these three features. No other primate is hairless; subcutaneous fat is virtually unknown among the anthropoid apes and never in the places where we deposit it most thickly, and sweating is also unique as a cooling technique – one so costly in terms of moisture loss rate that one Professor of Dermatology (Montagna, not referenced) terms it a 'major biological blunder'. Under desert conditions at 100°F a naked man would sweat up to 28 litres a day and lose 10–15 per cent body-weight by salt-loss. As an adaptation to the supposedly hot arid savannah environment this is an odd move, to say the least! Conventional accounts seem to imply a rather tortuous sequence of events in which hair was first lost in order to keep the hominid cool, then fat added so that it could keep warm, then sweating evolved to cool it down again, 'all to cope with a climate which every other savannah dweller had accepted without turning, or shedding, a hair' (p. 46). Wheeler (1984), mentioned earlier, must be considered seriously to weaken this argument, however.

Loss of hair itself is disadvantageous in several respects, for example skin gets cut easier, and neonates cannot cling so easily – or be held on to. Two theories other than 'cooling' are offered for hairlessness: sexual attractiveness and neoteny. Darwin himself favoured sexual selection, but as Morgan points out, 'Characters gained through sexual selection evolve by a process of exaggerating some feature which is *characteristic of the species* . . . But features which are *uncharacteristic* of the species are more likely to repel than attract' (pp. 32–3). She also rejects the arguments often linked with this regarding its role in 'making sex sexier' and facilitating pair-bonding, or encouraging males to return to base to share their meat, etc. In general increased sexiness does not, Morgan claims from comparative data, enhance monogamous bonding, while she shares the currently widespread scepticism about these male hunting scenarios. She might perhaps have added that male hairlessness is *not* necessarily considered sexy.

As for neoteny, the fact of the matter is that at 6 months the human foetus is covered with a coating of hair known as 'lanugo', which is occasionally retained by a neonate (a condition called hypertrichosis). It was the patterning of this lanugo, as we have said, that reminded F. Wood Jones (1929) of the passage of water over a swimming body, stimulating Hardy's initial speculation. 'It is manifestly contrary to the whole notion of neoteny that a human foetus should first acquire the coat of hair once appropriate to an adult primate, and then subsequently abandon it' (p. 36). Apart from the large pachyderms (for whom there is also some evidence of aquatic links) on the terrestrial side and polar fauna such as sea-lions on the aquatic side, hairlessness is an aquatic mammal feature (though some smaller ones such as otters and beavers evolved water-resistant fur). The longer a species has been aquatic, the greater the hair loss. In dolphins it is virtually complete. Hairlessness greatly enhances efficiency of swimming, and even in humans shaving off of body hair can have a measurable effect on speed. Subcutaneous fat evolves to offset the loss of insulation arising from depilation, but in addition it increases buoyancy and produces a streamlined hydrodynamically efficient body form. Such a layer is universal in sea-mammals, from whales and manatees to porpoises and sea-lions. It is suggested that retention of scalp hair, especially in women, is for children to cling to, and there is some anthropological evidence that this practice is still known. Being the only aquatic mammals descended from the primates, with prehensile hands, we could expect this to be unique to the aquatic ape. Subcutaneous fat is laid down late in gestation and accounts for the relatively overweight human neonate's smooth and chubby appearance, by contrast with the 'cadaverous' looking chimpanzee baby.

The aquatic theory offers an explanation of both our hairlessness, the pattern of such hair as we do have, and our possession of a subcutaneous fat layer. What of sweating? This of course would be unnecessary for an aquatic animal, but would be a natural development for a mammal returning to land and continuing to live in close proximity to water, so that the high rate of moisture loss associated with this cooling technique was immaterial. The bare skin plus subcutaneous fat combination cannot be easily reversed; there would be no immediate pay-off in altering the hair-fat ratio, but the homeostatic sweating mechanism solves the problem nicely. The onus must surely be on Morgan's opponents to provide an alternative explanation for this set of aquatic mammalian traits which is more plausible than her convergent evolution theory.

2 *Bipedalism*

As we have seen, the origins of bipedalism are a mystery. It is increasingly recognised (e.g. by Lovejoy, 1981) that as a savannah adaptation it makes little immediate sense; it is intrinsically unstable, rendering us liable to trip and fall or be knocked over, nor is it more energy-efficient than quadrupedalism, while chimpanzees manage to carry things quite effectively with knuckle-walking and use quadrupedal locomotion for speed. 'A long dreary catalogue of physical disorders – muscular strains, prolapses, hernias, backaches and disorders of the legs and feet reflects part of the price we pay for walking upright' (p. 56). On the savannah the slight initial advantages for improved vision would hardly have outweighed the drawbacks. 'Except for man, no mammal, carnivore or herbivore, on the savannah or off it, has chosen to walk and run habitually with its spine at right angles to the earth' (p. 57). 'A pattern of mammal behaviour that emerges only once in the whole history of life on earth takes a great deal of explaining' (p. 54).

Again, once viewed in the context of aquatic life rather than savannah-dwelling, the bipedal, or at any rate upright, posture, ceases to be quite so odd. It would initially be useful in wading, and once afloat the upright 'treading water' position is biomechanically natural. Seals, otters, dugongs, manatees and dolphins all commonly exhibit it. But, more importantly, even while swimming the spine and lower limbs remain aligned, the lower limbs no longer hang at right-angles to the spine as in quadrupeds, but continue in a straight line behind it. Instability, of course, is no problem; you cannot trip over while swimming. Such a situation has typical consequences for the pelvic girdle arrangement. 'If this modification of the skeleton – a pre-adaptation to bipedalism – took place in the water, it would not be unique; it would be a common and natural development' (p. 60). (The only other animal with a humanoid bipedal walk is the penguin, although the beaver too adopts a walking gait when carrying.) On return to the land, the aquatic ape would already have the anatomical adaptations which rendered erect posture easy, and been immediately able to exploit some of its secondary advantages.

One aspect of bipedalism is the need for a greatly developed sense of balance. The only other animals which can match human abilities in this respect, being indeed superior, are sea-lions. Aquatic life would require, presumably, an enhanced ability to orient oneself via internal balancing mechanisms in the absence of gravity. The spinal flexibility of humans, as evidenced by gymnasts, is also unique among primates, while typical of aquatic mammals. While in themselves not conclusive, such features which are shared with aquatic mammals but never found in either savannah or woodland dwellers or primates must lend cumulative weight to Morgan's thesis that physiologically we exhibit a recognisable aquatic syndrome.

But why did we not evolve flippers, as Gowlett crushingly inquires? The answer is twofold and fairly simple. Firstly, being descended from primates who had already evolved highly versatile leg-movements (unlike the typical rear legs of quadruped plains dwellers) our initial swimming movements would more likely have resembled frog-like kicking, than flipper-like paddling (and even now neonates placed front-down use their limbs in a sort of breast-stroke style, especially when in water). Shortening of the hind-limbs is not a necessary aquatic adaptation as, indeed, frogs prove, and the swimming style of a newly aquatic ape would be quite different from that of a newly aquatic dog (the ancestral line of the pinnipeds). The current state of play in the debate over early australopithecine gait and foot morphology sheds no light one way or the other on this, and Stern and Susman's comparison of the Laetoli footprint makers' gait as akin to walking in flippers refers to side-to-side placement of left and right footprint trails, not foot-size itself. Morgan introduces a further line of argument on the matter, concerning the fact that 7 per cent of the population actually do have some webbing of the feet (fewer for the hands). Of this she observes:

> Congenital abnormality can take many different forms. Most commonly it is the absence, or distortion, or incomplete development, or reduplication, of some item in the standard pattern

> of the human physical structure. It is extremely rare for congenital abnormality to take the form of adding a feature . . . which is usually believed to have been absent from our own species and from *our whole biological order* (the Primates) throughout its evolutionary history. (p. 82, italics E.M.)

Gowlett must come up with something a bit more convincing. This brings us on to human swimming and water adaptations generally.

3 *Swimming and drowning*

It is pertinent here to note that humans continue to display a high degree of adaptiveness to water unique among primates other than the swamp-dwelling proboscis monkey of Borneo and the talapoin, which lives by the Congo and leaps into the water to escape danger. Our fellow great apes loathe water by and large and can hardly be induced to immerse their faces in the stuff. If liking for water were simply a matter of 'regression to the womb', it ought to be shared by most placental mammals. In fact it isn't, and the parallels between the two situations are slender. A foetus in the womb is held firmly, rotates slowly, and is curled up; the baby in the water is free-floating, can thrash about and is also in a quite different auditory milieu. Morgan devotes a whole chapter to the water-adaptiveness of babies and cites evidence that even aquatic parturition can be trauma-free, though she does not believe it was ever normal. The presence of a mid-wife is, however, typical of some aquatic mammals, such as dolphins. Neonates are able to control their breathing so as not to inhale water; they float contentedly and without distress. It is possible that prior to ten months a baby will learn to swim without tuition, though at present most evidence in this area is anecdotal. The whole topic of neonate affinities for water is a fascinating one of considerable significance for the credibility of the Aquatic Ape theory/hypothesis. If the recapitulation argument is considered respectable enough to be given a scientific airing in relation to infant food-sharing, it deserves at least the same attention in relation to their swimming before they can walk.

Even aside from tales of neonatal piscatorialism, humans, some anyway, are remarkably good swimmers and divers. We share with other diving animals the so-called 'diving reflex' (provided our faces are unmasked) in which the phenomenon known as 'bradycardia' occurs: a dramatic reduction in heart-rate (from around 70 to the low 30s in humans) reduces the rate of oxygen consumption. Humans can actually dive deeper than otters and beavers, and even deeper than some varieties of porpoise and dolphin. Certain features of our strange nasal apparatus also fall into place: we possess musculature capable of flaring our nostrils – a 'tentative step' in the direction of being able to actually close them as aquatic mammals can – and the specifically evolved protruding nose functions to deflect water when diving, a feature shared with the curious-looking proboscis monkey, which also swims. On the other hand, Carey and Steegman (1981) have found a correlation between nasal protrusion and lower humidity, which is greatest in cold, dry, climates. Morgan draws frequently on Wind (1976b), though not always in agreement throughout her discussion of this topic.

The standard counter-argument to this seems to be that such behaviours are learned and not universal. This is less than adequate. Many behaviours such as speech itself require practice and 'learning' to be perfected, but are nevertheless part of the species' natural repertoire. Most current models of ontogenetic development (e.g. Lewontin, 1983) would firmly reject such an argument as an oversimplification. And surely a behaviour does not have to be universal to qualify as genetically rooted; if it did then heterosexual mating would be ruled out! Our physiological equipment is such that we *can*, should we choose, acquire and perfect behaviour of particular kinds (such as swimming) and this physiological pre-adaptation for the behaviour must have been acquired somewhere during our evolutionary history. Its universality in the repertoire would naturally have declined if it was last mandatory several million years ago, though as just noted some argue that swimming *would* be universal if everyone was introduced to the water prior to ten months.

Perhaps the most dramatic evidence of an aquatic adaptation has come to light quite recently with the discovery that some people can survive long periods of total immersion (up to half an hour or more) without damage. Although they lose consciousness, the brain does not appear to be permanently affected after resuscitation. The most publicised case occurred in 1977 when an 18-year-old youth was brought back to consciousness after 38 minutes under water in Lake Michigan. Again, although such evidence may verge on anecdotal, the total picture of the relationship between *Homo sapiens* and water is of a swimming-pool building, beach-gathering, swimming, diving, showering, splashing and wading animal who, even if a desert dweller, creates fountains and pools where possible and whose idea of luxury is often to wallow in a scented bath. What other land animal commonly washes itself by immersion or in cascading water? Why do we not lick ourselves clean? Such tastes are no easier to reconcile with an arid savannah-bound heritage than the physiological oddities already discussed.

4 *Speech*

Chimpanzees can make a wide range of noises and their vocal communication system is now known to be surprisingly elaborate. Their primary channels of communication, though, are visual, gestural and tactile. Our sound-dominated communication system is curious in two principal respects: (a) why did this particular channel come to gain such importance; and (b) how did we acquire the uniquely complex system of breath control required for human speech (and indeed, song)? Morgan's suggestion is that the aquatic phase pre-adapted us for these evolutionary innovations. As far as the dominance of the sound channel is concerned, it is fairly clear that in an aquatic environment the gestural and tactile channels will be less effective, while the smell channel, already in decline in arboreal primates, accelerated its decline in the hominids – a rather unexpected phenomenon if only a return to the ground had occurred, for this should, logically, have stabilised the decline. Other aquatic mammals, dolphins and whales especially, rely on sound and make utterances possibly comparable in complexity to our own. The control over breathing required for fully developed speech follows directly from the aquatic swimming and diving life-mode, as indeed do certain strictly physiological developments of the laryngeal system and the ability to block the nasal passages with the velum. She also raises the question of whether on return to the land humans did not evolve their unique axillary and urogenital scent 'organs' to amplify or revive the nearly atrophied scent channel.

The role of the aquatic phase in the origins of speech is envisaged as a laying of the necessary physiological and psychophysical bases for the behaviour: conscious control of breathing 'a feature of, and essential to, all diving mammals' (p. 102) and the reliance on the auditory channel required by marine conditions where traditional visual channels are less efficient.

5 *Other 'aquatic' traits*

The traits we have been looking at are the most important, but not the only, indications of an aquatic ancestral phase. Other features which Morgan incorporates into her theory include rather general factors such as our relatively slower biological clock; our precocial pattern of child-bearing (or 'K-strategy'); and our larger brain size, shared with dolphins and whales and, she argues, related to the more complex control and adjustment required in aquatic conditions. This last point is frankly weak, and does not tally with the evolution of brain-size data. There are also a number of more specific behavioural and physical features such as face-to-face copulation and forwardly directed vaginal channel, and weeping of salt tears both as a response to irritants and as a stress reaction.

Elaine Morgan's case clearly requires an answer from physical anthropologists and deserves better than Gowlett's offhand dismissal. If we dignify it with the fully fledged status of a 'theory', it should be possible to derive some potentially falsifying hypotheses in true Popperian fashion. As it is we seem to share, to a considerable degree, a constellation of traits typical of aquatic mammals. The alternative explanations offered are frequently inadequate, usually *ad hoc*, accounts of single traits, and fail to clarify why, if our evolutionary course was simply from trees to savannah, albeit as now favoured via plusher woodland, we do not more closely resemble other animals which took a similar route. Some of our features, such as sweating and precocial child-bearing, seem positively disadvantageous for a savannah dweller. Given current moves away from 'savannah' theories, acceptance of a fragmented Miocene environment as the speciation-favouring scene of our debut, continuing bewilderment over the origins of bipedalism and over the degree of morphological divergence we show from our genetically similar pongid cousins, Morgan is only taking current lines of thought a stage further. A peculiar animal presumably requires a peculiar evolutionary history and the tree-sea-land sequence meets this requirement. The serious difficulty concerns its timing, for where in the record can we fit the requisite time-interval of perhaps as much as 1–2 million years? (Hardy (1977) suggests 20 myr of semi-aquaticism, with 5–6 hours a day in the water, but Morgan does not follow him in this.)

The aquatic phase would presumably have to come *after* the chimp-hominid split but before the Hadar and Laetoli australopithecines. In recent years the split has, as we have seen, been brought forward to the 5–6 myr range, while the hominid fossil record is nudging the 4 myr mark. Even so, no one could claim that palaeochronology was as yet remotely approaching definitive resolution as far as events in protohominid evolution are concerned, and if the 'aquatic phase' evidence is otherwise weighty enough the onus would be on palaeochronology to fit it in, rather than reject it because it can't.

Reviewing the status of the model in 1984 (Morgan, 1984), Elaine Morgan felt that it was standing up rather well, except in the matter of brain size. It must be stressed that Morgan does not see her theory as a rival to the savannah theories for the period since Lucy, but as accounting for the initial emergence of the unique hominid physiology and behavioural traits. Even during the 'savannah' period it now seems likely that the hominids lived in proximity to water, in lake or riverside environments.

6 *The Agaiumbu*

A curious piece of anthropological evidence, not mentioned by Morgan, and little known in the literature, concerns the now extinct 'duck-footed' Agaiumbu of New Guinea described by C. A. W. Monckton (1920). These people lived in a lagoon on the north-east coast which linked the River Musa to the sea. They were extremely timid and nearly helpless on dry land, living in a village half a mile from the shore, built on poles. From our present point of view the interesting fact about them is that although they had presumably only been driven into their aquatic environment within the last few thousand years at the most, and still sometimes purchased brides from land-dwelling tribes nearby, they had already acquired some physiological adaptations to the aquatic life. Monckton describes them in comparison to their neighbours, the Baruga:

> Placing an Agaiumbu man alongside a Baruga native of the same height, one found that his hip-joints were three or four inches lower than those of the Baruga; one also found that his chest measurement was at least on average three inches greater, while his chest expansion ran to as much again. The nostrils of the Agaiumbu were twice the size of those of any native I have ever seen; they appeared to dilate and contract like those of a racehorse. Above the knee on the inside of the leg was a large mass of muscle; on the leg below the knee there was no calf whatsoever, but on the shin bone in front there was a protuberance of a sinewy nature. The knee-joints were very wrinkly, with a scale-like appearance; the feet were flat as pancakes, with practically no instep, and the toes long, flaccid, and straggling. Walking on hard ground or dry reeds, the Agaiumbu moved with the hoppity gait of a cockatoo. Across the loins, instead of curving in fine lines as most natives do, there was a mass of corrugated skin and muscle. The skin of the feet was as tender as wet blotting-paper, and they bled freely as they crawled about upon the reeds and marshy ground of our camp. They had a slight epidermal growth between the toes, but nothing resembling webbing, as alleged by the Baruga; the term 'duck-footed' therefore, had only meant tender-footed . . . or more literally, 'water-bird footed.' (1936 ed., p. 185)

They were of course extraordinarily adept swimmers. Monckton's account was confirmed by his companion Sir Francis Winter in a letter to the Australian Governor-General. The encounter, in 1902, was followed within a few years by a massacre of the Agaiumbu by a neighbouring tribe. They were even then but a remnant of their original numbers, having suffered an epidemic before Monckton met them. It is impossible to disentangle at this remove how far their physiological peculiarities were genetic or resulted from the peculiar ontogenetic circumstances of

Agaiumbu maturation. The aquatic ape envisaged by Morgan was of course a much smaller animal than modern humans, while the Agaiumbu were having to adapt to the water from current human size. The Agaiumbu evidence is relevant, I think, because it does suggest how rapidly some sort of aquatic adaptation can be made, physiologically, even by modern *H.s.sapiens*.

Whatever the eventual fate of Morgan's theory, the aquatic factor as such must surely be incorporated in future models of hominid evolution.

Conclusion

Given that the area is rapidly changing and effectively in its early stages, I have by and large refrained from attempting to evaluate the positions I have been discussing other than to draw attention to their controversial aspects or note specific features I see as problematic.

For behavioural scientists the most interesting are Parker and Gibson's and Wynn's neo-Piagetian models and Holloway's comprehensive structuralist approach with its blend of socio-psychological and anatomical perspectives. At the level of empirical research one is fascinated by the positively Holmesian micro-wear and bone-damage work of Shipman, Keeley, Bunn, Potts and others. Morgan's theory, just discussed at length, though considered outlandish, awaits a serious appraisal by a non-dogmatic physical anthropologist.

By and large, varied though the theories discussed have been, and rare as total agreement is, there are really no clear fundamental cleavages in the area, no basic theoretical divides of such a profound kind that their positions are mutually exclusive and irreconcilable. This cannot be said for the topics which concern us in the next chapter, where we focus on three particularly crucial issues: the relationship between physical and cultural evolution – 'genes and culture', altruism, and the origins of language.

CHAPTER 5

Genes and culture, kindness and speech

Introduction

The dynamics of the behavioural evolutionary processes dealt with in the previous chapter eventually, possibly with the advent of *Homo erectus*,[1] generated an apparently novel and unprecedented form of life-style – that which we loosely term 'culture'. The hallmarks of this include complex social organisation, technology, language and 'art', though they did not all appear simultaneously, nor, individually, did they appear overnight. 'Art' in particular does not appear until relatively late in the day, while the timing of the origin of language is anyone's guess. There is broad agreement that this form of social life radically altered the evolutionary situation, and that henceforth a process of change of a new kind enters the picture which some, but not all, have been prepared to call 'social evolution' (or 'biocultural evolution'). The profoundest effect of this appears to have been the cessation, then reversal, of the normal evolutionary pattern which follows the advent of a new genus-adaptive radiation or divergent evolution. Typically a new genus diversifies into various species as it spreads through ecological niches presenting different selection pressures. The australopithecines, immediately prior to *Homo*, were still apparently diverging in this fashion as the existence of both robust and gracile forms in close proximity testifies. From *H.erectus* onwards our lineage has not only failed to diversify in this fashion, but has moved into the reverse direction – convergent evolution. Our genetic diversity has progressively reduced (the extinction of the Neanderthals being a dramatic case in point, but the vanishing of the native Tasmanians in the last century is another), and continues to do so (Dobzhansky, 1944; Kitihara-Frisch, 1980; Neel, 1983). Kitihara-Frisch argues that this unprecedented course of events is related directly to the adoption of the new 'cultural' life-style, based, in her view, on our newly acquired capacity for symbolis-

[1] Though in the present writer's view later for reasons to emerge during the chapter.

ation. Other consequences of the onset of social evolution are: acceleration of the rate at which change can occur by enabling the direct cultural transmission of innovative behaviour and technology across generations and between groups in a non-genetic fashion; elaboration of social organisation and diversification of behaviour to a point where economic, sociological and social psychological forces come into play for the first time; and (optimistically) that the evaluation of change is brought under rational control and direction.

In this chapter I will first review the 'social evolution' issue itself, then move on to consider the topic of 'altruism', and finally the thorny problem of the evolution of language, which is probably the single most significant facet of the human cultural adaptation.

A Social evolution

The notion that a process of social or cultural evolution has operated to elevate our species from savagery to a highly civilised present is well established in western thought. Prior to Darwin, Comte had outlined the stages of human thought from magic via religion and philosophy to science. In Darwin's wake (see Chapter 2) came the ideas of L. H. Morgan and of popular writers such as Benjamin Kidd (1894), who gave the phrase 'social evolution' a wide currency. It is worth considering for a while the connotations this phrase carries. Its principal implication is that the modern human life-style has arisen by an orderly 'natural' process in some way related to, or akin to, biological evolution. 'Progress' is thereby legitimated as an objective feature of the world and the course of, first, our prehistory, then our history, represents its continuance. Biological evolution produced humans as its highest achievement and in doing so raised the rules of the evolutionary game to a qualitatively superior level. The extent to which the term 'evolution' is intended literally or metaphorically varies between writers, but the effect of its adoption (intended or not) is nearly always to imply a continuation, in and by the human species, of a natural cosmic process. Modern civilisation is justified as the fulfilment, or a step towards the fulfilment, of transpersonal forces. Such justification of the present order is, as was described in Chapter 1, one of the principal roles of creation myths.

This connotation remains even when a writer explicitly eschews all notions of teleology or conscious ideological purpose. There is an inevitable tension between this disclaimer, entailed as it is by espousal of a deterministic scientific stance, and the very notion of social evolution. D. T. Campbell's major paper of 1975 is a case in point, and in it he actually laments the lack of English terms for change which do *not* imply progress. Even if prehuman biological evolution was essentially directionless, social evolution is not, indeed the introduction of conscious purpose on to the scene is one facet of its revolutionary character. But it is hard to avoid a further retrospective move of construing this eventuality as itself the outcome of *unconscious* purpose even in the preceding biological phase. If biological evolution produced us how could it have been directionless? We are the direction in which it was heading!

This in turn raises what is perhaps one of the toughest problems now facing us; how to appraise our species' past, and by implication its present, without *a priori* assumptions regarding whether it is, in some objective sense, 'normal' or 'pathological' in character. If this is in part what we are trying to find out, to assess it in either way in advance is to beg the question. The third possibility, that such notions are anthropocentric, reflecting only our subjective criteria, merely leaves the problem to someone else. This is an issue which is only now surfacing clearly. There has been a rapid change of tenor between the still fairly anthropolatrous orientation of the immediate post-World War Two years and current writers such as Neel (1983) who are fundamentally pessimistic. Hence, although the course of 'social evolution' as gauged from archaeological and palaeoanthropological findings does seem to have been something of an orderly unfolding or development through successive stages of a widespread, if not universal, regularity, and although too a quasi-evolutionary process of selection and adaptation of technical, social institutional and ideological phenomena seems to have occurred, whether or not all this amounts to progress in any but totally anthropocentric terms is one of the biggest begged questions of all time. To a cynic, or a woolly mammoth, we are about as progressive as Dutch elm disease.

This is not to suggest that accounts of social evolution have been in any way homogenous; on the contrary they are extremely diverse, a diversity hinging on the manner in which the relationship between the sociocultural and genetic levels is conceived (this all indeed being an aspect of the old nature–nurture controversy).

Comprehensively summarising current theories and knowledge in this area is particularly difficult. Historically, it has been one of those issues which perennially attract discussion from widely differing directions, while being the clear province of no particular discipline at all. The upshot is that until the last decade or so, for all the spilt ink, it is hard to identify a coherent body of cumulatively developing scientific or academic literature on the subject. Too often what we have is a final chapter in which the author of a text unwinds, following the hard scientific slog, and waxes philosophical about the human condition. All too frequently this amounts to little more than a recitation of clichés and platitudes. Commonest among these during the post-war decades was what I would term 'The Great Choice' cliché. This arises from the deep-rooted need to preserve some elevated status for humanity, combined with the perception by such writers that there is precious little evidence for such a notion in our species' current behaviour. A relatively late version of this is at the end of Thorpe (1974), where the tenor is getting just a mite desperate:

> Thus the task of mankind is perhaps greater and his activities for good or ill more momentous now than ever before in the history of the world. It is tragic that such multitudes are oppressed and rendered impotent and ineffective by a failure to find meaning in their lives when in fact the potential fullness of human life has never been greater or more obvious for those who have eyes to see. It is indeed both a joy and a terror to be living in times such as these – joy at the fullness of life and the opportunities for greater enlightenment; terror at the danger and disasters threatened by evil forces – for never has the human race had greater opportunity than now to rise above itself and to bring the human spirit to new levels of transcendence. (p. 384)

Sociologists, anthropologists, psychologists, historians, philosophers and polymaths such as Lewis Mumford have all contributed to this confusing mélange. In the post-1970 period the social evolution question has become a facet of much broader discussions about the nature of evolutionary processes and the possibility of using evolutionary principles as a basis for a general 'evolutionary epistemology', a framework for a comprehensive philosophy of knowledge. This position is discussed further below, but for the moment we need only note that it sees the central evolutionary principle of 'blind variation and systematic selective retention' as applicable across the board as a model for human learning, conceptual change in the sciences, technological change and the development of social institutions and social evolution generally (Campbell, 1975). ('Sociobiology' may be seen as a school of thought within this wider evolutionary epistemology movement, which contains many who are highly critical of it.)

It is then no easy task to tease out from the presently available literature any concrete models of human social evolution since the Middle Palaeolithic. It is somewhat easier to identify a number of diverse positions regarding the genes-culture relationship which have been advanced since 1950, and these will be my main concern here. Even so some further preliminaries are in order.

One topic which is taxing the evolutionary epistemologists is how to conceptualise the unit in question when discussing social evolutionary processes. Even in biological evolution there is a conceptual difficulty here regarding the notion of 'species'. Are 'species' the units of biological evolution? They after all consist on the one hand of populations of varied individual members and on the other hand represent but one moment in time of a genealogical lineage extending back to the origins of life. On both counts the identification of a species as a coherent entity becomes blurred in principle. With social evolution the problem is far more serious. We may consider biological evolution as involving (a) reproducers (some, but often not all, species members), (b) traits or alleles which they transmit from one generation to the next, (c) the transmission process itself (e.g. sexual reproduction) and (d) environmental selection pressure. To formally equate social with biological evolution then we need to identify equivalents of (a) reproducing entities, (b) traits, (c) transmission and (d) environmental selection processes. We also though require a formal equivalent to 'species', i.e. a *taxonomy* of types of society (or whatever the things are we are supposing evolve). The difficulties on all these counts are legion. Some authorities have felt that they can be overcome fairly easily (Cohen, 1981); others consider them sufficiently recalcitrant to undermine the whole analogy (Dickemann, 1981). Others again believe that a solution to the 'species' problem at least can be reached by a radical reappraisal of the concept itself (Hull, 1982).

Finally there is a fact to which amazingly little critical attention has been given in the study of behavioural (including social) evolution. The traits which physiological evolution involves are, if not permanent throughout an organism's life-span, at least present physically for extended periods of time. Behaviour, by contrast, is intrinsically transient, that is, however conditional upon genetics and however 'wired-in' in terms of the movements involved, a behavioural trait is only visible in particular circumstances. More specifically we cannot know what an organism *cannot* do simply by watching it. Its behavioural capacities can never be fully explored, at least not without intense and prolonged study of captive groups in naturalistic surroundings, and even then it is probably debatable. It is the capacity for enacting the behaviour which is inherited, not the behaviour itself, as is routinely noted. But this situation means that the nature of the expression of behavioural traits is fundamentally different from the situation for physical ones. It is at one remove, and more intimately bound up with environmental conditions. Since humans have the capacity for radical behavioural *innovation* also, the physiology/behaviour correspondence is rendered even more difficult to unravel than in species where only limited behavioural innovation is possible. This gulf between the genetic-cum-physiological level and the cultural behavioural level is too wide to permit simple linkages between the two, but as we will see some contemporary writers, particularly the sociobiologists, feel they can in principle at least bridge the gap.

So, a dilemma confronts us. At some point along the way, our lineage shifted to this novel form of life-style – 'culture'. Henceforth, as Holloway and others have understood, this cultural arena produces its own selection pressures on those within it autonomously from those generated by the physical natural world. Indeed it may be seen as buffering us from these

(Kitihara-Frisch, 1980). A new kind of dynamic seems to have arrived. This 'culture' is itself an entity over and above its members. From *H.erectus* onwards the story of human evolution becomes increasingly the story of how *culture* changed and developed. The physical evolution story does not cease, but increasingly the changes involved (in brain size and form, in general robustness) seem to be comprehensible only by incorporating the cultural dimension. It converges on adaptations suiting those who possess them to increased efficiency in social organisation, communication and flexibility, rather than diverging along the lines required by varied habitats (though a few features such as the epicanthic fold which characterises orientals may represent this latter more orthodox process).

As far as data are concerned, these cultural developments are easier to track than the physiological ones, since for every fossil there are thousands of artefacts plus hearth sites, living floors and food remains. But are we looking at 'evolution' any more? Cultural phenomena are not tied to genetic transmission; they can spread rapidly from one human group to another by cultural contact. It is not the makers who evolve so much as the objects they make, but a flint cleaver does not make another flint cleaver, the tool is not a species of reproducing entities. Tools, so far, need us to make them, but *we* might be being selected for our ability to do so. The cultural object, the flint cleaver, the basket, the fire-drill, is a factor in *our* evolution. By now, though, there are several levels in play: the individual's biologically based behavioural capacities and general 'fitness', particular cultural/social phenomena (tools, customs, words), and the social or cultural tradition as a whole to which we give such proper names as Mousterian, Magdelanian, Aztec or Egyptian. The concept 'social evolution' somehow embraces each of these, even though they vary enormously in the mechanisms by which change is governed and the rate at which it occurs. Part of the diversity of the literature on the topic arises from the fact that writers have concerned themselves with different levels. When we are referring to 'human evolution' then we are, for the post-*H.erectus* period, talking about the evolution of at least three different kinds of phenomena – our physical and behavioural traits, our societies, and the particular products of our physical and mental labour.

We will now consider the various positions taken on the genes-society relationship of which I am afraid there are no fewer than nine, with even more sub-variations (though I do not pretend that this classification is definitive!).

1 *Physical evolution supplanted by social evolution*

(a) Social evolution is strictly analogous to biological evolution; only the mode of transmission has changed. This is the classic 'Social Darwinist' model in which civilisation emerges from a struggle for survival between peoples, individuals, ideas, etc., the 'best' being selected. The most common current form of this is the sociobiology perspective, but this more properly belongs in position 5, below, since it does not consider physical evolution to have stopped. Some earlier writers, however, did believe that humans had reached physical perfection, that no further advance on the physical plane was possible, hence in part the appearance of social evolution. There are few strict adherents to this position now and it is included primarily for the sake of completeness, its mantle having passed to the sociobiologists.

(b) Human society is inherently determined by social forces operating in a scientifically lawful fashion, but these are of a radically new kind. Whether or not these are termed 'evolutionary' varies from theorist to theorist. A classic example of such an approach is the Marxist model, where socio-economic laws which govern the nature of the relationship between labour, product and social structure supersede physical evolutionary ones. Engels (1884), influenced by L. H. Morgan is the key statement of this position. In this century the archaeologist V. G. Childe (1963 [1951]) was its most influential proponent. Another version is Kitihara-Frisch (1980) who sees the advent of symbolising capacities, which in her view underly both tool-making and language, as initiating a process of 'convergent' evolution in which social organisational and cultural factors come to over-ride the normal primacy of physiological adaptation to the natural environment. Culture 'buffers' the individual against the natural world by technological means such as clothes, tools, artificial dwellings, etc. A somewhat similar orientation is adopted by the anthropologist W. Goldschmidt (1976), who although accepting some similarities between biological and cultural evolution (both involving continuity, variation, selection and extinction) considers that the latter hinges primarily on learning processes rather than 'genetic programming' and is hence quite different in character – being teleological, 'Lamarckian', and very rapid, as well as involving a different unit, the community or social group, rather than the individual. In all such positions the genes-culture relationship is considered at best tenuous. Our biological evolution is substantially complete and we have moved into new realms altogether.

Perhaps Mumford's invocations of the radical impact of consciousness and creativity belong here too, since he considers physical evolution to have been perfected and the stage to have been set for entirely new developments (Mumford, 1967).

(c) The extremer environmentalist position would have us liberated entirely from either genetic/biological *or* transpersonal social forces. We can create (and have) of ourselves what we will. A variety of rather differing accounts cluster here as it is the most explicitly anti-evolutionary stance and can be reached from several directions. Since the radical

Behaviorism of J. B. Watson earlier this century it has been adopted as something of a central dogma of much academic psychology that the individual's own environment is the paramount determinant of its behaviour via its unique learning history. The cultural relativism of anthropology after Boas and the similar commitment to environmentalism of post-World War Two US social psychology meant that the evolutionary perspective was eclipsed in both Psychology and Anthropology. Around the mid-century, then, ideological factors played a large part in downplaying evolution in western liberal thought, since it was felt to be dangerously conducive to racism and anti-humanistic determinism (a view not without justification). In practice this meant that humans were seen as psychologically almost infinitely flexible; genes played little part in affairs at the behavioural and cultural levels other than the part they might take in determining the individual's gross morphology, thereby rendering them susceptible to particular kinds of social psychological process (such as stereotyping). The gene's role, such as it was in these cases, was thus always mediated by cultural factors. Curiously this extreme environmentalism provided a rationale both for highly positivistic behaviourist social engineers *and* anti-reductionist humanistic psychology, positions otherwise poles apart.

From another direction, such a position would be adopted by many Christian thinkers for whom the constraints on our free-will and on society at large are, if they exist at all, 'spiritual' in character, or derived from some overall plan to which God alone is privy. The very notion of seeing ourselves in an evolutionary perspective is deeply offensive, be it physical or social in kind. Why it is felt to be so offensive is an interesting psychological question into which we cannot enter here. This may of course verge into extreme Fundamentalism. In most orthodox Christian cosmologies human history is an unfolding of a divine plan, not an evolutionary process, and they may thus be included either here or under 1b.

In general extreme environmentalist positions are now adopted primarily on the grounds that other positions, particularly evolutionary ones, are ideologically suspect. This is an oversimplification, and being replaced by more sophisticated models.

(d) A somewhat more complex position was adopted by one of the most eminent geneticists, Theodosius Dobzhansky (1955, 1962) in which social evolution *plus* biological evolution add up to a uniquely 'human' evolutionary process;

> Biological evolution has produced the genetic endowment which has made culture and freedom of choice possible. But from then on, human evolution has become in part a new and unprecedented kind of evolution – evolution of culture and of freedom. This certainly

> does not mean that the biological evolution of man has come to a halt, as some writers like to suppose. The two kinds of evolution, biological and cultural, are combined in a new and unique process which is human evolution. (1955, pp. 374–5)

> Human evolution has two components, the biological or organic, and the cultural or superorganic. These components are neither mutually exclusive nor independent, but interrelated and interdependent. Human evolution cannot be understood as a purely biological process, nor can it be adequately described as a history of culture. It is the interaction of biology and culture. (1962, p. 18)

Even if biological evolution has not ceased, its classic mode of operation has thus been left behind. The high educability and behavioural versatility of humans has resulted from the selection of genetically controlled *plasticity* of traits, particularly mental ones, and there is no reason to suppose this has ceased. It is interesting that Dobzhansky joined with one of the leading anti-hereditarian anthropologists, Ashley Montagu, in one exposition of this (Dobzhansky and Montagu, 1947). In its hard version, the notion that physical evolution has actually stopped and been replaced totally by social evolution is now rarely expounded.

The positions just identified are not all mutually exclusive of others discussed below, and writers often shift emphasis from one image to another in the course of their work.

2 *Aspects of a single evolutionary process*

This is the 'evolutionary epistemology' model which constitutes the most influential recent development in evolutionary theorising. The central concept is that all biological systems, from the gene-level to social structure and individual learning, constitute systems of knowledge acquisition. As a doctrine it has been coalescing for a number of years, since the late 1940s, from a number of different strands as varied as Piaget's developmental psychology, Popper's philosophy of science, the rise of cybernetics and systems analysis, and the varied genetic theories of Waddington, G. C. Williams, Stephen J. Gould and Richard Lewontin. Plotkin (1982), now one of its leading advocates, characterises it thus:

> The argument is that a theory that will encompass all biological phenomena, including those of socio-cultural origin, will have to be based on multiple selection processes and multiple units that are selected. (p. 7)

> (evolutionary epistemology) . . . maintains that the adaptive organization that exists at multiple levels, from genetic and

> biochemical substrates through the morphological and behavioural traits of the phenotype and on to its ecological and socio-cultural interactions, is all part-product of processes of information gain. (ibid.)

Thus it is moving beyond a reductionist commitment to ultimate explanations purely at the molecular level. A historical review of the origins of evolutionary epistemology can be found in D. T. Campbell (1974). The 'nesting' of different levels of knowledge gain on Plotkin's own model was given in Fig. 3.10. This represents the most comprehensive account so far of how the genes-culture (and intervening level) relationship can be integrated into a single general evolutionary model, though Lumsden and Wilson (1981) run it a close second, as we will see. It must be stressed that Plotkin endorses D. T. Campbell's 'blind-variation-selective-retention' commitment, as the basic routine by which knowledge is gained. 'All forms of knowledge gain in any biological system, and within any level of any system, appear to be achieved in the same way.' (1982, p. 8)

An influential figure for this school is G. C. Williams, who in the mid-1960s provided a penetrating analysis of a number of important issues in the natural selection model of evolution. His account of the notion of 'progress' is particularly relevant in the light of my introductory remarks, for he cast an icy douche on the whole idea, stating outright that

> I would maintain . . . that there is nothing in the basic structure of the theory of natural selection that would suggest the idea of cumulative progress. (in Plotkin (1982), p. 47)

> The concept of progress must have arisen from an anthropocentric consideration of the data bearing on the history of life. (ibid.)

> I doubt that many biologists subscribe to the view of evolution as a deterministic progression towards man, but there is a widespread belief in some form of aesthetically acceptable progress as an inevitable outcome of organic evolution. (ibid., p. 40)

He then proceeds to cast doubt on all the candidates that have been proposed by biologists to sustain some notion of evolutionary progression: accumulation of genetic information, increasing morphological complexity, physiological differentiation, increased effectiveness and the tendency to see particular species as goals towards which their ancestors were heading are all refuted or seriously questioned. On *H.sapiens* he sees the habit of depicting the primates with tree shrews at the beginning and humans at the end as making 'it easy to imply that progress toward man is a recognized evolutionary principle that has operated throughout the history of the primates'. 'It is mainly when biologists become self-consciously philosophical, as they often do when they address nontechnical audiences that they begin to stress such concepts as evolutionary progress' (p. 58).

Within psychology the strongest advocate of this 'evolutionary epistemology' has been D. T. Campbell, notably in his A. P. A. Presidential Address of 1975, but the actual structure of the model he proposed there renders him more appropriately placed in the next group.

Although the evolutionary epistemology perspective is hard-headedly deterministic in its picture of evolutionary processes, its advocates are far from sharing a common account of the ways in which these operate with regard to social evolution. The extremer sociobiology camp constitutes only one group within evolutionary epistemology which may also be said to include Lewontin, one of its leading opponents (Rose, Kamin and Lewontin, 1984). The picture of social evolution emerging from Plotkin's account would be that the advent of human culture added another layer, or perhaps several layers, of information-gaining system to the genetic level. Nevertheless the genetic level remains in play and the whole ensemble constitutes a single complex information-gaining-and-selecting process. Even so, it is ultimately 'directionless', the result of blind variation and selection mechanisms.

My own principal query here is about the notion of 'blind variation'. This has come under challenge already within the evolutionary epistemology camp (Lewontin, 1983; Ho, 1984; Ho and Saunders, 1982, 1984). Notwithstanding the ingenuity of learning theorists to blur the boundary, there remains a difference between the way in which behavioural innovation is achieved by a cat in a puzzle-box, which panics (generates random behaviour) until it hits, by chance, on an 'adaptive' response, and insight behaviour or behaviour based on cognitive analysis of a situation (as can occur in chimpanzees and humans, among others). The 'trial and error' routine is constrained by a number of factors: (a) the size of the organism's existing response repertoire, especially the way in which this is itself constrained within physiological structural limits; (b) the organism's ability to give weighting to various responses in terms of their probability of success; and (c) the organism's 'insights' into the mechanisms governing the phenomena (to) which it is trying to adapt. There is a radical difference between responding to familiar systems in unfamiliar states and responding to unfamiliar systems *per se*. At any level, we are, in the light of (a), obliged to analyse the way in which the available response repertoire has itself been generated. In human individuals it seems, as I will discuss later, that a key feature in initiating the 'cultural' move was a dramatic amplification of the response, or behavioural, repertoire involving a strategy for enabling the organism to expand this continuously. *Psychologically* this is perhaps the key turning point, as will be argued in due course. Thus I am unsure what is meant by 'blind' here; if it means 'random' this is surely untenable, since it can only be random within certain parameters, and how did *they* come about? A snail is unlikely to hit on coughing as a way of discreetly attracting your attention. If it means 'without knowledge of outcome', this ignores the fact that the organism is likely to have some expectations regarding the relative probabilities of outcomes of different behaviours, and move from most-likely-to-succeed responses to long-shots. (There is even Mr Spock's paradox – rationally concluding that since there is no rational solution to the problem at hand, the solution must be irrational.) As far as behavioural and social evolution are concerned, there are, in short, two qualifications to the 'blind variation' notice: the generation of variation itself and the weighting of items within this range.

Although this evolutionary epistemology approach has provided us with some promising theoretical frameworks for tackling social evolution, aside from D. T. Campbell's paper (discussed below) there has been little in the way of direct discussion of the phase we are currently concerned with, it having so far been applied mainly to history of science and history of ideas topics. A recent, brief, formal statement of the 'single evolutionary process' position is de Winter (1984).

3 *Social and biological evolution in conflict*

It has long been a commonplace to suggest that we possess some intrinsic tension in our natures between a primitive, bestial, 'dark' side and a rational civilised one. Unlike the previous position, this generally implies that there *is* something intrinsically pathological in our situation. Three variations on this can be distinguished.

(a) The 'strong' version – A classic instance of this which continues to have adherents is the Freudian model, classified by one recent commentator (Maddi, 1976) as a 'psychosocial conflict model'. The requirements of social life are seen as essentially opposed to the individual's instinctive drives. Social evolution is the record of the progressive subordination of the biological to the social in a constructive, but only partly successful fashion. This is at heart a tragic vision of our predicament.

Though not couched in Freudian terms, the same basic picture has emerged more recently in Sagan's and Maclean's interpretations of brain morphology (Sagan, 1977; Maclean, 1982) where the high-powered neocortex, vehicle of rational faculties and desires, is locked in battle with a primitive reptilian brain, source or repository of savage instincts. Both of these represent a persistence of the older nineteenth-century image of a psyche split between the evil savage and the noble man of reason.

(b) A 'weak' version (e.g. Washburn and Howell, 1959) is that there is not so much a split as a lag. Social evolution has 'advanced' beyond biological evolution, due to its rapid character by comparison with genetically based evolutionary change. Clearly the notion of 'lag' carries an implication of 'direction'. This weak version has become commonplace as an explanation of the patent discrepancies between our technological skills and the manner of their application. The paradox of such 'lag' models is how a species can create an environment to which it is not adapted; how indeed it can be adapted to create such an environment! This kind of picture does not provide explanations for the state of affairs it describes, at least in the way of detailed models of the evolutionary process itself, or how its relationship to physical evolution has developed.

(c) D. T. Campbell (1975), proposed the outlines of a model of the nature of social evolution and its relationship to physical evolution which requires examination in rather more depth. The principal aim of this paper was to affirm the importance to Psychology of the evolutionary perspective and suggest that traditional belief systems, features of social organisation, morality and social custom ought not to be rejected as irrational and maladaptive simply because they appear so to contemporary psychologists. Rather they deserve respect as adaptive products of thousands of years of sociocultural evolution and need to be evaluated, though not uncritically, in that light.

> Just as human and octopus eyes have a functional wisdom that none of the participating cells or genes have ever had self-conscious awareness of, so in social evolution we can contemplate a process in which adaptive belief systems could be accumulated which none of the innovators, transmitters, or participants properly understood, a tradition wiser than any of the persons transmitting it. (p. 1107)

At the outset he makes it clear that,

> I speak from a scientific, physicalistic (materialistic) world view. The evolutionary theory I employ is a hard-line neo-Darwinian one for both biological and social evolution, the slogan being 'blind variation and systematic selective retention'. (p. 1104)

Sociocultural evolution is thus another arena, alongside the biological one, in which general evolutionary principles play themselves out, the 'evolutionary epistemology' position. He describes sociocultural evolution thus:

> By sociocultural evolution we mean, at a minimum, a *selective* cumulation of skills, technologies, recipes, beliefs, customs, organizational structures, and the like, retained through purely social modes of transmission, rather than in the genes. (p. 1104)

> If there are general principles of organizational effectiveness as in the division of labor, then quite independent streams of social

> cumulation may be shaped by this common selective system so that streams converge on similar structures moving from simple social systems to complex social systems along parallel routes. (ibid.)

Of course, past adaptiveness does not guarantee present adaptiveness and there will in any case be 'noise' in the system;

> For a natural-selection type of socio-cultural evolution to work, the retention system must be capable of perpetuating uncomprehended functional recipes. The retention system must operate, as in biological evolution, by perpetuating everything it receives from the edited past. Inevitably this includes a lot of noise, maladaptive mutations and chaff, along with selected kernals of wisdom. (p. 1107)

> . . . the wisdom produced by evolutionary processes (biological or social) is wisdom about past worlds. (p. 1104)

In moving on to analyse in more detail how the biological and social levels are interrelated Campbell focuses particularly on the altruism and morality issue. It is this analysis which qualifies him for inclusion here, since he holds not only that modern urban civilisation is a product of social, rather than biological, evolution but that

> This social evolution has had to counter individual selfish tendencies which biological evolution has contrived to select as a result of the genetic competition among the co-operators. (p. 1115)

> On the one hand, there is biological evolution optimizing an individual person and gene-frequency system. On the other hand, there is a social-organizational-level social evolution optimizing social system functioning. For many behavioral dispositions the two systems support each other. *For others, the two are in conflict, and curb each other.* (p. 1116, my emphasis)

To elucidate this further he proposes a 'two system analysis'. On this model we assume that biologically the individual tends towards behaviour *X* (say selfishness) while the socially optimum requirement for the population is a much lower level of *X* than the individual is inclined to exhibit and a higher level of *non-X* (say altruism). The 'moral preachments' of culture will tend then to insist on total commitment to *non-X*. In doing so unambiguously and simply these moral preachments act as a 'low-cost, one-sided homeostat setting'. Since their absolutist demands can never actually be met, they do nevertheless ensure that behaviour shifts from *X* towards *non-X* sufficiently to stabilise in practice around the 'biosocial' optimum. The strength of the moral demand indicates the opposing strength of the biological drive. 'Any recurrent single-pole moral preachment becomes an indicator of a biological bias away from the biosocial optimum in the opposite direction' (p. 118). In other cases, as in opposing proverbial injunctions ('look before you leap', 'nothing ventured, nothing gained' are given as an example of this, though they are not I think a good one – 'Many hands make light work' and 'Too many cooks spoil the broth' is clearer), the traditional wisdom strives to maintain behaviour at an optimum mean level from which deviation in either direction could become dysfunctional if persisted in. (Surely, though, such proverbs are situation-labels, not general injunctions?)

Although Campbell is offering this analysis in a speculative spirit to open up the topic, he clearly views social evolution as having resulted in a variety of homeostatic mechanisms, such as systems of moral precepts and ideologies, by which the individual biological character of society's members can be held in check and socially required learned behaviours, sometimes quite contrary to these, maintained. Given his deterministic stance, we ought perhaps to stress at this point that the school of thought to which Campbell belongs cannot ever construe social evolution as differing from biological evolution by virtue of the advent of some novel purposive principle. Social evolution must be as blind and as rooted in natural selection as biological evolution; it is only the mechanisms of transmission and storage which have changed. In Campbell, too, there is also this underlying suspicion that the social is wiser than the individual, the social system as a whole transcends its units – a direct analogy between body cells and individual members of society is apparent in our first quotation above. He is not, though, complacent about the present, which he views as a 'runaway positive feedback' situation in which stabilising mechanisms have failed. This raises again the question as to whether or not the present situation is somehow objectively pathological. How *can* an evolutionary process 'go wrong' if one views it all from a 'hard-line deterministic' position?

Campbell's address had a mixed reception, especially from anthropologists, much of the criticism being directed at his treatment of altruism (Wispé and Thompson (eds), 1976). Some of this has been superseded and discussion of altruism as such is reserved for the next part of the chapter. I would like to draw attention to two pertinent points that were raised within the evolutionary perspective. Firstly, R. Cohen took issue with the conservative drift of Campbell's paper, arguing that in view of the very rapid changes occurring in the human situation at present, there was selection *for* discontinuity and rejection of traditional injunctions. Secondly, G. Levine again brought things down to earth by observing

> Our present traditions may have had to pass tests to survive. But there is little reason to conclude that our traditions have been tested for benefits to successive societies. It is reasonable to assume that they

have been tested for benefits to some of the selfish interests that rise to leadership in most societies. (Wispé and Thompson (eds), p. 377)

But the unresolved paradox behind Campbell's position is surely that unless social evolution *has* generated a situation in which rational insightful evaluation of the requirements of society can transcend 'blind' evolutionary mechanism, there is no point in us striving to enhance such transcendence anyway. Our very analyses of social evolution are but blind responses to circumstances which will owe their effectiveness or success not to their correctness but to their *incidental* adaptive value. This, though, conjures up some much broader enigmas in the philosophy of science which we cannot enter into.

4 *Social and biological evolution in parallel*

This is really a milder version of 1(b), in which the continued operation of biological evolution in relation to, for example, immunological resistance, is accepted, but the dominant social forces are accepted as a non-biological realm. Some modern geneticists (e.g. Neel, 1983) are concerned with the incidental genetic consequences of current changes in human society: the breakdown of isolation between breeding groups (entailing the rapid spread of alleles and breakdown of co-adaptational trait clusters), and the effects of medical science in perpetuating genotypic maladaptations which were hitherto constantly held in check since they were lethal or prevented reproduction (e.g. phenylketonuria). This is hardly a comprehensive position on the whole social evolution issue, rather a focusing on one particular aspect, but at least one influential textbook (Bodmer and Cavalli-Sforza, 1976) deals almost exclusively with such issues in its discussion of social and cultural aspects of evolution.

5 *Social evolution subordinated to biological evolution*

This line has been advocated with some fervour by C. D. Darlington from the 1950s on, culminating in his two books *The Evolution of Man and Society* (1969) and *The Little Universe of Man* (1978). His principal target was J. Huxley's 'psychosocial evolution', a position akin to Dobzhansky's (Darlington, 1979). The most influential versions at present do not, however, appear to stem so much from Darlington as from ethology. A number of books appeared during the 1960s, some extremely successful, which sought to show how far human cultural behaviour was in fact biologically rooted. (Desmond Morris's *Naked Ape*, Lorenz's *On Aggression* and Ardrey's *The Territorial Imperative* and *The Social Contract*

1 Nineteenth-century French artist's impression of flint-tool manufacture at Pressigny. From Figuier (1882), illustration by Emile Bayard. One cannot help observing that the man centre foreground is somewhat exposed to danger from flying debitage!

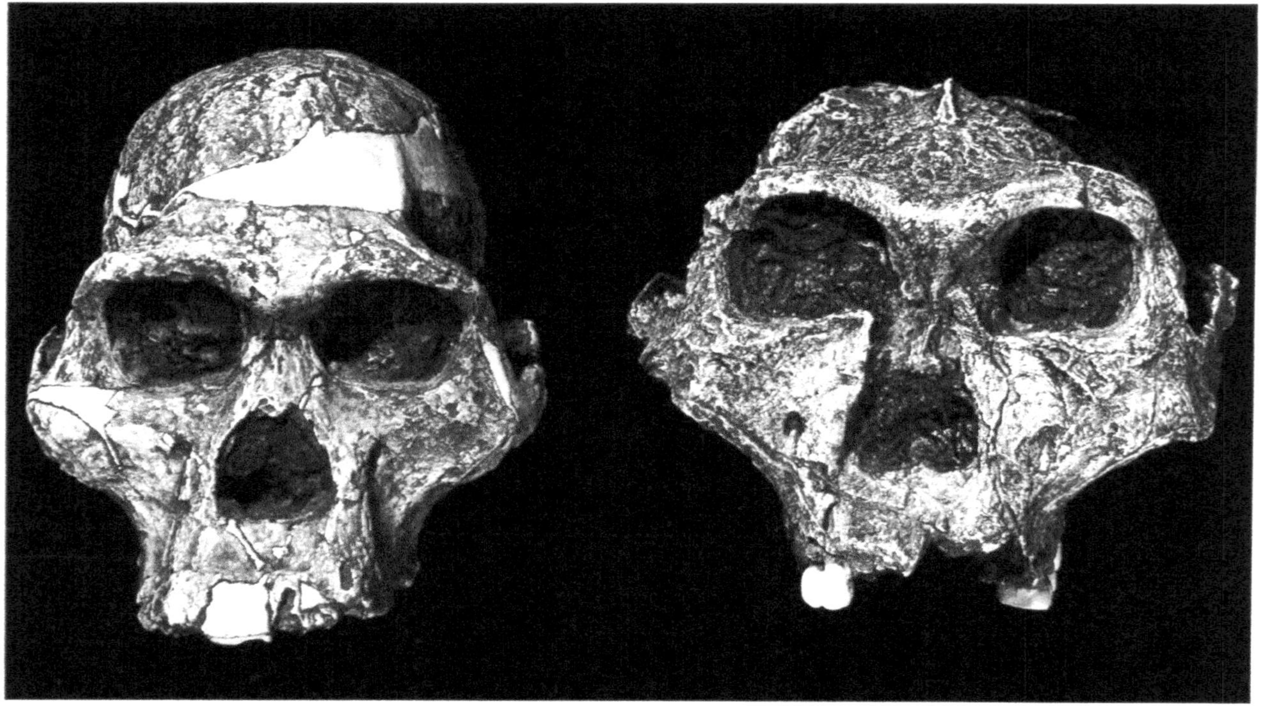

2 Two South African australopithecines. *Left* St. 5 (from Sterkfontein), a gracile *A. africanus* (originally this specimen was named 'Plesianthropus'); *right*, the robust Sk. 48 from Swartkrans originally given the name 'Paranthropus' by Broom, a term returning to favour for *A. robustus* generally. *By courtesy of the British Museum (Natural History).*

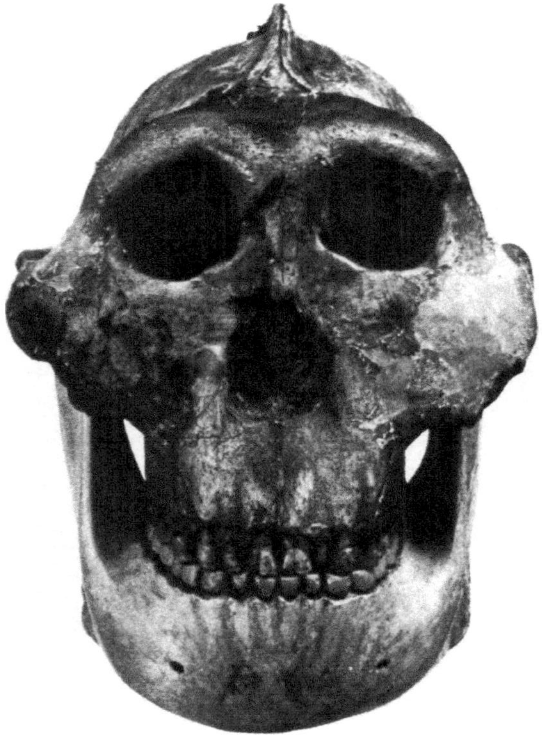

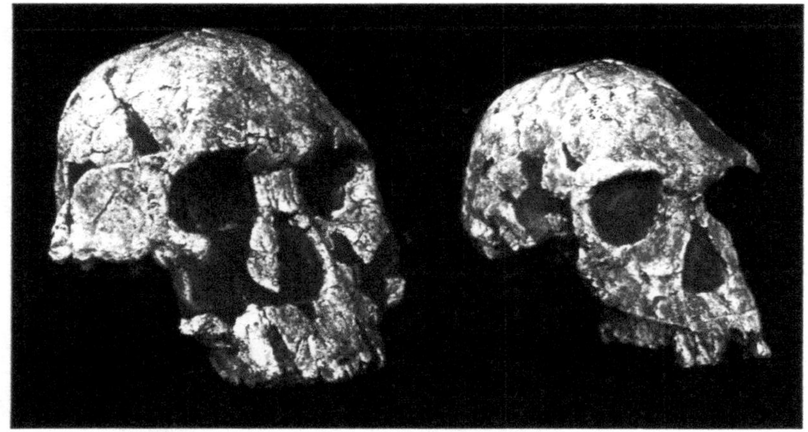

3 (*left*) '*Zinjanthropus*' or 'Nutcracker Man', as this famous specimen was called by Louis Leakey, is now classified as *A. boisei*, a subspecies of *A. robustus*, though if *Paranthropus* is successfully revived as a taxonomic category this will presumably become *P. boisei*. *By courtesy of the British Museum (Natural History)*.

4 Two *Homo habilis* skulls from Lake Turkana. *Left* KNM-ER 1470 found in 1972, one of the most famous fossil hominids; on the *right* KNM-ER 1813 whose classification as *H. habilis* is now rather insecure (Stringer, 1986). *By courtesy of the British Museum (Natural History)*.

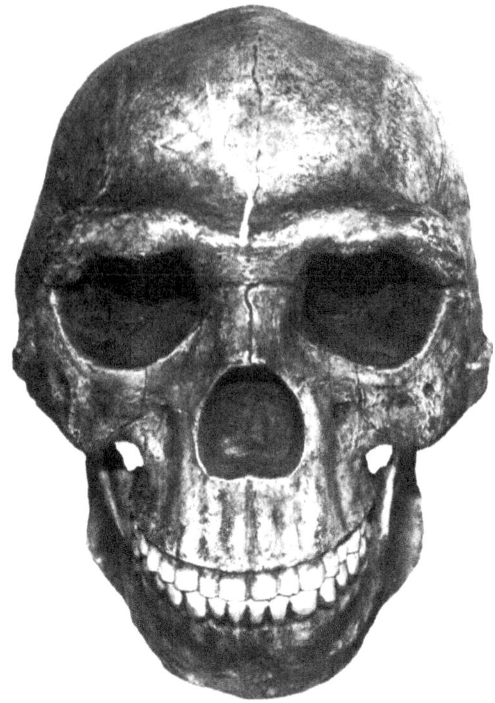

5 Restoration by Weidenreich of female *H. erectus* from Peking Man site at Choukoutien, originally classified as *Sinanthropus pekinensis*. *By courtesy of the British Museum (Natural History).*

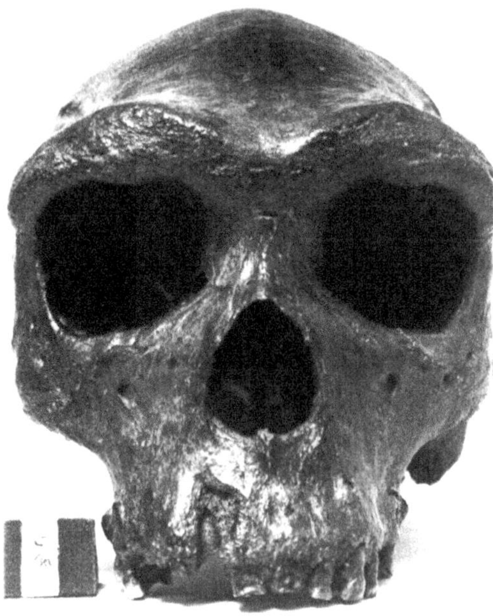

6 The 'Broken Hill' skull long known as 'Rhodesia Man'. Once dated at around 40,000 BP and considered as an African Neanderthaloid, it is now placed at around 125,000 and belongs in the 'Archaic *Homo sapiens*' group. *By courtesy of the British Museum (Natural History).*

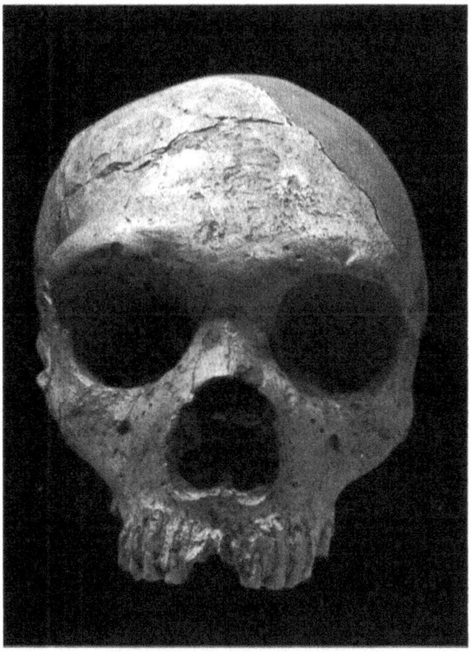

7 The Gibraltar Skull, found as long ago as 1848. Dating from around 50,000 BP, this is a typical European Neanderthal. *By courtesy of the British Museum (Natural History).*

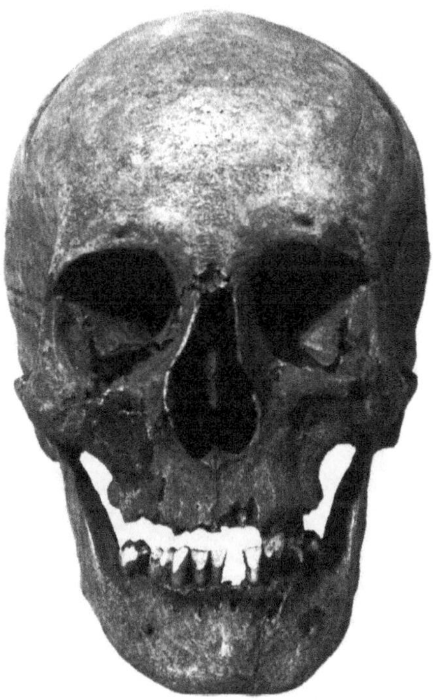

8 A 'modern' *H. sapiens*, brow-ridges virtually gone. This particular man was possibly a centurion and dates from the sixth century *AD* (British). *By courtesy of the British Museum (National History).*

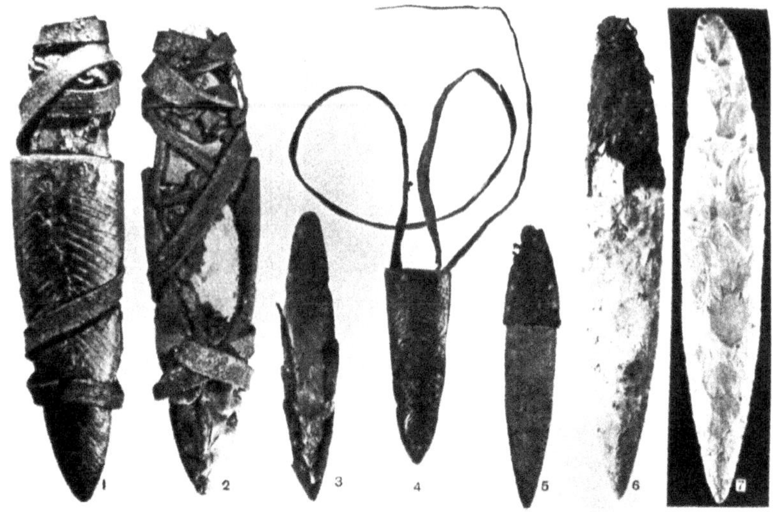

9 Neolithic dagger photographed at different stages of disassembly to show method of construction. Reported in the *Proceedings of the Prehistoric Society* (1937) vol. III, pp. 178–9 (Plate XII). *By courtesy of the Prehistoric Society.*

10 Replica of Acheulean hand-axe by Mark Newcomer showing relationship to debitage. Experimental replication of this kind is proving invaluable to prehistoric archaeologists. When the author visited Dr Newcomer he was preparing for the experimental butchery of a hippopotamus (late of London Zoo). *Photo: Mark Newcomer.*

11 Some have argued that the invention of carrying technology was a key step in hominid evolution. This photo of a young New Guinea mother returning from chopping firewood seemed to me to provide an irresistible illustration of the point. *Copyright 1969 National Geographic Society. Photo: Malcolm Kirk.*

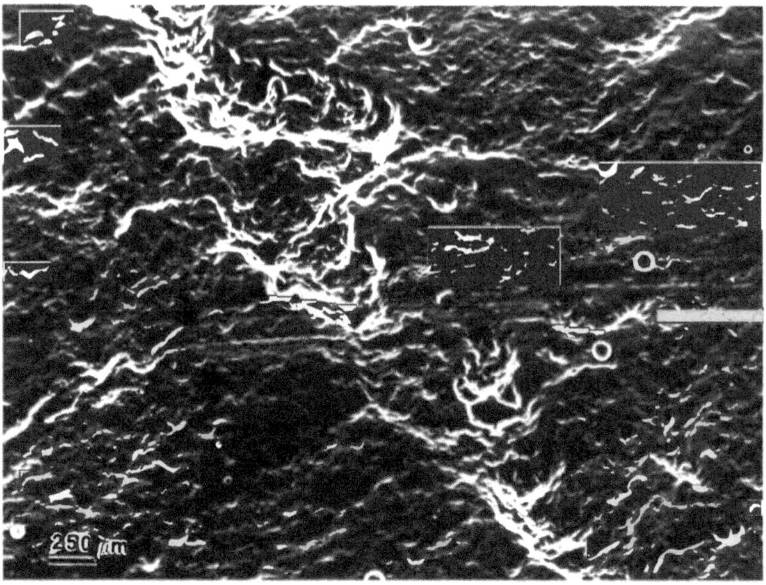

12 Pat Shipman's photo proving hominid scavenging at Olduvai 1.7–1.9 myr. It shows a broad carnivore tooth scratch (diagonally top to bottom) overlaid by a narrower slicing mark (left to right) made with a stone tool. *Photo: Pat Shipman.*

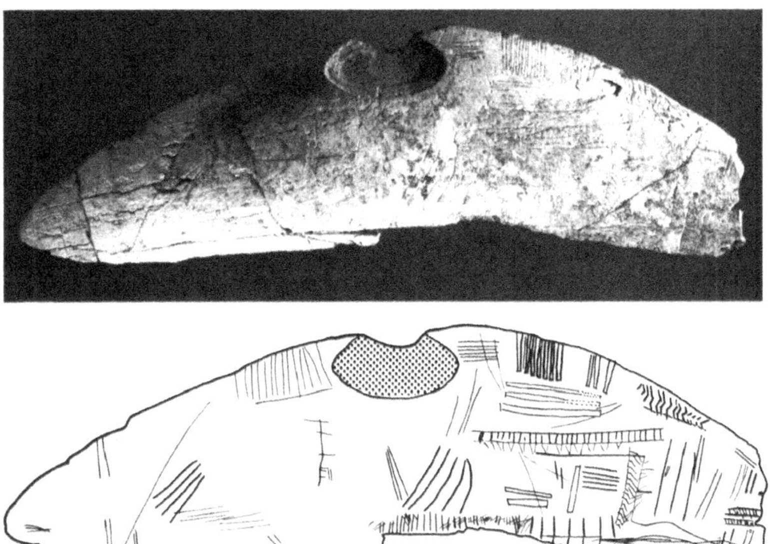

13 An Upper Palaeolithic mammoth ivory plaque from Eliseyevichi in the Soviet Union with a variety of incised motifs. Alexander Marshack has made an extensive study of these objects, and shown that, far from being chaotic, such motifs occur time and again, and constituted a developed 'symbol system'. (See Chapter 6). *Photo: Alexander Marshack.*

being the best known examples of this genre.) Darlington's *magnum opus* concentrated primarily on the significance of genetic factors during the post-Mesolithic period. These works by contrast drew on the new findings of post-war ethological studies. During the 1970s this tradition gave rise to 'sociobiology', and the most ambitious attempt yet at formulating a sociobiological model of the genes-culture relationship is Lumsden and Wilson's *Genes, Mind, and Culture* (1981). This is a highly technical treatise relying much on detailed mathematical modelling. In it the authors claim to have provided the outline of a complete account of how the genes-culture-genes circuit operates, a process they christen 'co-evolution'.

This is not nearly such an adamantly reductionist thesis as some earlier sociobiological treatments of the issue, although, in Wilson's phrase, genes 'hold culture on a leash'. Nevertheless it is still inspired by the possibility of not only integrating the biological and social sciences but by the belief that they could, together, eventually become truly predictive, at least in a broad sense. The 'genes-culture' half of the circuit is stressed far more than the 'culture-genes' half.

This account's central features, for our purposes, may be identified as (a) the crucial role given to 'epigenetic rules' as the link between genes and culture and (b) the concept of the 'culturgen' as a label for the units of analysis at the cultural level. The following passage provides a fairly good summary of the overall picture as well as a flavour of the book's style:

> We have perceived culture as the product of a myriad of personal cognitive acts that are channeled by the innate epigenetic rules. The 'invisible hand' in this marketplace of culturgens has been made visible by characterizing the epigenetic rules at the level of the person and translating them upward to the social level through the procedures of statistical mechanics. Gene-culture co-evolutionary analysis runs counter to the organicist conception of many social scientists, which views culture as a virtually independent entity that grows, proliferates, and bends the members of the society to its own imperatives. (Lumsden and Wilson, 1981, p. 176)

And another, slightly less technical:

> The structures of mind and culture are most effectively understood as developmental processes, underwritten by genes whose frequencies are the product of the protracted interaction of social behaviour and selection forces working from the environment. Thus to understand culture fully is not just to perceive the rich detail of which it is composed, but to follow through each step in the

> evolutionary circuit – from psychological time and cultural evolution, to evolutionary time and genetic evolution, and back again. (p. 237)

We need, then, to consider what these key terms, 'epigenetic rule' and 'culturgen' mean.

(a) Epigenetic rules (i)

Put simply, these are the rules governing the developmental processes, the 'unfolding' of the genetic programme through time in the course of the organism's growth. These serve to channel development in particular directions, render it differentially sensitive to various kinds of environmental influence, and so on. Lumsden and Wilson distinguish between 'primary' and 'secondary' epigenetic rules, drawing on a considerable range of psychological work in the process. Primary epigenetic rules are 'the more automatic processes that lead from sensory filtering to perception' (p. 36). These include our differential sensitivity to different light wavelengths (which underlies cultural colour-classification systems), and apparently built-in food preferences demonstrable even in infants. The secondary epigenetic rules concern the way in which sensory stimuli are evaluated by memory, emotion and cognitive processes. An example appearing early in life is the 'fear of strangers' response during the period from around 6 to 18 months. Other factors entering here include the limited processing capacity of short-term memory, processing times of various stages in the cognitive system and visual pattern preferences (e.g. for faces over non-faces in infants). Such behaviours as non-verbal communication and incest-avoidance are also considered to belong here. For Lumsden and Wilson these epigenetic rules represent the basic behavioural level at which genes are expressed, and they underly and constrain whatever cultural superstructures are subsequently raised. Their mathematical analyses of the effect of these on culture lead them to conclude that far from being blurred or lost as one ascends to the cultural level, the effects of epigenetic rules are actually amplified 'in the translation process' (p. 118).

> Our principal conclusion about gene-culture translation (is) that relatively small changes in the epigenetic rules can force profound changes in the overlying cultural patterns. (p. 110)
>
> Only a small amount of innate bias in favour of a culturgen, or a genetic or environmentally induced alteration in the sensitivity of the individual to the choice of culturgens made by others, can create large shifts in the ethnographic curve.* (p. 124)
>
> (*i.e. the distribution of 'culturgens' in the culture -GR).

It is necessary to stress that these 'profound changes' have yet to be empirically demonstrated.

(b) Culturgens (i)

We might begin by quoting its coiner's own definition, although it is somewhat cumbersome:

> . . . a culturgen is a relatively homogeneous set of artifacts, behaviors, or mentifacts (mental constructs having little or no direct correspondence with reality) that either share without exception one or more attribute states selected for their functional importance or at least share a consistently recurrent range of such attribute states within a given polythetic set. (p. 27)

Culture therefore is to be conceived as composed of 'culturgens', which may vary from a tool like a stone axe or bicycle or sword or fire-drill, through to a rule about who can marry whom, a style of building houses, or an institution such as kingship. It is on these culturgens that, at the individual level, the epigenetic rules come to operate, via the individual's tendencies to prefer some of the available culturgens over others – e.g. a preference for bows and arrows over slings and pebbles; for slinging babies on the hip over carrying them on the back; or for vegetable food over meat. In articulating the relationship between epigenetic rules and culturgens, attention is focused on long-term memory, 'the node-link structures of long term memory' are 'the ultimate biological theater of interplay between genes and culturgens' (p. 238). The storage and processing capacities of long-term memory are among those features most directly inherited, along with the epigenetic rules, while culturgens constitute the actual content or input with which it deals, evaluating and learning the outcomes of adopting the various culturgens available. A change in bias at this level will, it is calculated, reverberate throughout the system at a surprisingly rapid rate:

> We have summarized our inferences by a thousand year rule: the alleles of epigenetic rules favoring more successful culturgens can largely replace competing alleles within as few as fifty generations, or on the order of one thousand years of human history. (p. 304)

For psychologists this work is particularly significant, since it attempts to synthesise a considerable amount of current work in developmental, cognitive and social psychology into the co-evolutionary model, particularly in the context of epigenetic rules. There are continuing elements also of the traditional effort to cast the present condition of *H.sapiens* in a role of cosmic significance: 'mankind has literally altered the form of organic evolution' (pp. 325–6). The advent of culture is a unique and major event in evolution, arising largely because our ancestors, through a lucky combination of circumstances – physiological, ecological and geographical – were in the right place with the right bodies at the right time.

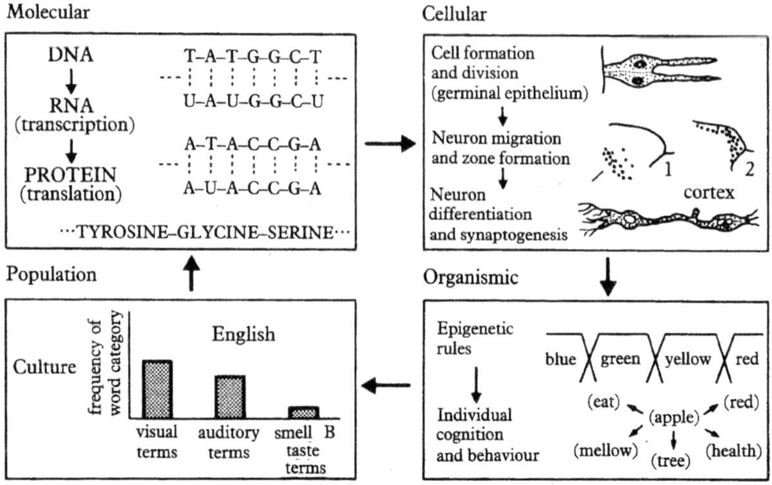

Figure 5.1 Genes and culture co-evolutionary circuit proposed by Lumsden and Wilson (1981). The full circuit of causation in gene-culture co-evolution, showing the grouping of steps within the four principal levels of biological organisation. The molecular, cellular, and organismic steps constitute epigenesis; the transition between the organismic and populational levels consists in gene-culture translation; while population phenomena affect gene frequencies through natural selection. *Reproduced by permission of Harvard University Press.*

The total operation of co-evolution is seen as a huge circuit (see Fig. 5.1) not unlike that proposed by Plotkin and Odling-Smee, although the latter depict a more complex network of feedback connections which actually renders their model quite different in character to Lumsden and Wilson's, since the culture-gene link *per se* is far less clearcut. Current knowledge of human evolution from the palaeo-disciplines is rarely used in the work, and a passing summary of it (p. 326) is already out of date in major respects (the 'hunting hypothesis' ceased to be, as they describe it, 'the prevailing hypothesis' regarding proto-hominid life-style well before their publication date).

Criticism of the model can I believe be levelled at the concepts of 'epigenetic rule' and 'culturgen' without getting involved in the complex mathematical modelling which Lumsden and Wilson adopt. If the basic concepts are flawed, their subsequent mathematical treatment, however rigorous, is somewhat beside the point, unless, like concepts in physics, the concepts are themselves fundamentally mathematical. This is not the case here. I am fully aware that it can be argued that some definitional incoherence is a virtual necessity in a theory's infant phase, and that Galileo would never have succeeded in refounding physics had he striven to meet such criteria (Feyerabend, 1975). I find it hard though to cast E. O. Wilson's sociobiology in the role of struggling and delicate theoretical neonate. Both of these concepts appear to me to be dubious in the way they are used in this work.

(c) Epigenetic rules (ii)

Firstly, the role allocated to these oscillates between 'rule' and 'constraint'. Much of their psychological and psychophysical data refer to the biologically governed parameters of behaviour, rather than its specific content. A shift in parameters is surely a somewhat different phenomenon than a shift in, say, timing of when a specific behavioural category develops, or determination of the specific form it takes.

Secondly, the net effect of their use of the expression is paradoxically to serve as a virtual *tabula rasa*. It covers so broad a range of phenomena that almost any variance in the way genes are translated into behaviour becomes possible by invoking 'epigenetic' processes. To elaborate on Wilson's own metaphor, the leash linking genes and behaviour can be looped, knotted and tangled to whatever degree is required in order to preserve the notion that a shake at the gene end is somehow the cause of a behavioural innovation. All these convolutions can be subsumed as 'epigenetic rules'. We will see a particular usage of this kind later on when discussing altruism. The risk here is that the 'epigenetic rule' loses any concrete predictive value, the situation is very similar indeed to the issue of the utility of the 'trait' concept in psychological theories of personality. It became progressively apparent through the late 1960s and early 1970s that knowing a person's score on a 'trait' measure added very little indeed to the experimenter's ability to predict their behaviour, since 'situation variables' and 'moderator variables' (age, sex, social class, etc.) were by and large sufficient (Mischel, 1973). The utility of trait concepts is actually rather different, and they are best not thought of as 'causes' of behaviour at all (Cochrane, 1984). 'Epigenetic rules' are not dissimilar to 'traits' in the role currently being assigned to them, nor in their sharing with them a lack of class homogeneity (such as Cochrane discusses with reference to traits).

Thirdly, Lewontin (1983), as cited in an earlier chapter, uses the flexibility of epigenetic processes to quite the opposite effect to Lumsden and Wilson, namely as evidence that the notion of a clear gene-phenotype linkage across all environments is unsustainable. His attack on the 'unfolding' metaphor in this paper has serious implications for the Lumsden and Wilson concept of 'epigenetic rules'. In genetics in general it is probably fair to say that the manner in which the DNA level translates to the molar morphological and behavioural levels is increasingly seen as very extended and probabilistic in character. Some present writers, far from seeing epigenetic processes as a translating link between genes and behaviour, see them as virtually an autonomous level altogether (see, for example, the radical critiques now appearing by Ho and Saunders). In short, the gaps between the boxes in Fig. 5.1, especially between Molecular and Cellular, and between Cellular and Organismic are possibly far wider than the authors give them credit for.

I am not here claiming that such a brief critique of the concept undermines the entire rationale for the complex model offered by Lumsden & Wilson, but it might serve to indicate how it is far more speculative in character than its style of exposition suggests.

(d) Culturgens (ii)

In order to model cultural processes mathematically one must postulate units, and the introduction of the term 'culturgen' serves this purpose. The question is whether or not such entities can really be identified, and if so, how broad a range of cultural phenomena they cover. I have three criticisms of this concept.

Firstly, can such units be isolated? A recent paper by Hull (1982) argues strongly for the necessity of viewing change in scientific thought in terms of social contexts of communication, etc. One cannot identify unitary 'ideas', discrete concepts, which survive or perish in a process of conceptual evolution. Similarly one might argue that in cultural evolution generally the meanings of the individual items by which a culture is expressed depend on entire contexts of meaning. Lumsden and Wilson do introduce a notion of 'culturgen-packing' to try to handle this but it is not adequate, being used only to account for the size of the available 'culturgen store' in quantitative terms. If we take as an example of a culturgen the word 'nigger', it is surely apparent that its demise as an acceptable term (except perhaps jocularly among some members of the class it was originally used to label) is due to a whole range of complex interacting factors. These include its identification with strong racist attitudes which were themselves becoming less acceptable, explicit positive self-assertiveness and self-definition by the group so labelled, scientific discrediting of the concept of 'race' and so on. The relative decline in the frequency of the verbal culturgen 'nigger' cannot be treated in isolation from these sociopolitical factors. Nor can I see how it relates in the least to 'epigenetic rules'. 'Culturgen-packing' is not simply a quantitative packing of items but a structuring of cultural meanings, a web of interlinking customs, mores, beliefs, artifacts and so on. To analyse co-evolution in terms of separable culturgen units thus strikes me as highly artificial.

Secondly, the model is so comprehensive that it seems to deny that there is *any* level of autonomous sociocultural change, governed by essentially non-genetic forces. There is only one leash to which culture is attached – genes. Yet in the realms of economics, and the immediate cultural consequences of demographic changes such as migration, population explosion, or decimation by disease and disaster, there is surely little need to invoke the individual epigenetic level. Of course such phenomena are interlinked in numerous other 'circuits', operating on various time-scales, and links back to genetic selection both before and after the event may no doubt be traceable. This does not, though, mean that this linkage, one among many, has any special explanatory primacy. The problem I think is that Lumsden and Wilson, unlike Plotkin, neglect the existence of subsidiary feed-back circuits operating between the levels of the overall circuit pictured in Fig. 5.1. An intense positive feedback loop between technological innovation, manufacturing processes and social structure presumably operated during the Industrial Revolution. It is not at all clear where the governing, or constraining, hand on the genetic leash enters the picture – at most one pictures genetic *consequences* trailing in the wake of such cultural upheavals.

Finally, the actual natural selection process as applied to culturgens is not entirely clear. Not much attention is given to culturgen *innovation*. Certainly there is no evidence that the major cultural innovators have had their personal inclusive fitness particularly enhanced (Newton and Kant probably died virgins, Michelangelo was gay . . .) nor with a few exceptions (such as the Bachs and Huxleys) do the offspring of the innovator appear more adapted than anyone else at utilising the new culturgen. Should adoption of a novel culturgen prove of sufficient selective advantage at the individual level, the nature of the subsequent selection process would be a weak selection against those unable to learn it. Even so, apart from the culture's language, there are few 'culturgens' which it is essential for *everyone* to choose for their survival, at least in cultures of any complexity. The sheer diversification of behaviour *within* cultures since urbanisation began around 7,000 BC suggests that a culturgen has to be positively extremely unfit indeed to become totally extinct. After all there are a number of expert hand-axe makers around even now. (Ironically it may sometimes be the ultra-valuable culturgens which are at most risk of being lost, since accessibility to them is restricted, and their transmission across generations hampered. The bronze-casting techniques of the Chinese Shang Yin period around 1000 BC were superior to those of the Renaissance, for instance (Willetts, 1958).) As Lumsden and Wilson describe the process, the individual is actively engaged, in the course of her or his epigenetic development, with processing the culturgens which they encounter. But this in itself can be construed as entailing that almost *any* physiologically viable genotype will be able to find or create a congenial niche for itself within an urban culture – or if not, the reasons do not lie in its individual genotypical character but in its location within that culture: the social class it happens to have been born in, or the economic-sociohistorical conditions which happen to be prevailing (e.g.

whether it is in a boom or slump economically), or other fortuitous factors unrelated to genetics (such as psychological crippling following an emotional catastrophe).

In conclusion, then, this model, although tantalisingly comprehensive, and drawing on a wide range of disciplines, must still be considered as highly speculative. Not all theoretical developments in genetic and evolutionary theory at present lend support to it, and its key concepts – 'epigenetic rule' and 'culturgen' are still problematical. True, there has been backtracking from previous extremes of sociobiological reductionism, and the preparedness to incorporate psychological findings and research into the modelling is welcome. They have not, though, unequivocally demonstrated that the leash on which genes hold culture is short enough ever to be pulled tight to visible effect. Nor, since the situation they propose is actually a circuit, is it entirely clear why the gene level, rather than the culture level, should be taken as the starting point. (See Midgley, 1984, for more on this.)

I have concentrated on the Lumsden and Wilson version of this position since it is the most influential, and a more recent sociobiological account (Lopreato, 1984) appeared too late for consideration (but see Medawar, 1984 for a damning review).

Earlier versions, notably those referred to above (Morris, Ardrey, etc.), tended to espouse a more strongly gene-dominated position. This led them to suggest that most cultural phenomena were direct expressions of instinctual imperatives, and that aspirations to change some basic features of human culture such as those pertaining to sex-roles, hierarchical social structuring and aggression, were doomed to failure, these being fixed by unalterable genetic programmes. Such conclusions elicited the wrath of many anthropologists (e.g. Ashley Montagu), psychologists and sociologists, who argued for the over-riding importance of environmental factors. This simplistic juxtaposition of genes and environment as opposed, separable sets of factors has at last begun to wane, though the limitations of overly gene-oriented approaches are still being ardently argued (Rose, Kamin and Lewontin, 1984). It is unfortunate that the strong ethological position came to be associated with right-wing ideological positions, since many of the insights this approach yields into human behaviour are salutary and stimulating. It is surely possible to see a link between, for example, advertising and territorial demarcation, without lapsing into strict biological reductionism.

6 *Cultural evolution dominant over physical evolution*

This notion is not incompatible with some of the other models we have been looking at since it is usually proposed not so much as a model of past gene-culture relationships (except in a loose sense in which culture is equated to 'domestication') as of the present and future relationship. That such a situation might be possible to achieve was central to the nineteenth-century Eugenics movement led by Galton. While Eugenics as such has ceased to be an ideological movement, current developments in genetic engineering re-open the issue from a technological direction. This extends even to the reproductive process itself: cloning, embryo-implantation, artificial insemination and surrogate mothering represent radical departures from conventional modes of sexual reproduction though as yet affecting only a minuscule proportion of the population. Both human and non-human physical evolution is being progressively subordinated to anthropocentrically perceived sociocultural requirements. Neel (1983) has reviewed some of the consequences of current developments from a geneticist's viewpoint. Since this perception of the genes-culture relationship is more to do with the future than the past we need not consider it at further length here. One might remark before moving on though that the scientific dynamic involved in advancing these developments has not been one whit inhibited by the fact that most conjectural attempts at envisioning its consequences (e.g. by science-fiction writers) are resolutely pessimistic.

7 *The social context is a novel setting for normal evolution processes*

This is, broadly, Holloway's position. It is not evolution as a process which has altered but the relationship between the evolving organism and its environment – which is now a sociocultural one – and this produces a novel pattern of selection pressures. This perspective is more concerned with physical evolution – convergent replacing divergent evolution since *H.erectus*, the sustaining of wide individual differences and the like – than with social evolution as such. The novel cultural context, bringing selection pressures into play, is itself a part of the environment to which persons have to adapt. Their adaptations enter in turn into the group repertoire, perhaps as cultural phenomena for others. Although addressing a facet of the total evolutionary situation, this position does not answer questions about the nature of social evolution in itself, but does provide a convenient position for those who wish to see social evolution as autonomous: Ashley Montagu (1962), for example, would appear to be opting for this insofar as he admits a biological level of things operating at all. In Holloway the feedback from physical to cultural levels is more positively formulated and the two less clearly separable conceptually (see the discussion of Holloway in Chapter 4).

The arena in which orthodox natural selection continues to work is in such areas as immunological resistance which are under fairly direct genetic control, but the effects on culture are tenuous. The key feature of this position is that it does not involve the invocation of novel evolutionary principles to explain the course of culture, it sees culture as some kind of outgrowth of increased behavioural complexity, an outgrowth which enters back into the evolutionary scenario as a new kind of environmental factor to which species members have to adapt, as ants have to adapt to life in antnests. Some writers, such as Parker (1982) see cultural forms as tracking the evolution of cognitive ability,

> The emergence of higher intelligence in *Homo sapiens* was associated with entry into a new niche which we can characterize as the niche of *abstract culture* (as opposed to the preceding niche of concrete culture) . . . access to the niche of abstract formal culture came through terminal addition of formal reasoning onto preexisting concrete operational reasoning abilities; ecological access through the challenge of a variety of habitats; and physical access came through the geographic mobility conferred by the preexisting concrete cultural niche. (p. 17)

Even so, there is nothing intrinsically mysterious about the evolution of culture. I will return below to a sense in which this position may provide another valuable angle from which to tackle some features of cultural evolution (see 9 below).

8 *Social evolution is an untenable concept*

There is only one sort of evolution, and the rest is history, which is essentially different in character from the evolutionary process. This position differs from 1(b) for example, in seeing the human past as *neither* reducible to sociohistorical or economic laws, *nor* under rational control to any significant degree. Understanding the past involves a complex unravelling of factors ranging from climate to the personalities of key individuals. Engaging in such a study is itself an activity of a different kind to orthodox science, even though it may involve using modern scientific technology as part of its methodological repertoire. Karl Popper's rejection of 'historicism' – the idea of 'historical laws' – would be a case in point (Popper, 1977). It differs from 1(c), too, in that it does not actually endorse a total 'free-will' position: cultural change *is* determined and we are not free to choose whatever future we want, but the determining mechanisms require detailed exegesis in each case. From this perspective the past of our species is more like a biography, a single unique life-story which is unpredictable and idiosyncratic, an interplay of uncontrollable circumstances (including genetic ones) and human plans and purposes, non-reducible because never repeating itself. This view is appealing in many respects, just as ideographic approaches are appealing in Psychology. The complexity of reality is given its dues and aspirations to making predictions about the future firmly held in check. But it is in some ways too extreme, ignoring the fact that for a substantial part of our prehistoric past we can detect regularities, rule-bound dynamic patterns and so on even at the cultural level. These may be temporally undirectional and generate a sequence of states each of which is unique, but this does not mean that history and prehistory are an intrinsically unstructured course of events. Some historians (e.g. Braudel, 1972) have opted for a highly synoptic approach in which they try to construct an overall vision of the past incorporating geographical and climatic factors at the baseline and tracing the emergence of the actual historical sequence within the context of these transpersonal forces. This is more congenial to the evolutionary perspective.

Historically, the extreme anti-social evolutionary position may have more to do with preservation of discipline territory and interdisciplinary rivalry than with understanding the past as such.

9 *The 'open behavioural repertoire' model*

This is the view currently espoused by the present writer. Our species is special only insofar as it has evolved a relationship with its environment which is ecologically unchecked – somewhat as if an organism was infected with a virus to which it had no immunity. 'Culture' is the overt form of this relationship, but is itself a consequence of a behavioural innovation which was *initially* fairly undramatic at the psychological level. This move, that is, was not in itself particularly special from an evolutionary point of view. Whether the subsequent course of sociocultural events warrants the name 'evolution' – with its connotations of orderliness and progress – is, as suggested earlier, simply a colossal begged question. That regularities and order may be discerned in it is accepted, but at the present historical juncture their value and significance is strictly undecidable.

The particular behavioural innovation in question refers to the acquisition, probably at the level of cortical neurological organisation, of a technique for *positively* generating behavioural novelty. Somewhere along the line (and where that somewhere was I will return to shortly) we acquired *an open-ended behavioural repertoire*. The crucial question is how an open-ended behavioural repertoire can be generated and sustained, how can an organism exhibit continuous behavioural novelty? This requires that a number of necessary preconditions have been met – an ability to discriminate between and store large numbers of stimuli along many different dimensions, physiological flexibility as such, and *a way of sustaining social coherence among group members in the light of their behav-*

ioural diversification. Even granted these conditions, a key problem still exists – where do the novel behaviours come from? How, as we say, did they get their ideas? This is such an elementary query that it is rarely addressed, but it is at this point that a specifically psychological thesis is required. I suggest that this innovation arose from acquisition of a *generalised* neurologically embodied 'programme' of the form 'To achieve X identify a known X-achiever and copy'. That isolated, discrete, imitational innovations of this kind are possible in other species is not denied, what is new is the adoption of this as a generalised strategy. While this does not seem especially dramatic, it has profound consequences since it involves the individual in a search within its environment for novel items to incorporate into its own behavioural repertoire. (This is, I discover, akin in some ways to Mumford's picture (Mumford, 1967), but without his misleading high drama.)

I hypothesise that this only became fully possible with the acquisition of lexical language as a basis for information storage, and after the basic locomotor repertoire was substantially complete. Put simply, I am saying that environmental phenomena became seen as potential 'ways of being', a store of behavioural models, to be drawn on when the individual wished to bring about an effect perceived as caused by such a 'way of being'. Now this has important secondary implications. In terms of social coherence it is potentially highly disruptive, for it could involve a rapid divergence of behaviours and broader psychological orientations – a loss, in a sense, of species identity. Adoption of animal-species modes of behaviour, for example, would be a major source of behavioural diversity, but also of conflict matching the perceived relationships between animals themselves. This potential fissioning of the species into an array of identifications with particular environmental phenomena was, I argue, held in check by an initially linguistic form of 'culture' in which phenomena are *collectively* labelled and evaluated, and individuals within the group similarly identified and provided with feedback regarding their identity in a social context.

This model has two especially important implications for the genes-culture issue.

(a) Individual differences in physiology, particularly morphological ones, acquire value as facilitating particular kinds of behavioural skill, so that the group as a whole possesses a collective behavioural repertoire greater than that of any individual within it. Insofar as natural selection pressures operate they shift from focusing on the individual's gross morphological competence to enact a species-specific behavioural repertoire, or adapt to a specific ecological niche, on to their ability to exploit their gross morphology as a resource for acquiring behaviour which is valued by the social group, thus even a blind person might survive if they can establish the value of their aural skills. The focus of selection is thus dramatically changed. Gross morphological diversity within the species can be sustained and is of positive social value, as underlying a corresponding diversification of the collective behavioural repertoire. This situation would also, incidentally, render the 'dominance' ranking more complex, since it becomes polydimensional. Thus although there is convergent evolution on features facilitating social life in *Homo* as a whole, it is highly, though not uniquely (cf. domestic dogs) polymorphic *within* that range. In any case phenotypic diversity, not genotypic diversity, is what concerns us here.

(b) 'Culture' therefore starts as a technique for maintaining species-identity or coherence in the face of behavioural amplification. It controls, maps and evaluates the broadening behavioural repertoire. This involves the basic step of formally differentiating humans from the external world from which they are drawing their behavioural repertoire – the ubiquitous 'Culture-Nature' distinction of anthropology. Once established, this situation possesses a dynamic of its own as recursive applications of the core behaviour-generating strategy get under way and permutations of items within it are explored. What I am suggesting here is best described, in Lévi-Strauss's term as 'physiomorphism'; the opposite of anthropomorphism. Instead of projecting human characteristics on to the physical world, we extracted them from, or recognised them in, it (Lévi-Strauss, 1966).

The question of 'timing' is a difficult one, and although the move I have described might constitute a 'catastrophe' in the technical sense, it would be preceded by a phase in which basic behavioural skills already approached current levels, I do not see it then as predating lithic technology. Since the dynamics of the process, once established, would seem to be fairly rapid, it would probably be wiser not to place it particularly early. The connection with lexical-syntactic language strikes me as very close, since the basic propositional form 'A is B' entails and incorporates the required reversible, reflexive application of environmental properties to the speaker. Although we will be looking at evolution of language later, I will risk saying here that I believe the *full* acquisition of human language, and the start of culture in the physiomorphic process just outlined, coincide, and that they probably occurred no earlier than the emergence of the *H.s.sapiens* grade itself, around 100,000 bp at the earliest. This is consistent with Lieberman's recent account (Lieberman, 1984), which is discussed below. Developments during the *H.erectus* phase just do not seem rapid enough to justify the presence of a positive behavioural innovation mechanism of the kind I envisage. From this viewpoint, 'culture' has basically been the exhaustive progressive assimilation of the physical environment to immediately perceived *H.sapiens* interests. By reflexively turning back on 'Nature' the properties derived from it we have, as we used happily to gloat, 'conquered' it. In the meantime we have progressively buffered ourselves even more successfully against 'natural selection' processes. But I must emphasise again that this stems from what was originally a relatively simple neurological 'wrinkle', a generalised programme of the form 'To achieve X identify a known X-achiever and copy' emerging from the already developed 'imitative' capacities of *H. erectus*.

I do not see this model as optimistic. It amounts to saying that we have succeeded in out-flanking all ecological control mechanisms and are now running destructively amok.

Conclusion to Part A

Accounts of 'social evolution' and the 'genes-culture' relationship have varied across the spectrum from optimistic to pessimistic, from emphasis on genes to emphasis on culture, from mutual dependency of various levels to virtual autonomy. None of these positions is devoid of ideological connotations in the broad sense, and this would seem to be impossible, notwithstanding Durant's wish that we stop playing the mythologising game (see Chapter 1). Current approaches in sociobiology attempt to integrate the different levels in models which place cultural phenomena ultimately under genetic control, even though feedback mechanisms may be admitted. Other current genetic theorists as well as anthropologists are more sceptical of this enterprise, and some are frankly condemnatory. The 'evolutionary epistemology' approach has shed some light on issues in the history of ideas and history of science but has not as yet been applied in any depth to the prehistoric phase of our species. I have tentatively proposed an account of why 'culture' arose in the first place, and the genetic consequences of this, which hinges on the achievement by our species of a way of generating behavioural novelty in an open-ended fashion. This is confessedly pessimistic.

A central feature of 'culture' as customarily conceived is the control exerted over human inter-relationships. It is commonly noted that human culture involves forms of mutual support between members far beyond anything which appears in the non-human world, outside the social insects. There are almost universal cultural injunctions towards 'morality' which usually involve encouragement of generosity, charity, kindness and similar virtues. The presence of such behaviour in our repertoire has given evolutionary theorists much to puzzle over. They constitute the problem of 'altruism' to which I now turn.

B Altruism

A central feature of human society seems to be preparedness to behave in ways where the needs of others take precedence over the needs of the individual in contexts far removed from such things as child-rearing or protection of kin. Although, like much else, this may have discernible animal precursors, it is generally considered that it has reached a quite unique level of development in human society, and is one of its hallmarks. It has also long been thought that this represents something of a paradox for orthodox evolutionary theory, since it appears to involve selection for preparedness to engage in 'self-sacrifice'. How can genes which give rise to this kind of behaviour ever gain ascendancy when those who possess them are, by definition, going to survive, or reproduce, less than those of others who do not behave in such ways? Darwin himself, as we saw in Chapter 2, first identified the problem that the most self-sacrificing would 'perish in larger numbers than other men', and proposed that it could be explained by the inheritance of helping behaviour which had initially been rationally perceived by the actor as advantageous to them also, to this he added a similar acquisition of love of praise and fear of shame. Both of these required some measure of inheritance of acquired characteristics. His formulation of the paradox, if not his attempt at resolving it, has survived by and large down to the present. It is now translated into technical terms, in a classic paper on the topic, thus:

> Altruistic behavior can be defined as behavior that benefits another organism not closely related, while being apparently detrimental to the organism performing the behavior, benefit and detriment being defined in terms of contribution to inclusive fitness. (Trivers, 1971, p. 35)

This definition is very important for our discussion since it has remained fundamentally unchallenged within sociobiological debate, while opponents of sociobiology have usually tackled it on a much broader front. The differentiation of 'individual fitness' from 'inclusive fitness' has become serious for evolutionary theorists, but does not affect the central point here, since the primary point is detrimentality to fitness as such. The Darwinian image of the problem is produced again, quite uninhibitedly, by E. O. Wilson (1978), who opens his chapter on the topic by invoking battlefield behaviour as the highest form of altruism, referring to men 'who threw themselves on top of grenades to shield comrades, aided the rescue of others from battle sites at the cost of certain death to themselves. . . . Such altruistic suicide is the ultimate act of courage . . .' (p. 149).

The present discussion will be highly critical of the sociobiological approach to altruism, more so than is perhaps appropriate in a textbook; nevertheless the conceptual difficulties which this definition presents, along with other aspects of the sociobiological account, are too serious to leave aside, and have not, to the present writer's knowledge, been adequately answered. Since Trivers's own account (dealt with later), positions have tended to consolidate around the following possibilities:

(a) On closer analysis apparently 'altruistic' behaviour turns out *not* to be detrimental to the organism in terms of either individual or inclusive fitness after all. Technically speaking then there is no such thing as altruism. The class of altruistic behaviour as defined by Trivers is empty.
(b) Altruism may exist now but it *originates* in behaviour which was not altruistic, and is sustained in human societies by social forces, etc. which over-ride the usual negative consequences. This is clearly the position in those versions of social evolution which see us in a situation of basic conflict, which see genetic selection as having effectively ceased, or which see it as having been augmented by some superordinate social forces.
(c) Opposing both of these is the view that altruism is not controlled at the genetic level anyway, and cannot therefore be subject to selection either for or against in any biological sense. The grounds for this rejection range from empirical to philosophical.

Before addressing Trivers's work directly, attention must be drawn to three preliminary conceptual difficulties with accepting his definition of altruism, just quoted.

1 There is persistent ambiguity as to how far altruism in the technical sense – as defined by Trivers – is equivalent to altruism in the everyday language sense. Sociobiologists frequently protest that their opponents have failed to appreciate the purely technical sense in which it is being used. This suggests that it is not meant then as a rigorous *redefinition* of the term analogous to defining 'burning' in terms of oxidisation, but has a different range of reference, a different meaning. But it is still unclear whether the class of behaviours termed altruistic in the technical sense is (a) a subset of those termed altruistic in everyday language or (b) an independent set with some overlapping (and if so, how much?). If they really wanted to steer clear of contention the sociobiologists ought to have chosen a less contentious term. In practice it is apparent that many sociobiologists *do* consider their analyses to have a bearing on altruism in the ordinary sense. (Interestingly they do not offer a technical definition of the polar term 'selfish', which is equally problematical in their usage.)

2 A second conceptual difficulty is that in defining altruism by its *consequences* for inclusive fitness (and note that it must be as these are judged by sociobiologists as the ones sufficiently expert to diagnose 'apparent' detrimentality) the question of actor's motives has been eliminated from the discussion. This is very odd, because it is normally considered that it is via motivation that the connections between the genetic level and the psychological and behavioural levels are mediated.

But even more seriously, this definition means that the apparent status of an act as altruistic or not can oscillate indefinitely as subsequent events enhance or curtail the actor's inclusive fitness: thus if you rescue someone from drowning and they reward you in such a way that your inclusive fitness is enhanced, it was not an altruistic act, but if at a later date they kill your family in a car accident then it becomes altruistic again! This seems absurd, yet the time-span following an act which is relevant to its altruistic character is very extended in sociobiological accounts themselves, and there is no way this absurdity can be dodged, granted Trivers's definition. It is clearly a quite unsatisfactory way of defining any behaviour, let alone one supposedly under genetic control, to pick on its consequences to the exclusion of its motivation. (Although some radically behaviourist theories *do* reject the need for motivational concepts, they are theories which would have little truck with 'altruism' either.)

The phrase 'apparently detrimental' could, possibly, mean apparently *only in the first instance*, prior to deeper investigation. This would enable Trivers to continue calling the behaviour altruistic even if he subsequently identifies a reason why it does, contrary to these first appearances, enhance inclusive fitness. But even this cannot really salvage the situation, since apparent detrimentality will still be subject to individual differences in judgment and criteria for identifying 'apparent detrimentality' cannot be unambiguously specified. What is apparent to someone depends on what they know already. We can in fact *never*, on this definition, be certain that a covert advantage for inclusive fitness does not exist, as yet unidentified, and thus we cannot ever know whether or not an altruistic act has occurred. Nor, though, can we be sure it has not, in the absence of total knowledge of the entire subsequent course of life of both parties and indeed the lives of their descendants. (Voorzanger (1984) has more conundrums of this kind.)

3 Two different kinds of act seem to be merged in the definition, those in which deliberate self-sacrifice (either of life itself or physical well-being) occurs and those in which a high risk is taken but which, if successful, do not involve the actor being seriously injured or losing her or his life. The former quite obviously appear to be detrimental to inclusive fitness since at their most extreme the actor cannot breed subsequently, nor can they assist in promoting the survival of their kin. On rare occasions the sacrifice itself might enhance the latter, but then it would no longer be altruistic in the Trivers sense. Such behaviours, i.e. self-sacrificial ones, are rare, elicited in idiosyncratic situations and are highly diverse in character. It is unlikely that they figure at all significantly in the genetic selection process. We are more concerned, surely, with the second type of apparent altruism – engaging in high-risk activities. They after all provide the classic paradigm used in Darwin, and have not been basically queried since. This prime example, of death in battle

by the brave, is an odd example in many ways, to which I will return later, but for the moment note that death in battle is here positively correlated with level of bravery and both with altruism. Yet there is no *prima facie* evidence that success and failure in high-risk 'altruistic' behaviours – battle or any other – of this second type is either random among those engaging in them (which would consistently select against engaging in them at all), or that *failure* is highest among the most altruistic (as the Darwin example assumes). On the contrary what is being selected against here, if anything, is not engaging in high-risk altruistic behaviour, but engaging in it *unsuccessfully*. What is perhaps being selected *for* is a complex of attributes such as accurate appraisal of risks, physical strength, sense of balance, fast reaction time, resourcefulness and so forth. In battle, even if we grant that this is an arena of altruism, which is dubious, the correlations are not between survival and cowardice or between death and bravery, but between survival and either winning or negotiating a cessation of hostilities. To win requires greater physical skill, greater intelligence in assessing the situation and greater rational control over the emotions, 'discretion' as the saying has it 'is the better part of valour'. One might cynically observe that battle culls the most bloodthirsty, savage and psychopathic young males, not the most altruistic and noble! (Of course twentieth-century armaments render this all obsolete; I am to be understood as referring to the kind of hand-to-hand fighting which would have characterised most of the human past.)

The wider upshot of this is that such high-risk altruistic acts as trying to save someone from drowning, rescuing someone from a burning building, etc., do not have the character, ascribed to them by Trivers, of being to the apparent detriment of the actor's fitness *unless* they are unsuccessful in some measure, i.e. the altruist dies or is injured. Failure is not random among those engaging in such behaviour, but it is biased towards the least skilled, not the most altruistic.

Several subsidiary points might be made before quitting this issue: engaging in high-risk altruistic acts is not so much a function of individual choice as a function of particular age/sex positions in the community (e.g. unmarried young males, though there is cross-situational variance in this); success in such high-risk endeavours is typically rewarded by the group in terms of enhanced status, etc., thereby augmenting individual fitness – and even in the case of failure, kin of the deceased may be treated in such a way as to offset the potential detriment to inclusive fitness incurred; interestingly, in some situations requiring high-risk acts which are judged to involve odds too long to be rationally acceptable, a common solution has been to select the actor by chance (e.g. drawing lots or tossing a coin), thereby neutralising any systematic consistent selection vector.

In general then, the first clause alone of Trivers's definition of altruism would be sufficient to define it in the everyday sense. Adding the technical clause only brings a number of conceptual difficulties in its train, which have, remarkably, gone largely undiscussed: the abandonment of motivational criteria in favour of apparent consequences renders the concept fundamentally incoherent; the relationship between everyday and technical senses of the term becomes quite obscure: the failure to differentiate between actions which are intrinsically detrimental and those which are so only if they fail, further confuses the role of 'detrimentality to inclusive fitness' in the debate.

Nevertheless the fact is that in the period since its appearance Trivers's paper has provided the framework in which a wealth of discussion of the issue has been conducted, even if this has not left it entirely unmodified. Though its premises appear flawed to the present writer it cannot be omitted from an account of the predominating ideas on altruism. In the remainder of Part B, I will discuss Trivers's paper at some length and then, more briefly, a paper representative of the current state of the debate. This is necessarily a very limited coverage of the massive literature on altruism now available.

The aim of Trivers's 'The Evolution of Reciprocal Altruism' is to demonstrate how altruism, in the sense defined earlier, or, more specifically, reciprocal altruism, can be naturally selected for under certain conditions, notwithstanding the apparent detrimentality of the act to the actor in the first instance. The paper falls into three principal sections: the first provides the formal theoretical demonstration of his case; the second considers some instances of reciprocal altruism in non-humans (notably cleaning-symbiosis in fish and alarm calls in birds); and the last tackles the human 'altruism system'.

An earlier evolutionary theorist, Hamilton (1963), had shown some altruism to be explicable in terms of 'kin-selection', which is to say that the detriment to the actor is more than offset by benefit to their gene-sharing kin – fitness of the shared genes is thus enhanced. Following a hint made long before by J. B. S. Haldane, it was held that inclination to altruism was a direct function of degree of relatedness between actor and recipient. Hamilton's concept of kin-selection is now broadly established in evolutionary theory. Trivers's move is to show how this does not provide a comprehensive picture of altruism – and indeed he excludes the kin-selection form of altruism from his definition (' . . . benefits another organism, *not closely related* . . .' (my emphasis)). His initial example of an altruistic act is saving someone from drowning. It can be formally demonstrated that provided the benefits of the act to the recipient sufficiently outweigh the costs to the actor, reciprocal altruism will come to be favoured by natural selection in species meeting various life-style criteria. Taking the drowning example, let us assume that the chance of drowning is reduced from 1 in 2 to 1 in 20 if someone attempts to rescue you. If the roles are subsequently reversed, i.e. if the act is later reciprocated, 'Each participant will have traded a one-half chance of dying for about a one-tenth chance'. It can also be formally proved that 'random altruism' as opposed to altruism conditional upon reciprocity, will result in *elimination* of 'altruistic genes' (assuming such to exist).

The major life-style criteria for the evolution of reciprocal altruism in a species are: relative longevity (enabling reciprocity to occur), a low dispersal rate (so that parties are likely to remain in contact) and a high degree of mutual dependence (so that opportunities for altruism are frequent). Broadly speaking we can, it seems to Trivers, translate the issue into games-theory terms and predict altruism to evolve when the long-term pay-offs for reciprocal altruism outweigh those of non-altruistic selfish or 'cheating' behaviour.

All this might be well and good, but the further elaboration of the case brings to light a number of difficulties and a covert shifting of ground. Even the 'altruism in animals' cases are not entirely clear. In the cleaning-symbiosis behaviour of certain fishes we find a situation where one species (e.g. the wrasse *Labroides dimidiatus*) cleans another (e.g. the grouper, *Epinephelus striatus*) 'sometimes entering the gill chambers and mouth of the "host" in order to do so' (p. 40). This is actually essential for the 'host' or 'client' fish's health as well as providing food for the 'cleaner'. But if the earlier analysis is kept in mind it is surely obscure how good an analogy it provides for the cleaning-symbiosis case since (a) how can the cleaner fish 'cheat' and (b) what risk or 'detriment' is operating on either party? (The fact that the client forgoes the chance of a meal is hardly sufficient here – it would in fact get ill if the cleaning was not done and is not presumably hard up for food if its teeth need cleaning!)

On moving to human altruism, it rapidly emerges that we are dealing with a complex 'altruism system' in which the participants engage in continual calculations and evaluations of one another's levels of reciprocity *and* of their motives. This Trivers surveys quite widely drawing on findings from Psychology and Anthropology. Humans, it appears, do indeed operate a 'reciprocity calculus' (as I will term it) – introducing sanctions against cheaters and under-reciprocators, using altruism to signal and consolidate friendship, etc. The trouble is that in all this the 'altruism genes' themselves have fallen from sight, we are no longer dealing with a single phenomenon called 'altruism' but skill at operating in, and understanding, a very complicated system of social relationships mediated by networks of mutual obligation. Altruism and selfishness are in everyone's repertoire and it is the situational pay-offs (as understood by the actors) which determine which are called into play. Trivers admits that

> Individuals will differ not in being altruists or cheaters but in the degree of altruism they will show and in the conditions under which they will cheat. (p. 48)
>
> . . . children in experimental situations do not divide bimodally into altruists and 'cheaters', but are distributed normally. (ibid.)

The 'technical' sense has also disappeared and we seem to be using altruism in a sense fairly close to its everyday language meaning. The existence of an 'authentic' altruistic *motive* is also now introduced as it is perception of the presence or absence of this which determines our judgment of others' behaviour. This raises a serious paradox as we will see. (We then of course have to find behavioural criteria for the presence of the motive aside from the supposedly altruistic behaviour itself – but the two levels, motivational and behavioural, are nevertheless still being differentiated. The behaviour by which we identify the presence of a particular motive is viewed as symptomatic, perhaps inadvertently 'betraying' that all is not as it otherwise seems.)

The difficulty arises, I think, because of a failure to distinguish between the following:

(a) The operation of the obligation network within a group of continually interacting people – by which group performance is controlled, relationship-structures maintained and modified and formally expressed, etc.
(b) The single altruistic act directed either to someone outside the obligation network or, if they are within it, directed in such a way that it is clear that its operation in that instance is to be considered as suspended.

The drowning example is typically, in our culture anyway, a case of the latter type, but most of Trivers's discussion refers to the former. Note that Trivers does not suggest that those who save the drowning consciously engage in calculations of odds before diving in (other than at the crude level of whether they stand a chance of succeeding or not at all), nor does he suggest how they could even know what the odds were. Indeed the actual location of these probabilities is obscure. In any real situation they are either 1.00 or 0.00. Each situation is unique. There is no *general* probability figure for likelihood of drowning, it is merely a mask for ignorance of the entirety of factors operating in a particular case. Even were there a general probability, the discrepancy between perceived and objective odds in most people's judgments is notorious – but it is the former which determine behaviour.

To return to the paradox raised by admitting an altruistic motivation, this may be expressed as follows: on Trivers's account altruism *is* of benefit to inclusive fitness, which is why it evolved, but it can only operate if we deny to ourselves that that is why we do it. If we admit to actually engaging in the calculations of the reciprocity calculus we are, by virtue of that very fact, no longer altruistic. A kind of utilitarian fiction has to be maintained – we best serve ourselves by deceiving ourselves that we are sacrificing our interests to those of others! (Indeed E. O. Wilson (1978, pp. 155–6), says just this.) This paradox arises from the failure to make the distinction just mentioned. The motivation in type (a) obligation networks is not altruistic (in the everyday sense) so much as anti-exploitative – the concern in the group is with genuine acceptance and recognition of the value of *balancing*, so far as possible, the obligational books. Luck may dictate that some are better or worse off than others, but this fate is not to be a result of their 'cheating' or being victims of 'cheating' in the reciprocal obligation system. On this, vast structures of jurisprudence are erected. This is a different motivation from that which might be termed 'genuine altruism' where the *denial* of expectation of later reciprocation is of the very essence (e.g. Sir Philip Sydney's dying donation to another of the water for which he was thirsting). This is, indeed, delicate since the recipient of such an act may find it difficult to shift entirely from the obligation-network perspective – hence the 'anonymous donor' syndrome. Even so the actor's own denial of expectation of reciprocity is not considered inauthentic – even if ritually sugared with some such phrase as 'some day you'll do the same for me'.

There is surely a sense in which the phrase 'reciprocal altruism' is itself a contradiction – if reciprocation is involved then the behaviour is not so much altruistic as co-operative.

Finally, the evaluation of 'costs' in inclusive fitness terms, crucial to the whole analysis, becomes very stretched when we consider the 'altruism system' in humans. I have earlier queried this in connection with high-risk behaviour. It is even more tenuous in such instances as gift-giving, low-risk assistance, etc. Consider Trivers's own introduction to the human altruism section of his paper:

> Any complete list of human altruism would contain the following types of altruistic behaviour: (1) helping in times of danger . . .; (2) sharing food; (3) helping the sick, the wounded, or the very young and old; (4) sharing implements; and (5) sharing knowledge. All these forms of behaviour *often meet* the criterion of small cost to the giver and great benefit to the taker. (My emphasis.) (p. 45)

(This is the criterion for altruism to evolve, not for identification of behaviour as altruistic.) But surely, one can go further; in many of these cases there is *no apparent cost at all* in terms of inclusive fitness (at least not apparent to me!). But the existence of *some* apparent detrimentality *is* central to Trivers's definition of altruism. Not all costs are costs in inclusive fitness terms, but these are what matter for the crucial definition with which Trivers begins his paper.

The appeal of games-theory analysis at the formal level nevertheless outweighed the conceptual weaknesses of the paper and, especially after Dawkins's *The Selfish Gene* (1976) the notion that *all* apparently altruistic behaviour was actually only enlightened self-interest, even if managed by self-deception (as in the paradox just described), gained greater authority than ever in sociobiological analyses of the issue. This use of the concept 'selfish' has been severely criticised on basic philosophical grounds by Midgley (1979a, 1979b, 1984). Trivers's own treatment of human altruism was not nearly so reductionist in temper as what followed. What is now happening, though, is an effort at re-incorporating all the obviously important situational and psychological variables into the picture under the guise of 'epigenetic processes'. This follows the Lumsden and Wilson model already considered at length. The role of the 'epigenetic processes' in current theoretical, biological and genetic thought is as was stressed a highly controversial one. A recent example of the application of the concept to altruism, by a psychologist sympathetic to sociobiology, is MacDonald (1984), whose paper 'An ethological-social learning theory of the development of altruism: implications for human sociobiology' we turn to next.

Here we are no longer concerned with altruism genes but with altruism as 'a biological system involving the integration of cognitive, affective and perceptual processes'. The paper is explicitly directed at sketching 'an evolutionary theory of some of the proximate genetic and cultural mechanisms or epigenetic rules (Lumsden and Wilson, 1981) involved in the development of human altruism' (p. 97). Admission of the complexity of the factors surrounding the individual's phenotypic, epigenetic, acquisition of altruism (centering on 'affective systems') requires a reformulation of the gene's role. It specifies 'not the behavioural phenotype, but the reaction of the organism to the environment'. What are inherited are 'systems' (altruism, attachment and affective or emotional systems).

> Such systems are highly flexible, since there need be no genetic correlation between the relevant aspects of the environment, such as parental behaviors, and the behavioral phenotype of the organism. As a result, environmental factors resulting from ecological contingencies such as the need for males to leave the family for extended periods of time . . . may have profound consequences on the development of children. Social controls such as could occur in a highly stratified, exploitative society, may be very insensitive to the genotypes of the individuals subjected to them and yet have important affective consequences within the family, as would occur if these controls resulted in inadequate resources for the family or resulted in the breakup of the family unit. (p. 107)

> The genetic basis for the affectional system does not code for some behavioral trait that is modified by the environment *simpliciter*. Rather, the affective response to environmental stimulation, which has been under natural selection, affects, among other things, what models are attended to, what expectations the child forms regarding others, and what generalizations he/she makes on the basis of experience. (ibid.)

> In a sense, the affective systems are nothing in themselves. It is only when they interact with other systems that they profoundly affect social and cognitive development. (ibid.)

MacDonald's paper certainly provides a useful review of the developmental factors affecting the altruism system (in the sense introduced by Trivers), and among other things gives infant-mother attachment mechanisms a central role. But in loosening the specificity of the genetic contribution as he does he, like Lumsden and Wilson, is in danger of leaving the genotype little explanatory work to do other than setting certain broad parameters on what is behaviourally possible. For the psychologist uncommited to the sociobiology approach, the study of the evolution of altruism, though abundant in quantity, fails to shed much new light. Clearly there is an evolutionary issue of major importance involved here somewhere, namely the manner in which humans established reciprocal obligation networks as a basis for group cohesiveness, as a medium for structuring and expressing interpersonal relationships and so on. Seeking a root for some of this in attachment mechanisms, as has been done as far back as Bowlby (1969) is reasonable too (though we are now a long way from death in battle!). These provide a prototype for the caretaking relationship which, it transpires, proves worthwhile maintaining into adult life, when its expression is a primary form of altruism. Indeed most altruism may be said to take the form of reinstating the caretaker-infant relationship temporarily between actor and recipient, and altruism-eliciting behaviour often closely resembles its infant prototypes. One exception to this of course is death in battle, which has mesmerised so many writers on the topic. Surely it was peculiar, though, to pick as a paradigm of altruism behaviour which involves trying to kill people? Certain cues – behavioural, postural – clearly elicit caretaking behaviour which similarly adopts fairly ritual forms of patting, murmuring, hugging and so on, resembling adult-infant behaviour interactions. Yet in some ways the reciprocal obligation network has more to do with economics than altruism (*sensu lato*). It represents the *suspension* of the unbalanced caretaker/dependant system and the substitution of a parity one.

A psychologist, making an initial analysis of factors affecting the likelihood of altruistic behaviour being exhibited – the individual's 'altruism threshold' we might call it – would incorporate numerous variables such as the individual's perception of a given situation as one in which altruism was expected for one in their social-role position, 'person-variables' (in the sense used by Mischel, 1973) and cognitive competence – ability to identify altruism-appropriate situations. One could generate a host of 'types' from this – from those who always see altruism as their responsibility, eagerly accept it, and often find opportunities of manifesting it, via those who often feel it is their responsibility but for some reason cannot manifest it very often (e.g. a handicapped young person perceiving there to be a role-demand on young people to be altruistic in high-risk situations who simply cannot, for physical reasons, engage in such actions) to the individual oblivious to the demands of altruism from within or without. The contextual and social role factors here would seem to render any explicit genetic contribution very obscure. When the social-role aspect and individual learning history factors are subtracted, and general cognitive competence allowed for, what is left for genetics aside from its much broader parameter-setting for personality development as a whole?

For the moment I can only tender the following conclusions on the 'Evolution of Altruism':

1 That what has been discussed has not primarily been altruism but reciprocal obligation systems.
2 That altruism in the everyday sense is *not* clearly paradoxical in the way Darwin and most of his successors have thought.
3 That altruism in the technical sense as defined by Trivers (1971), and repeatedly accepted ever since, is a totally incoherent concept.
4 That there is little *prima facie* case for allocating a central role to the genotype level in determining individual likelihoods of being altruistic in particular situations – although of course the capacities for such behaviours and the emotions underlying them are biologically rooted.
5 In the light of wide divergencies in the current literature on the nature and extent of DNA control over epigenesis, the sociobiological modelling of altruism, as of other cultural and social behaviours, must be considered purely speculative.

Hard evidence on prehistoric altruism is in any case elusive! The present writer would be happier with 'softer' ethological approaches such as those adopted by Eibl-Eibesfeldt and Bowlby, in which the evolutionary roots and prototypes of social and cultural behaviour may be traced without excessive reductionism, and holding explicit genetic theorising in abeyance. This is how most human evolution theorists have been using primate behaviour studies, for example. Tracing altruism to attachment behaviours is no more reductionist than tracing lungs to the balancing air-sacs of fishes. But finally:

6 One must question whether the *perception* of there being an apparent altruism paradox is not itself an expression of a fundamental contradiction in US and, if less acutely, western culture in general, between an official Christian ideology of compassion, love, and humanitarian values and an economic ideology based on unbridled self-interest. To provide an analysis, as many sociobiologists have done, in which altruism turns out to be selfishness, and selfishness, in turn, turns out to be the best way to serve everyone's interests – hence the highest altruism – provides a nice, albeit Orwellian, way of reconciling the irreconcilable. Selfishness is Altruism. Under pressure they can always retreat into saying they mean these terms in a scientific technical sense only, but unfortunately they have not provided us with an intelligible definition of what this is!

C The evolution of language

Speculation on the origin of language dates back to antiquity. The earliest research reported is Herodotus' story of the Egyptian pharaoh Psammetichus's attempt at having two children raised by a shepherd who was to never communicate with them. Eventually they allegedly spoke the word 'becos', Phrygian for 'bread', thereby convincing the regal experimenter of the superior antiquity of the Phrygians over the Egyptians. Similar anecdotes are told of a number of later despots. But although it is one of our longest standing intellectual inquiries, progress has, until recently, been minimal. Two principal factors have hindered investigation: firstly – and most obvious – language does not fossilise, so there can be no *direct* evidence as to either its existence or forms in prehistory; secondly, familiarity has long led us to underestimate the sheer complexity of the phenomenon of human language. Only this century, and especially since World War II, has the latter become apparent, notably as a result of the work of linguistic philosophers (Wittgenstein being the foremost) as well as psycholinguists. Indeed the latter are in some respects only now catching up with the philosophers, as they wrestle with the problems of computer simulation (see, for example, Winograd, 1980).

The evident futility of investigating the topic led the French Société de Linguistique de Paris to formulate, as Article II of its bylaws (approved in 1866): 'The Society will accept no communication dealing with either the origin of language or the creation of a universal language' (Stam, 1976). This sentiment was shared by the Philological Society of London's President, Alexander J. Ellis. Eighteenth- and nineteenth-century linguists and philologists nevertheless produced a clutch of theories on the matter jocularly referred to as the 'bow-wow', 'pooh-pooh', and 'ding-dong' theories, to which 'yo-heave-ho' and 'sing-song' versions were subsequently added (see Table 5.1) each ascribing the origin of language to a different key mechanism. Wundt's view was that language grew from 'the cries and expressive movements of primitive man and his animal forebears' (Warden, 1932), a version of the 'pooh-pooh' theory, while Sir Richard Paget proposed a model in which the tongue and other speech apparatus 'pantomimed' the referents of words or the hand-gestures elicited by them:

> . . . the great discovery was made that if while making a gesture with the tongue and lips, air was blown through the oral or nasal cavities, the gesture became audible as a whispered speech sound. If, while pantomiming with tongue, lips and jaw our ancestors sang, roared or grunted – in order to draw attention to what they were doing – a still louder and more remarkable effect was produced, namely, what we call voiced speech. (Paget, 1930, p. 133)

TABLE 5.1 *Early theories of the origin of language*

Theory	*Main proponents*	*Mechanism*
'bow-wow'	J. G. Herder (1770), C. G. Voigtmann (1865)	'Onomatopoeia', imitation of natural sounds, in Voigtmann birds especially.
'pooh-pooh'	H. Wedgwood (1866)	Emotionally expressive exclamations.
'ding-dong'	F. M. Müller (1861)	Noises made by objects when struck. Müller abandoned this theory, which he also ascribed to the *Naturphilosophie* advocate Oken.
'yo-heave-ho'	L. Noiré (1877) M. Moncalm (1905)	Natural sounds emitted during collective group effort.
'sing-song'	O. Jespersen (1894)	Vocal play, differentiation of language out of wholistic 'musical' utterances.
Pantomime	R. Paget (1930)	Mimicking with vocal apparatus (see main text).

Sources: M. Moncalm (1905), W. Paget (1930), J. H. Stam (1976), C. J. Warden (1932).

The development of the issue up to the late nineteenth century is examined in depth by Stam (1976) who discusses the roots of the modern study of the question in the early eighteenth century, Herder's famous *Treatise on the Origin of Language* of 1770, the views of the extraordinary Lord Monboddo and nineteenth-century figures such as Schlegel and the eminent linguists Whitney and Müller, culminating in the aforementioned effort at 'annihilating the question', as he puts it, in the 1860s and 1870s. (A briefer account may be found in Aarsleff, 1976.)

The question could not of course be annihilated by *diktat* and answers to it continued to appear from a variety of directions. All the same, it long remained a topic which was not entirely academically respectable, and discussions of it generally arose as speculative by-products of other linguistic research (such as Paget's studies of phonology). Two works in the 1950s – Du Brul's *Evolution of the Speech Apparatus* (1958) and Revesz's *Origin and Prehistory of Language* (1956) – signalled its revival as a scientifically acceptable issue. The broader advances being made in human evolution research thereafter continued to stimulate interest in the matter and its latest phase may be said to have properly 'arrived' in the mid-1970s with Lieberman's *On the Origins of Language* (1975) and the massive New York Academy of Sciences Symposium of the following year (Harnad et al., 1976), *Origins and Evolution of Language and Speech*. Influential accounts appearing during the 1970s variously linked language to tool-use and gesturing (Hewes), cerebral organisation (Holloway, Passingham, Falk), 'sound-marking' of territory (Jerison), bipedalism (Falk) and the 'feedback circuits' of varying complexity involving brain and behaviour which now began to flourish (Tobias, 1979; Wind, 1976a). The most recent major work is Lieberman (1984) *The Biology and Evolution of Language*. It is with these that we will be primarily concerned in the remainder of this chapter.

There are two principal routes by which we may approach the problem. We can, for a start, consider the anatomical evidence pertaining to brain organisation and the vocal apparatus to identify if, and when, the necessary physiological prerequisites for human language production are met. Secondly, from a theoretical direction, we may find grounds for inferring the operation of language from evidence of other behaviour (e.g. patterned tool-making) which we might believe could not be achieved without it. Statements as to the timing or nature of the origin of language might, that is, be necessary corollaries of broader models of our behavioural evolution. From the interaction of these empirical and theoretical perspectives some insight into the genesis of human language may yet be gleaned.

Much of the background debate has focused on the degree of similarity between human language and the communication systems of other species, a discussion influenced to a considerable degree by a now classic analysis of 'design features' of language by Hockett (1960). In recent years this topic has been dominated by the various programmes aimed at teaching language to higher primates (see Chapter 3). Somehow or other our ancestors presumably made an unaided transition from a typical higher primate call system, such as that of chimpanzees described by Marler (1976), to our fully fledged syntactically structured, lexical, language. Though even to say this is to imply a continuity between the two behaviours which some would deny.

Before turning to the current state of play on the topic we ought to ponder, if not for long, on what human language *is*. This is a deceptively simple query. We tend to imagine that, being able to use language, a moment's reflection would be sufficient to answer it, but actually this is no more valid than supposing that the ability to see is sufficient for an understanding of perception. Such obvious answers as that language is for transferring information, or classifying the environment, or is based on 'naming' are far from self-evidently true. It is salutary to quote Wittgenstein here.

> . . . the *speaking* of language is part of an activity, or of a form of life.
>
> ***
>
> Giving orders and obeying them –
> Describing the appearance of an object. . . .
> Constructing an object from a description (a drawing) –
> Reporting an event –
> Speculating about an event –
> . . .
>
> Play-acting –
> Making a joke; telling it –
> . . .
>
> Asking, thanking, cursing, greeting, praying.
>
> – It is interesting to compare the multiplicity of the tools in language and of the ways they are used, the multiplicity of kinds of word and sentence, with what logicians have said about the structure of language. (1967, 11e–12e)

For 'logician' here read 'theorist on the origins of language' and for 'structure' read 'function'! This multi-functional character of modern language has often been lost sight of, and the assumption that just one of these functions is logically central too easily made. It is here that the primate-human gulf yawns widest. Language is not just, or even basically, naming, referring, making propositions. It is the medium in which speakers live and in terms of which they structure their relationships, articulate their motives and evaluate their lives and those of others. In short, it is 'a form of life'.

Such a situation did *not*, we may plausibly infer, arise overnight. The transition from primate to human language most likely involved numerous intermediate phases and forms which are now nowhere represented.

The major areas of concern today relate to the functional roots of the evolution of language – its relationship to other 'hominising' trends such as increasing tool-use, encephalisation, and bipedalism – and the timing of the appearance of fully-fledged human language (broadly, whether it was early, late or very late). No extant group, however 'stone age' its life-style, lacks a fully grammatical language, nor has one even been reliably recorded. Earlier travellers sometimes perceived native languages as primitive concatenations of crude grunts and squeaks, but this was due entirely to their ethnocentric perspective. Those who later mastered such tongues discovered on the contrary that their sheer grammatical complexity was often greater than that of, for example, English. As to the inter-relationship between language and thought, the extent to which what is 'thinkable' is determined by language (the Whorfian hypothesis) as opposed to the extent to which all languages can be made to express any thought, we must sidestep the issue here. No serious student can avoid tackling it sooner or later, though.

I turn now first to the anatomical evidence and then to the theoretical models of language-origin which have been proposed.

1 *Anatomical evidence*

This naturally concerns two main areas; the physiology of the vocal apparatus itself, and the brain areas believed to govern language. Although gradually increasing in both quality and quantity the raw evidence on both is still hardly more than sketchy.

(a) Vocal apparatus

Since our vocal apparatus primarily involves soft-tissue (the larynx itself, tongue, epiglottis, pharynx), palaeontologists have had to try and develop techniques for reconstructing hominid vocal regions inferentially from the morphology of the basal region of the skull (an approach opened up by Lieberman, 1975). Such reconstructions are rarely uncontroversial, a reconstruction of the oral and pharyngeal cavities of Neanderthals by Lieberman and Crelin (1971), for example, seems to have called down the wrath of nearly everyone else in the field, with Falk actually claiming that such a vocal tract was incompatible with breathing (Falk, 1975). Lieberman and Crelin concluded that Neanderthals could not have possessed modern human speech, but Wind (1976a, 1978) has taken issue with this. Lieberman's most recent defence of his position is discussed later (Lieberman, 1984). The fossil evidence on Neanderthals does not as yet permit a full reconstruction of their vocal tracts, since vital details of the basicranium are still unknown, but that it differed from that of adult modern humans is clear, and Lieberman now suggests that it was similar to that of modern human children. Laitman and Heimbuch (1982) have proposed detailed reconstructions of earlier hominid and protohominid vocal apparatus and they believe these indicate a vocalising capacity far more restricted than ours, with smaller, and hence less resonant, oral, pharyngeal and nasal cavities and higher placed tongues.

Wind's argument is that high levels of vocalisation, though not so high as among *H.sapiens*, are present among primates generally and that the significance of the vocal tract *per se* as evidence of language evolution is limited. Evidence from human pathology shows that even if we lack substantial parts of our vocal equipment we are capable of re-acquiring speech by employing other sound-making techniques. The system has, he says, 'considerable functional redundance' (1976a). But he goes even further to suggest that 'if surgery would reach the level where a chimpanzee larynx could successfully be grafted into an otherwise normal human being, such a person would be able to acquire a speech hardly discernible from the normal' (1976a, p. 626). In Wind's view the vocal apparatus was pre-adapted to vocalisation, and the evolutionary acquisition of human language primarily involved changes at the neurological encoding level and in the elaboration of brain-vocal tract channels. Once the *behaviour* changed, i.e. once our ancestors began to use the vocal communication channel in a more elaborated fashion, the morphological changes in the vocal tract region followed,

> . . . rather than vocal tract novelties, it seems to have been cerebral reorganization that has been decisive for the origin of speech-like communication, such as an increased ability to form cross-modal associations and an increased memory. (1976a, p. 628)

> . . . the primates, including their peripheral vocal and auditory organs, have been pre-adapted to a large extent for speech long before *Homo sapiens sapiens* evolved . . . it was cerebral reorganisation that triggered speech-like communication. (1978, p. 90)

Wind notwithstanding, the more recent Laitman and Heimbuch work (1982) does suggest that australopithecines would have had problems integrating eating and vocalisation behaviours, as well as having a more restricted noise repertoire. That vocal tract morphology follows in the wake of speech development is to be expected from the general principle that physical change follows behavioural change rather than vice versa, since it is only a behavioural change which renders the physical shift selectively advantageous. On the other hand, against Wind, if a major physical shift *does* occur, then it hardly suggests that the previous morphology was already fairly adequate, and this indeed seems to be the case with the *Homo sapiens sapiens* larynx (Lieberman, 1984). We turn then to the brain evidence.

(b) Speech areas of the brain

As discussed in Chapter 4, there are fairly strong grounds for believing that lateralisation of function has quite early beginnings in the hominid lineage, and this feature is assumed to be strongly associated with language. This is expressed in an expansion of the Broca's and Wernicke's areas in the left hemisphere (see Fig. 4.2), the former in the frontal cortex and the latter in the temporal and parietal cortex. Broca's area is thought to govern speech production, while Wernicke's area governs comprehension, though this is a considerable oversimplification of the often far from unambiguous anatomical evidence. (Noback, 1982, provides a summary of neurological control of speech.) Cases are also known where the right hemisphere possesses considerable linguistic competence (Gazzaniga et al., 1984). Passingham, Falk, Holloway and others working on this, although not always in agreement in detail, all tend to conclude that there is at least some development of these regions as far back as the australopithecines. Falk (1980) argued that early expansion of areas homologous with Broca's area in humans was a major factor in facilitating lateralisation of motor-control (right-handedness), as the adjacent motor regions were affected. This is an interesting reversal of the popular theory, advocated by Hewes in particular (Hewes, 1973) that language evolved from gesture. Falk's sequence is bipedalism – vocalisation increase – right-handedness. Lieberman argues that syntax is governed by the evolutionarily more basic general cognitive system, and that Broca's and Wernicke's areas deal with the peripheral encoding and production functions specific to speech as such, but *not* linguistic ability as a whole. Contra Chomsky he disputes the notion of a single 'language organ' in the brain. Language is, for Lieberman, rooted deep in the cortical systems governing thought and problem solving.

By the time the *Homo* grades are reached, cerebral asymmetry is clearly evident, and brain-size in *H.erectus* approaching that of moderns. We thus have a situation where the cerebral evidence, such as it is, is of long-standing expansion of regions *now* utilised for production and comprehension of speech, while the evidence regarding the speech apparatus itself is far more equivocal – implying a more limited facility in vocalisation until relatively late in the day. The possibilities, other than total misinterpretation of the data (which is always a possibility, though by now decreasingly likely) would appear to be:

(i) Wind's position. Protohominid vocalisation, though different from ours, was already structured in a complex way (though less complex than ours); the laryngeal and basicranial physiological evidence underestimates the versatility that could be attained with relatively rudimentary speech organs.

(ii) Regions of the brain now specialised for language were being used for other kinds of oral control (e.g. usage of mouth as a supplementary hand) and their early enhancement deceptively overplays the development of language itself. Lieberman's 1984 position would seem to go some way towards the idea that the growth of these regions was not necessarily originally related to linguistic behaviour.

(iii) Some language functions were developed early, e.g. complex emotional-state signalling related to the more complex life-style postulated for early australopithecines and involving, as seen in Chapter 4, limbic system changes also. These functions could be achieved by an essentially primate vocal apparatus combined with a limited expansion of the speech areas of the cortex which Holloway and La Coste-Lareymondie (1982) suspect may be present. Other language functions, culminating in elaborate lexical classification and syntactic structuring, accrued gradually and in the familiar 'positive feedback circuit' relationship to physiological changes at the speech apparatus level. The principal factors involved here would be those to do with encoding speed and perceptual discrimination and classification of environmental phenomena, combined, as Wind argues, with development of the brain-vocal apparatus channel rather than the brain-face/limbs channel. An extremer version of this would be Jerison's model (1982b) dealt with later.

Present knowledge of the physiological side of language evolution permits us to do little more than set some rather wide parameters for theorising on the matter. There is great latitude within these limits for a variety of very different, even opposed, models.

2 *Theoretical models*

Views of the timing and nature of language evolution now occupying the arena of academic debate vary widely. In what follows I will take four examples of what is currently 'on offer' to illustrate the different kinds of thinking and direction of approach which mark this topic. Before doing this, though, we might turn again to the much publicised primate language research referred to earlier. How significant is this for the present topic? The linguistic capacities of chimpanzees and gorillas, at least in the gestural mode, have been shown to far exceed what were once thought to be the limits of what non-humans could achieve. It is still, early optimism notwithstanding, unclear how far a capacity for *syntactic* structuring is necessary to account for their gestural 'utterances', and if anything the pendulum has swung back against this. (Herman et al. (1984) have, though, shown that comprehension of syntactic rules is within the capabilities of bottle-nosed dolphins while Lieberman (1984) roundly attacks the sceptics.) This work has figured largely in many accounts of the human-animal link (e.g. Thorpe, 1974) and by implication is supposedly highly relevant to the evolution of human language. In fact that relevance is not entirely clear. Certainly the findings clarify the maximum level of linguistic capacity possessed by our last common ancestor, though assuming that even in chimpanzees the language capacity has undergone some progressive evolution since our paths parted, such an ancestor's capacity is likely to have been inferior to that of modern *Pan*. The fact that chimps can also be taught how to pour out cups of tea is not, in any case, thought to shed light on the origins of human tea drinking. Nevertheless, the fact that primate language performance *can* be so demonstrably enhanced does compel us to focus more closely on the specific factors which might have led our language to develop as it has. Why did *we* extend these capacities to their present more elaborate level, while our closest relatives failed to do so? Furthermore, we can certainly no longer use possession of human language as a unique and clearcut way of demarcating us from non-humans. Wide as the gulf is, too many aspects of language are already present, at least as potential, among higher primates for language possession *per se* to be a viable criterion for qualitatively differentiating us from other fauna. But did human language evolve from non-human vocalisation? Is there a continuum between higher primate vocalisation and gesture to Olivier performing Richard III? Noback (1982) identifies three broad schools of thought on this: (a) that human language evolved from animal vocalisation – the commonest view and one which stems for Hockett's original analysis of 'design features'; (b) that it evolved from gesture or tool-use; (c) that its evolution was *de novo*, with little relationship at all to animal vocalisation (Myers, 1976). From a behavioural-psychological perspective a more convenient classification of accounts would be the following, which I am adopting: (a) gestural models (Hewes, Parker and Gibson); (b) social models (Holloway, Marshack); (c) cognitive models (Jerison, Falk); (d) physiological-cognitive model (Lieberman). In none of these is the primate language evidence *central*, but it is invariably present in the background as a factor governing the terms in which the issue as a whole is being conceptualised.

(a) Gestural models

Gordon Hewes, one of the leading advocates of this interpretation, has provided a venerable ancestry for such theories (Hewes, 1976), versions of which were widespread in the sixteenth and seventeenth centuries. There are variations in the extent to which modern 'gesture-theorists' assume the gestural precursor of vocal language to have evolved prior to the shift to that mode, and also in their views of the duration of the transition period, as well as its timing. All agree though that the gestural channel of communication is more fundamental, that it preceded the primacy of the vocal channel, and was the source from which spoken language arose. In this section I will discuss primarily Hewes, and Parker and Gibson (1979, 1982).

Powerful arguments can be made in support of the gestural-origins case, though as we will see, these are not decisive and sometimes weaken under closer examination. Explanations are also given of the subsequent transition to the vocal-auditory channel, and these two issues will be taken in turn.

(i) ARGUMENTS FOR GESTURAL ORIGIN

1 The highly evolved communication capacity of this channel in the higher primates, especially as revealed in modern primate language research, contrasts with their relatively stereotyped and limited use of the vocal-auditory channel and their lack of an evolved apparatus for vocal communication. The last common ancestor presumably 'displayed a capacity for referential communication at least as great as that of chimpanzees and gorillas' (Parker and Gibson, 1979). Gesture would thus have been the initial dominant communication channel between hominids prior to the physiological changes facilitating speech (which occurred much later for reasons only then to become operative).

2 Hewes (1976) argues that the neurological data do *not* indicate continuity between primate and human vocalisation control areas in the brain. Also, by and large, human speech control is autonomous from the limbic system, permitting 'decontextualisation', i.e. separation of message quality itself from the emotive content of the message, though the latter can be intonationally incorporated. (But see above, Chapter 4, on the role of the limbic system.) This contrasts with primate vocalisation which is fundamentally expressive in character and closely linked to the limbic system. Initially gestural language would have been accompanied by this kind of primate vocalisation, but it did not serve as the direct antecedent of spoken language (rather it could perhaps be argued that it still functions in the original form in grunts, screams and the like).

3 Hewes (and both Holloway and Jerison, though neither are gestural theorists) sees a formal parallel between tool-making, tool-use and (in Hewes's case) large-carcase butchery on the one hand and linguistic structuring on the other. These activities are the formal prototypes of language use: the user-tool relationship and the sequencing of complex behaviour laying the basis for syntax and reference. This argument may be better labelled a 'tool-use' model, although it still requires the primacy of gesture. For Hewes the two processes of gesture and tool-use in any case evolve alongside one another in the protohominids.

4 The gestural channel is still available, being used by the deaf and by others in environments prohibiting usage of the sound channel. It is also the first to appear developmentally, a factor which Parker and Gibson place great stress on in their recapitulationist model:

> Protolanguage involves gesture that expresses all or part of a particular meaning. In fact, the emergence of a 'gestural complex' between nine and thirteen months predicts and precedes the emergence of the first words. This gestural complex is comprised of referential pointing, object showing, object giving, and a gestural request for objects . . . (1979, p. 373)

They too see gestural language as related to tool-use:

> The intellectual prerequisites for this sort of problem solving (i.e. co-ordination of gestures for referential use) already existed in connection with object manipulation for tool use and only needed to be extended into the area of gestures. (ibid., p. 374)

Additionally, Hewes notes that gestural accompaniment to spoken language has been shown fractionally to precede or occur simultaneously with the associated utterance, but never follows it, as we might expect if the gestural aspect were a subsidiary elaboration evolving after a primary vocal channel.

Both the continued availability of the gestural channel and the intimacy of its involvement with spoken language suggest that, at the very least, it is not a mere late elaboration on the spoken word, but shares with it an almost equal facility for signal complexity – however rarely this capacity is now fully exploited.

5 The more nearly universal character of gestural language and its lower ambiguity have also been interpreted as evidence for its fundamental role. Unlike vocal languages, gesture could be construed as a species-specific (as opposed to culture-specific) communication mode rooted far more deeply than arbitrarily encoded spoken tongues. A final piece of evidence cited by Hewes in a frankly speculative vein is the unique (among primates) lack of pigmentation on all human palmar and finger-tip surfaces: 'I propose that this uniquely human feature . . . arose through natural selection during a long era of gestural communication, to maximize the efficiency of visual language . . .' (1975, p. 498).

From a number of different directions then, the evidence for a gestural origin of language appears to be fairly weighty, if far from overwhelming. Some of these points regarding the versatility of gestural language, and so on, could be accounted for by Lieberman's view that there is an underlying cognitive neural substrate which is not mode-specific – rather than a linear emergence of speech from gesture, a redirectioning of this underlying capacity would be involved. Comparative evidence certainly indicates that gesture is the more developed channel in our closest relatives. Developmentally gesture precedes speech, while the introduction of tool-use and tool-making into the picture provides us with a rationale for envisaging a selection pressure on the channel towards greater sophistication and elaboration among early hominids. Even if this is all granted there is still considerable lee-way for theorists to differ when considering the further issue of the transition from gesture to speech.

(ii) GESTURE-SPEECH TRANSITION

Two alternative pictures emerge from the accounts given by Hewes (1973, 1976) and Parker and Gibson (1979, 1982).

1 Hewes's model

As mentioned already, Hewes places tool-use and tool-making in a central role:

> The visual, kinaesthetic, and cognitive pathways employed in tool-making and tool-using coincide with those which would have been required for a gestural language system. Speech, on the other hand, utilizing the vocal-auditory channel, implied the surmounting of a neurological barrier – that of the cross-modal transfer of learning, which I think could only have taken place as a result of long-established natural selective pressures on the central nervous system, unlikely to have been completed in the early phase of hominization. (1973, p. 103)

A long period of 'general cognitive enhancement' ensues accompanying the slow but continual development of tool-use and gestural language alongside one another. For Hewes cerebral lateralisation may have far more to do with the demands of tool-usage on neurological control and organisation than with spoken language *per se*, in particular the need for preferential handedness.

Only rather late in the day does the final transfer from gestural to vocal channels occur. Several factors contributed to this: firstly a growing contradiction between use of hands for tool-use and for gestural communication as the two categories of behaviour each increased in effectiveness and scope. 'The way out of the impasse would then be a switch to an underemployed sensory modality: hearing' (p. 112). Although this is important, it was supplemented by other factors – superiority of sound in the dark, enhanced auditory acuity in monitoring environmental sounds, and some aspects of tool-related behaviour itself in which the mouth comes to be incorporated, such as controlled blowing, while the 'mouth can serve as a holder, for cutting, shredding, softening as well as for moistening . . .' (p. 113). Additionally, Hewes believes there may have been a usage of spoken language for territorial demarcation, comparable to bird-song use. Groups would evolve different 'dialects' and eventually different languages altogether, which would promote dispersal of the population and maintain social coherence of the group, giving an 'identity'.

The picture then is of an initial gestural repertoire being brought into conjunction with increased tool-use, due to proto-hominid life-style changes (of the kinds discussed in Chapter 4). A long slow evolution of these two interacting categories of behaviour sets in, promoting cerebral lateralisation and other forms of neurological re-organisation and expansion to enhance cognitive skills, manual co-ordination and cross-modal transfer. Finally, perhaps very late in the day, the pressures on manual behaviour become too abundant and complicated, and a shift of communicatory functions to the auditory channel gets under way, helped by the intrinsic advantages of that channel for non-directional communication, communication in the dark and group identification. Hewes tentatively locates the timing of the transition around the end of the Acheulean 'hand-axe' cultures, since the subsequent more rapid diversification of tool-types suggests improved transmission of motor-skills. Although verbal instruction is not actually crucial to teaching manual skills, we have noted elsewhere that single-word or phrase encouragement or guidance is typically involved.

2 Parker and Gibson's model

Parker and Gibson's account of language origins emerges directly from their recapitulationist model of cognitive evolution generally, a position which, as we saw previously, uses a Piagetian conceptual framework to structure comparative analyses of human and non-human behaviour. Their account is now the most fully elaborated one available, and in their 1982 restatement of their position they stress the necessity for some such over-riding theoretical framework:

> We cannot overemphasize the importance of theory in hominid reconstructions. . . . The value of theories lies in their heuristic power, in their ability to generate coherent and testable hypotheses which would never have emerged in their absence. (1982, p. 60)

Their model is an exception to our earlier statement regarding the centrality of primate language studies; they do indeed incorporate these as a central source of evidence. It will be recalled that for these authors the central factor in hominid intellectual evolution was the adoption of extractive foraging as a primary survival strategy. Since this involves sustained visual attention, it would be advantageous for such an animal to adopt vocal rather than visual communicating – which would be distractive. This scenario, by implication, places the origin of spoken language rather further back than Hewes would. They also consider the imitational component to be important, whereas for Hewes it is only of secondary importance in occasionally being involved in word etymologies. Nevertheless they claim

> . . . these comparative data support the theory that the first form of hominid referential communication was gestural (Hewes, 1973), and imply that our common ancestor with the great apes and hence the earliest hominids displayed a referential gestural complex similar to that of young children . . . and great apes. (1982, p. 49)

If we conceive of language as 'a symbolic structure for regulating joint action and joint attention', the question then arises 'what resource utilization patterns would have favored these regulations or, to put it another way, these manipulations of the labor potential of conspecifics' (1982, p. 58). One crucial difference between chimpanzee and human vocalisations is that ours are directed at specific individuals. The chimpanzee finding a food bonanza will loudly advertise it, but the situation is different for humans and protohominids:

> It seems likely that directed forms of referential communication are favored in situations where kin groups are competing for scarce dispersed seasonal foods. (1979, p. 374)

The precise origins of vocal language are less clear, though:

> Language in the vocal modality probably first arose as an imitational supplement to gestural language, then gradually replaced it due to the two advantages of vocal/auditory communication over gestural/visual communication: First, vocal communication is multidirectional . . . Second, the vocal channel is more energy-efficient . . . (ibid).

Developmental and comparative data on great ape language use leads them to propose that 'gestural protolanguage arose as an adaptation for food sharing among close kin in our omnivorous extractive tool using ancestors' (1982, p. 60). The developmental data do though show a continuity between gestural and vocal language, and the evolutionary course of events is obscured because of the lack of extant intermediate forms. To some extent they see the transition issue itself as less problematical than Hewes does (as he noted in his own Peer Commentary on the 1979 paper). The key difference between the two models is the highly specific character of the Parker and Gibson proposal, which contrasts indeed with most other current accounts. Extractive foraging and food-sharing are the lynch-pin for language development as for behavioural evolution at large.

When all is said and done, the crux of the gestural case against the 'continuity with primate vocalisation' model is perhaps encapsulated in Hewes's statement, 'If spoken language had evolved directly out of the primate call system, one would not expect it to have become localized in an entirely different portion of the cortex' (1976, p. 490). But has it? As yet the gestural theory or model of language origin has failed to win the day, and the most recent full-length work on the topic (Lieberman, 1984) remains opposed to it, precisely on physiological grounds, as we will see.

(b) Social models

The central role of tool-use and manual dexterity is strongly disputed by several authorities, although a fully worked out 'social origin' theory has not yet appeared. Among those who would seem to espouse this position are Holloway and Marshack. The major argument that can be invoked for this downgrading of the manual is that the *primary* function of vocal communication is in the context of social relationships, and it is this aspect of the life-style which was central in the initial phase of hominisation (as we saw in several of the models described in the previous chapter). (In truth though the boundary line is not so clearcut in the Parker and Gibson theory since 'food-sharing' is as much social as manual.) The information-exchanging and instrumental roles of language are seen as a later development. A number of auxiliary points can be invoked to lend support to this, Marshack, for example, notes:

> Tool making and tool use are not learned through language, and the skills of the hunt are learned by example and participation, not by linguistic description . . . . The most effective 'linguistic' or protolinguistic component of subsistence activity (whether vocalized or gestural) is often simple 'affect marking', indicating approval, affirmation, support, negation, warning, request for cooperation or delay, none of which requires a complex use of language or any significant increase in evolved linguistic capacity. Such affect marking forms a large part of communication during human subsistence activity even today. (1979a, p. 395)

For Marshack it is the cultural, social, context which is the basic factor in bringing about changes in linguistic behaviour, a view he subtly backs up with the chimpanzee evidence:

> Clearly, in the case of the chimpanzee, the potential capacity for protolinguistic behavior in the laboratory is not dependent on a change in either brain volume or brain structure, but is elicited by a change in context – in cultural, relational and behavioral contexts that contain periodic, culturally maintained feedback and reinforcement. (ibid.).

Language then is primarily to do with interpersonal relationships, and thus with emotional life as much as, and more fundamentally than, with cognitive life. The persistent involvement of the limbic system with language testifies to this. Such an interpretation of the evidence is at variance with Hewes's emphasis on precisely the opposite move to the evolution of language – 'decontextualisation' and autonomy from the limbic system!

Holloway (1981a), as discussed in Chapter 4, is adamant regarding the logical primacy of the social:

> . . . it was not the tools themselves that were the key factors in successful evolutionary coping. Rather, the associated social, behavioral and cultural processes, directing such activities as tool-making, hunting and gathering, were basic. (1981a, p. 288)

The developmental data too might suggest that there is more involved in language than gestural accounts of its origin could provide; the first area of intense verbal communication is between parent (usually mother) and baby, and one could mount a strong argument relating it to attachment mechanisms. (After all in popular language we refer to our 'mother tongue'!). Hewes's comeback to this is that lexical, semantically structured, language is *not* directly evolved from this kind of vocal communication, which he would not deny occurs. On the other side, the stress by Holloway and Marshack on the cultural character of language, the necessity of a particular cultural-social context to supply a *raison d'être* for the behaviour cannot be so readily countered.

The reader might at this juncture be beginning to feel a sense of frustration, in that the two schools are to some extent apparently at cross-purposes, differing not so much in their actual picture of the process but at the level of phenomena with which they are concerned, the immediate technical level of how language evolved or the broader situation which rendered its evolution adaptive. There are, though, substantial issues at stake, notably regarding the continuity between primate vocalisation and human language, and regarding the degree to which the affective, emotional, aspects of language are functionally central. For gesturalists the whole point is that modern language represents an emancipation of the channel from the simple affective-expressive role; for social theories this persists as its underlying central task, even if not, now, the only one.

From a psychological viewpoint this debate would also seem to relate to the differentiation between verbal and spatial 'intelligence'. Although correlated, these are not synonymous. As I suggested in the previous chapter, part of the difficulty here is that 'intelligence' as a single concept could refer to a phenomenon with two or more evolutionary roots. This is downplayed in Parker and Gibson who are keener to note the 'existence of parallel stages or levels of intellectual functioning in many different domains' (1982, p. 49), but the psychometric, rather than Piagetian, tradition would caution us against over-assimilation of abilities under the single heading 'intelligence'. Given the extraordinary range of jobs language encompasses some compromise image might be feasible, perhaps in which language is seen as 'laminated', that is it lacks a single 'origin' but has developed by a successive incorporation and integration of different behavioural faculties – the intonational, expressive dimension of speech is essential to its effective use (since it establishes the user-receiver relationship), but may have had a longer evolutionary history than the fine-tuned lexical dimension required for information transfer.

Social theories do not in themselves lead their adherents to a clear position regarding the timing of the advent of language, for in this matter Holloway would place it early, while Marshack would place it in the Middle-Upper Palaeolithic, with the archaeological advent of symbolism.

(c) Cognitive models

The doyen of brain evolution studies, Jerison, proposes a model which we may for convenience refer to as 'cognitive'. In it he stresses the cognitive role of language as a method of improving the individual's ability to rationally evaluate its environment, 'human language evolved in response to an environmental demand for additional cognitive capacity in early hominids. Not specifically for a new and better communication skill' (1982, p. 763).

Jerison's model is unique so far as I know, and quite original. He traces language to the loss of olfactory senses and the attendant means of olfactory marking used in e.g. wolves. Protohominids found themselves 'trying to make a wolf-like living without adequate sensory machinery' (ibid., p. 764). Their solution to this was to substitute the auditory-vocal system for the normal olfactory scent-marking system as a way of marking. The needs met, in social carnivores, by scent-marking, were now met, in these new hominid social carnivores by 'associating' key territorial and trail markers with a sound, enabling them to store increased environmental information and enhance 'mapping', necessary for the carnivore life-style.

> . . . a new kind of communication was an unexpected bonus that resulted from the evolution of the language system, primarily because of the design features inherent in the auditory-vocal channel . . . conspecifics will share the auditory-vocal map with one another both in its construction (naming) and in its reconstruction (repeating or remembering a name). Communication would occur . . . as sharing of consciousness. In communicating by this strange system, the reality of one individual becomes the reality of others that use the same language, at least to the extent that this channel contributes to the constructed reality of an individual. (ibid.)

Differences between the self's and other's vocalisations in turn generate different such realities: 'A self is placed as a peculiar object in the constructed real world' (ibid.). His interpretation of the brain evidence is quite contrary to Hewes's:

> Had human language evolved from gestural systems, it would very likely not have evolved with as extensive a neocortical and hippocampal involvement as is present, nor would the language areas of the brain be as intimately associated with the neocortical fields for the localization of hearing and the motor control of mouth and tongue. Wernicke's area and Broca's area would be elsewhere relative to the 'homunculus' that is mapped in the human brain, perhaps near the hand and thumb areas. (ibid., p. 765).

Falk (1980c) is also doubtful that the brain anatomy evidence is compatible with a gestural hypothesis. She argues, contrary to Hewes again, that there *are* homologous vocalisation and vocal communication areas in monkeys and humans. She thus supports the view that human speech is directly evolved from primate vocalisation. Although less specific than Jerison, her model might also be termed cognitive in that she considers the shift from arboreal to terrestrial habitats to have placed pressures on the existing, already vocal, primate communication systems, ' . . . vocal communication systems had become more complex under the selective pressures that *led to* bipedalism' (1980c, p. 76). It is the information processing strain which seems, on her account, to be crucial. The tool evidence is symptomatic of the emergence of proto-culture and elaborated communication (here she endorses Holloway), but tools are not themselves *sources* of language. With bipedalism, the use of forelimbs as an auxiliary communication channel became possible, as did increased tool-usage and manufacture. This involves a pressure for lateralisation. As mentioned earlier, Falk's interpretation of the order of events is unusual. She sees the neocortical regions involving vocalisation to have already become lateralised in the left hemisphere in primates, but under increasing pressure for a wider variety of communication, plus the evolving manual lateralisation, the manual control areas in the left hemisphere come to join up with the enlarging speech area, particularly the association areas bordering the oral control region. Thus gestural communication becomes possible *after* speech. 'Voices were "free" long before hands were' (p. 76). Language was selected for, then, by the processes which led to bipedalism; handedness and gestural language came *after*.

Both of these 'cognitive' models seem to entail an early origin of language, by which I mean they envisage that vocal language began to develop as far back as the first australopithecines. Jerison's 'wolf-niche' idea is somewhat dubious in the light of recent models of the early hominisation process, which stress gatherer-scavenging, 'extractive foraging' and a broadly omnivorous diet. But the need for terrain mapping might be as acute for gatherer-scavengers as for hunter-predators. Falk does not see her model as incompatible with Holloway's, but rather as contrary to Hewes's.

The distinctive features which differentiate these two approaches from the others are then: focus on cognitive processing pressures in protohominids, interpretation of neocortical arrangements as inconsistent with gestural theories, continuity of language with primate vocalisation (shared with social theories) and a generally early placement of the emergence of language.

(d) Lieberman's physiological-cognitive model

Philip Lieberman, whose *On the Origin of Language* (1975) played a major role in reopening the topic of language evolution, has now brought his account up-to-date in *The Biology and Evolution of Language* (1984). Lieberman's grasp of the complexity of the phenomenon of language is impressive, and he would not claim its emergence to have been a single discrete event. In outline he proposes that although there was a degree of enhancement of the vocal channel very early, this did not immediately lead to spoken language. Rather, from the australopithecines on, there was a steady growth of general cognitive ability, manifesting itself at the central neurological level. A positive feedback circuit is envisaged between behavioural competency and neurological organisation, with continuous, but mosaic, improvement in the analysis of sensory input. While brain evolution is accentuated by anatomical changes related to bipedalism and manual dexterity, there is little evidence of any change in the basic pattern of supralaryngeal ('vocal tract') physiology during the period up to the demise of *H.erectus*. While vocalisation may have been playing an important part, it could not, on this evidence, have remotely matched modern speech in the range of sounds which could be produced. The region still conformed to the basic prehuman mammalian pattern of obligate nose-breathing, long oral cavity and high-placed larynx, these facilitating chewing, ability to breathe while drinking, and the virtual impossibility of choking, as the larynx connects directly to the nasal cavity, with epiglottis and soft palate sealing off the back of the mouth from the trachea.

At this stage Lieberman envisages what he calls a 'functional branch-point' arising within the hominid population. Increasing cognitive evolution and use of the vocal-auditory channel now gives at least some of the hominid population a selective advantage for any anatomical changes that improve vocalisation. The *Homo* lineage splits (though it was probably not a simple bifurcation) into those opting, as it were, for speech development and those going for robust physiological adaptations for greater strength – including strength of biting and chewing. The former constitutes the '*sapiens sapiens*' line, while the latter included the Neanderthals. Lieberman does not, though, deny Neanderthals some level of linguistic facility and he fully recognises their cultural achievements.

> Obviously language enhances and indeed may be ultimately responsible for the adaptive value of this cultural pattern. I therefore find it hard to believe that Neanderthal hominids did not also have a well-developed language, particularly given the linguistic and cognitive ability of modern chimpanzees . . . (1984, p. 323).

Nevertheless, on the basis of Neanderthal cranial anatomy, they cannot, he believes, have possibly talked in the modern sense, for their vocal tract is still too similar to the mammalian 'standard pattern' (Fig. 5.2).

Our current speech apparatus is explicable only in terms of a selective adaptation for speech, and has been achieved at some considerable cost. It has involved a dramatic lowering of the larynx, to such an extent that we are at constant risk of choking during eating, while the required shortening of our mandible has crammed our teeth into a smaller space, generating both a tendency to impacted wisdom teeth and impaired masticatory efficiency.

> The picture that emerges with regard to the vegetative functions of breathing, swallowing, and chewing is that the adult human supralaryngeal airway, and its associated skeletal structure, is less effective in these functions than the nonhuman standard-plan arrangement. (ibid., p. 283)

This human pattern is absent in the neonate, appearing around three months and deviating progressively further from the normal primate pattern from then on. This, incidently, provides Lieberman with a strong case against neoteny models of human evolution (e.g. Gould, 1977; Gould and Eldredge, 1977). We are precisely *not* neotenous in this crucially important area relating to our most characteristically 'human' attribute – spoken language. It is in other primates that the neonate morphology of the vocal tract is retained into adulthood. If efficient speech could, as Wind claims, be achieved with the 'standard-plan' physiology evident in Neanderthals, why such an otherwise costly change in *H.s.sapiens*?

The full acquisition of spoken language in our line probably happened around 40,000 bp, with the attainment of a crucial processing rate, facilitating syntactic organisation:

> The presence of a fully encoded speech system in recent hominids may also have more directly contributed to the development of complex syntactic organization in human languages. The rapid data rate of human speech allows us to transmit a long sequence of words within a short interval. We can take the words that constitute a complex sentence into short-term memory and effect a syntactic and semantic analysis. . . . Given the same constraints on short-term memory that are evident in modern *Homo sapiens*, a speech rate that

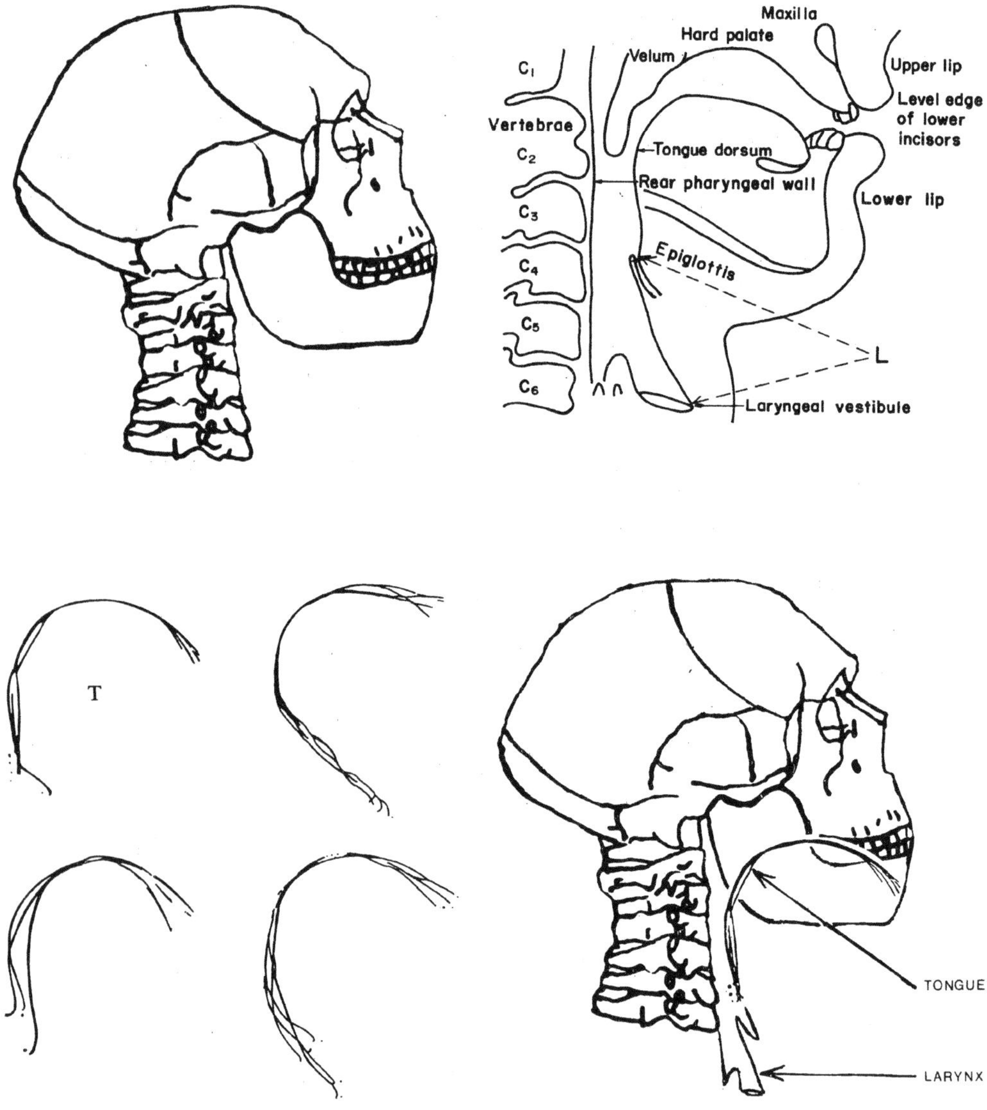

Figure 5.2 Lieberman's argument against Neanderthals possessing modern speech. From P. Lieberman, 1984. *Reproduced by permission of Harvard University Press.*

> was one-tenth that of modern human speech would limit vocal communication to very simple syntactic structures. (p. 325)

This in turn has the effect of enabling 'collective insight' to operate, that is, collective evaluation and problem solving. When a gradual quantitative change reaches a critical point it may bring about a radical, qualitative change in the nature of the system. This phenomenon, familiar in 'catastrophe theory', underlies Lieberman's 'functional branch-line' model of evolution, accounting for the apparently 'punctuated' nature of evolutionary change (Eldredge and Gould, 1972). The advent of collective processing is another instance of this, and he provides what I think is an interesting and helpful analogy:

> If the mode of linguistic communication is rich enough to communicate the set of parameters that describe a problem, then more than one person can apply insight to the solution of that problem. This process . . . may be one of the factors responsible for the rapid advance in human culture . . . in the last forty thousand years. In a similar manner we can see the process of collective insight at work with regard to the advancement of science since the seventeenth century, when the crucial step of scientific publication took place. Publishing the results of experiments and scientific theories in journals that were accessible to comparatively large numbers of people perhaps was *the* crucial step in the development of the scientific age. It allowed collective scientific insight to be applied to problems. . . . It is possible that a similar though more general step occurred some forty thousand years or so ago when the rapid rate of speech communication that is typical of the linguistic system of anatomically modern *Homo sapiens* became general throughout the hominid population. (pp. 327–8)

From here on new levels of social and cultural organisation are possible, which enabled our ancestors to oust the otherwise physically superior Neanderthals. Technology and art now developed rapidly, bringing us to our present extraordinary pass within a mere forty millennia.

The model is, in a sense, a two-stage one. Firstly, a phase of general cognitive enhancement associated with increasing tool-use and protocultural development. This involves no species-specific physiological adaptations, only an extension of features already in the primate repertoire – bipedalism, manual dexterity, and their concomitant neural substrates. This broad spectrum phase is maintained by a positive feedback between cultural and physical factors, more or less as in Holloway's model. The second phase possibly occurs in only one part of the post-*Homo erectus* lineage. This involves a highly species-specific set of modifications to the supralaryngeal tract, explicable only as an adaptation for elaborating the vocalisation ability of these forebears. From around 400,000 bp this branch diverges from the others. Finally this vocal-auditory channel – and the relevant neurological controls – become sufficiently refined for modern 'speech' to appear, and hominisation may be said to have become complete. Other *Homo* sub-species (of which we know only the Neanderthals for certain) cannot compete and become extinct around 32,000 bp.

Lieberman's reconstruction of the Neanderthal vocal tract has been the most controversial aspect of his empirical work and hinges on interpretations of the morphology of the base of the skull (the 'basicranial line' especially), and the scope of the tongue to modify the air column form during vocalisation. A modern human supralaryngeal system just could not fit in a Neanderthal without placing the pharynx absurdly low – as far down as the chest. Their level of phonological performance would, he believes, have been in the range found in human neonates. This would have been sufficient to provide some vocal language, as their cultural level seems to require, but they did not, Lieberman believes, possess a fully developed language. The very fact that *H.s.sapiens did* oust them, inspite of their ostensibly greater physical prowess, points to an advantage at the intellectual or cognitive level, probably related to social organisational skills. This strongly suggests language, and concomitant 'collective processing', as the factor in question (see also Krantz, 1980).

Lieberman's restatement and recapitulation of his case (originally proposed in Lieberman and Crelin, 1971) places the onus quite firmly on his critics (Falk, 1975; LeMay, 1975; and Du Brul, 1977 being the foremost), even bearing in mind Wind's aforementioned down-playing of the weight to be attached to this order of evidence. Whatever wider necessary conditions had to be met, the final achievement of modern language is therefore intimately bound up with a physiological restructuring of the vocal tract. The costs this entailed are themselves evidence of the advantages which such changes must have possessed in order to outweigh them. The prelude to this is a sustained elaboration of general cognitive abilities governed by neural substrates *not* localised in the 'speech areas', and it is these which make syntactical structuring possible.

I cannot see Lieberman's account as intrinsically incompatible with Holloway's, indeed both place weight on positive feedback systems involving culture, and in terms of brain anatomy, Lieberman does not deny an earlier enhancement of the vocal channel such as might account for the putative enlargements in the left hemisphere of areas now related to speech control evident in endocasts. It gibes less well with Falk's and Jerison's models and is very at odds indeed with gestural theories:

> . . . it is probable that these systems (i.e. modern sign-languages) work at a level comparable to vocal language only because we already

> have the neural substrate that evolved to structure vocal language. It is likely that our earliest hominid ancestors started from a syntactic and cognitive base that was only somewhat more advanced than that displayed by modern pongids. It is also likely that their communications were from the start vocal or that they at least made use of a wide range of vocal signals.
>
> . . . One of the selective advantages of fully vocal language is that it enhances our ability to use tools, carry things, and indeed any of the tasks that involve using our hands. Vocal language thus continues the trend set by upright bipedal locomotion. (p. 326)

Some aspects of Lieberman's model, notably the profound revolutionary sequelae of attaining 'collective insight' and the late timing of the appearance of modern speech, lend some weight to another model, the last which I will consider – that of Julian Jaynes (1976a, 1976b). This is perhaps the most heterodox and iconoclastic view yet proposed, being launched more or less simultaneously at the 1976 New York Academy of Sciences Symposium (Harnad et al., 1976) and in a controversial book, *The Origin of Consciousness in the Breakdown of the Bicameral Mind*, a best-seller in the US though it fared less well in the UK.

For Jaynes, language is a very recent development, owing its emergence to the peculiar conditions in which Ice-Age hominids found themselves, especially in northern regions. Only a dramatic change in environmental circumstances can, Jaynes argues, account for the success of an equally dramatic change in behaviour. Normally species are well adapted to their environments in a variety of complex and subtle counterbalancing ways, and it is highly unlikely that a major behavioural innovation would be other than disastrous.

In essence what Jaynes proposes is a six-stage model, the events in question beginning during the last Ice Age, and not being fully completed until as late as the Mesolithic. He does though consider that 'language developed somewhat differently and at a different pace in different ecologies and races' (1976a, p. 315). To summarise these stages briefly, indicating what they are rather than providing Jaynes's arguments for them, they are as follows:

(i) *Intentionalisation of vocalisation* This is more or less the same as Hewes's 'decontextualisation'. Vocalisation is brought under voluntary, cortical, control.

(ii) *Age of modifiers* Modification of cries by simple differentiation of intensity and way the final phoneme was uttered. (This lasts up to c. 46,000 BC). This constitutes a first step towards syntactic language, taken further by separating these final phonemes off as modifiers:

> Thus we might expect that a small lexicon of modifiers like near, far, at the river, behind the hill, etc. was developed in this way long before the specific referents themselves could be differentially indicated. (p. 317)

(iii) *Age of commands* 'Modifiers once separated from the cries they modified become commands' (ibid.). He sees this as especially useful in hunting, a view now considered highly dubious – hunting being generally a very silent activity except in certain expressive phases during the chase. Interrogation and negation also, he considers, begin here. This phase lasts down to about 25,000 BC. It is best conceived as involving the interpersonal elaboration of intonation to signal speaker-listener relationships, and also involves verbally mediated *self*-control.

(iv) *Life nouns* Only when these preliminaries are stabilised do we get the *noun*-type 'words', naming life-forms of particular salience to the group – prey animals and predators primarily. This coincides with the period in Europe, from 25,000 – 15,000 BC, when animal pictures and carvings make their debut.

(v) *Thing nouns* Extension from the previous phase, corresponding to the Magdelanian, with its expansion of technological objects – harpoons, bone carvings, spearheads, cordage, ornaments, etc.

(vi) *Names* Only in the Mesolithic do personal names become universal, though preceded by occasional naming of special individuals. One index of this is the elaboration of individuated grave-goods and burial markers.

Jaynes presents this only as 'a logical analysis to which a kind of commonality of the historical realities might approximate' (p. 319).

It must be mentioned in passing that Jaynes's paper (1976a) contains some straight errors, e.g. he has *H.habilis* alive at 400,000 bp and an interpretation of 'premesolithic' campsite locations as indicating a fear of water, shared with other primates. This latter is simply not so! (Terra Amata is one obvious counter-example.)

The reception of Jaynes's theory has been sceptical, but the timing is not perhaps quite so outrageous as it originally seemed, even if a Mesolithic date for proper-naming still strikes one as a bit steep! The consequences of language possession are a factor on which Jaynes places considerable stress:

> Words are of such huge moment in the life of men that the acquisition of them and the ability to organize them into sentences that convey meaning must have resulted in very real behavioural changes; and these changes must be reflected in the artifacts left behind. (ibid., p. 313)

> . . . I am making language not merely one among several of the remarkable evolutionary achievements of the late Pleistocene, but the very pioneer and promoter of the rest. (p. 322)
>
> If man had had words for these matters a million years earlier, I think it would have been inevitable that agriculture would have begun then instead of only around 8,000 B.C. (pp. 322–3)

This clearly raises a central issue which has not yet been properly evaluated. Are Jaynes's intuitions on this point sound? Would it be possible for a culture to survive for millennia with modern language but a world-view that did not require them to progress technologically beyond the Acheulean or Levalloisian? The supposition of widespread independent evolution of language is also problematical and would involve the whole issue of the timing of ethnic divergencies, touched on elsewhere (see Chapters 3 and 6). By and large Jaynes's ideas have been ignored by those centrally engaged on the topic (e.g. Lieberman (1984) makes no mention of him at all) – mainly because it is too clearly speculative, and because few are prepared to accept such a recent timing for the culmination of full linguistic competence. Additionally, hunter-gatherer tribes possess fully developed languages and proper names without ever having achieved agricultural or 'Mesolithic' cultural levels. Yet Lieberman's own account lends credence to the notion of an Upper Palaeolithic arrival of modern speech, and it is unlikely to have developed to maturity overnight. Jaynes's is one of the few attempts at evolutionary theorising so far to have emanated from within academic Psychology, and, if nothing else, Jaynes shows, both here and in his book, where he tackles the wider question of consciousness itself, how psychological data and concepts can be creatively deployed in this context.

The true course of the evolution of human language is thus still confused, but things are far clearer than they were two decades ago (see Table 5.2). By way of rounding off this topic I will summarise what appear to be the principal points which any future comprehensive theory on the matter must take into account, and the questions which need to be addressed.

(i) Relatively late appearance of full spoken language.
The weight of the evidence must surely now be read as pointing this way; Lieberman's physiological case backs up the logical and psychological arguments made by people as diverse as Hewes, Marshack and Jaynes.

(ii) The intensity of the social function of vocalisation.
The ontological roots of vocal behaviour in the attachment situation, plus the essentially social character of vocal behaviour suggest that one cannot explain it entirely, or even

TABLE 5.2 *Different views of the origin of language*

	Early/Late/Very Late[1]	*Cognitive/Social*	*Gestural/Vocal*
Falk, D.	Early	Cog.	Voc.
Hewes, G.	Early	Cog.	Gest.
Holloway, R.	Early	Social	Voc.
Jaynes, J.	Very Late	Cog.	Voc.
Jerison, H.	Early	Cog.	Voc.
Lieberman, P.	Late	Cog.	Voc.
Marshack, A.	Late	Social	Voc.
Parker S. & Gibson, K. R.	Early	Cog.[2]	Gest.

[1] Early = pre-*Homo erectus*, Late = post-Neanderthal, Very Late = post-Palaeolithic. This is timing for full acquisition of language, however it is not always clear what degree of language acquisition is envisaged to have occurred early by some of the 'early' theorists, thus these positions are not quite so distinguishable as the table implies.

[2] The cognitive and social functions would almost certainly be seen as inextricably interwoven by Parker and Gibson, but their Piagetian emphasis suggests that this is the more appropriate of the two categories to put them in for classification purposes.

predominantly, as a manual behaviour or gestural spin-off. Though differing in their scenarios otherwise, both Holloway and Marshack make this clear. The question is, however, how does full modern speech relate to vocalisation as such? I would feel that affective coding is pervasive, though not so much lexical as intonational.

(iii) Lieberman's 'general neural substrate' governing cognition as the basis for both verbal and manual operations would seem to outflank simple gestural arguments and carries the study of the neurological basis of language to a new stage.

(iv) The necessity for identifying the logical sequence governing language acquisition and development is vital. So far the most promising model is Parker and Gibson's Piagetian one, though Jaynes also provides one and others, not discussed here, appear in the literature (e.g. Fox, 1979). Such a model, establishing the logical priorities of language features, is essential if a coherent evolutionary account is to be achieved. Parker and Gibson (1982) are clearly right about this.

(v) The need for ethological analysis of the circumstances in which adaptations involving vocal-channel enhancement can occur. This is a central feature of Parker and Gibson's work and also of Jerison's theory. It is a tangled topic, suffering a surfeit of plausible possibilities, but is vital to sort out if both the initial origins and the subsequent timings of key evolutionary phases are to be clarified.

The questions which need addressing pertain to four issues:

(i) General linguistic capacity. Regardless of channel, there is the background question of the nature of our communication abilities as such. This is highlighted by Lieberman's general cognitive neural substrate idea.
(ii) Specific linguistic channels. Relationships between gesture and speech, timing of appearance of modern vocal apparatus and degree of mosaicism in its evolution.
(iii) Adaptational advantages of enhancing language. Many of the models looked at deal in depth with this question, and somehow their insights need to be integrated.
(iv) Stages involved in this enhancement. The Parker and Gibson account and others focus on this. What is necessary of course is that each phase in the process possesses its own adaptational value in its own right, not simply as a preliminary to a subsequent one. How acceptable are recapitulationist and comparative approaches to this? And what alternatives are there?

The sheer complexity and profundity of human language is such that it eludes any single user, and studies of its evolution have served to raise this fact to consciousness. As in the study of personality in mainstream Psychology, the grand theories each turn out to be not so much false as only parts of a vaster jigsaw beyond the individual's capacity to master singlehandedly. Gradually the different perspectives enable us to, as it were, triangulate the dimensions of the problem. Like the building of cathedrals, its solution is a multi-generation business.

D Conclusion

In tackling the cultural aspects of human evolution this chapter has looked at three main issues: the notion of social evolution itself, the problem of altruism, and the evolution of language.

It is quite evident that on none of these is there anything remotely resembling agreement, but the debates have nevertheless moved on considerably from where they were a decade or so ago, so one might even, optimistically, claim that they had progressed. Regarding social evolution, the great variety of positions will perhaps thin out somewhat in future, as a more sustained and serious study of the topic is mounted. It is likely that the empirical work of palaeoanthropologists and archaeologists will come to occupy a more central role than hitherto, with some ousting of grand theoretical systems of the Lumsden and Wilson variety. The necessity of including a 'cultural' dimension to account for behavioural evolution generally has been increasingly apparent in the work of many not otherwise primarily concerned with large-scale modelling of 'social evolution'. The nature-nurture controversy underlying much of the controversy in this area must surely now be declining in severity, though since its demise has been being confidently proclaimed since the 1920s one hesitates to write its epitaph even yet. The understanding of social evolution can surely only benefit from a fuller incorporation of psychological perspective.

Altruism has received a vast amount of attention, but proves on examination to be an extraordinary muddle. This stems principally from the attempt by those in the area to give the term a technical definition which is itself incoherent. The paradigm examples of 'Altruism' in the literature (death in battle, rescuing from drowning) also turn out to be highly misleading. The much advertised 'Altruism paradox' is in fact a myth.

Finally, the area where most success is evident, is probably the evolution of language. Various lines of empirical research have been opened up here such as comparative studies of speech physiology, non-human linguistic performance and speech areas of the brain. The data stemming from these do provide a common reference point for those studying the topic, something lacking in the case of attempts at dealing with social evolution as a whole. At the same time the conceptualisation of the issues involved in studying language – an appreciation, at heart, of the subtlety of the phenomenon – is also developing in a complementary fashion.

CHAPTER 6
From *Erectus* to *Sapiens*

Introduction

This final chapter concerns the course of events from the appearance, at least 1½ myr, and perhaps half a million years before that, of *Homo erectus* to the arrival of *Homo sapiens sapiens*. Much of this has been touched on already, and the basic data was reviewed in Chapter 3. The present task is to pull together the various themes relating to three principal issues around which the chapter will be organised. Part A will elaborate on the problem of *H.erectus* and its relationship to modern humans, along with what we know of their life-style. Part B will cover the current state of play in the Neanderthal debate and again, what we know of how they lived. In Part C we turn to *H.s.sapiens*, considering our spread to America and Australia, and then the rather specific question of the problem of interpreting European Upper Palaeolithic cave art. In addition to these issues it occurred to me that it would be useful to pull together the earliest dates for a range of behavioural and cultural categories such as domestication of animals, use and control of fire and so forth, even though many of these have been mentioned in passing already.

Before embarking on this, a couple of preliminary points must be made.

(a) The bulk of the data for the post-*H.erectus* period is European or Middle Eastern, partly for historical reasons. This gives a 'Euro-centred' cast to the picture which it almost certainly does not warrant, although things are not quite so bad as they were when two English popular writers on *Everyday Life in the Stone Age* could urge on their readers that they had 'reason to be proud that the British Commonwealth has provided three of the world's most celebrated prehistoric men' (Quennell and Quennell, 1926)! Nevertheless there are reasons for supposing this region to have been genuinely crucial in the later stages of hominisation. It has been subjected to continuous climatic and attendant geographical change since the first of the Quaternary Ice Ages. As Figure 3.1 indicated, there have been four major (and numerous minor) oscillations between glaciations and 'interstadials', and the ecological effects of these are increasingly well understood. The last interstadial and subsequent Ice Age (the Würm or Devensian) appear to have been the most momentous from a human evolutionary viewpoint. As far as our ancestors were concerned, such fluctuations required not only that they manage to cope with severely cold climatic conditions, but that they undertake a continuous, if slow, adjustment to the changing nature of their food supplies. This seems to have led to increasing regional variation in subsistence practices as time went on. The situation in Africa during the Ice Ages and interstadials is less well known due to lack of information from many areas, but it seems likely that climatic fluctuations were not so dramatic or far-reaching, resulting certainly in changes in the relative sizes of different biotypic regions but not their region-wide appearance or disappearance. Cautious though one must be, it is possible that crucial later changes in life-style and technological innovation were accomplished outside Africa.

(b) The present state of research in many areas under consideration here is in a greater state of flux now than at any time since the Second World War. This is due to a combination of (i) new data, (ii) new methods (e.g. of microscopic analysis of archaeological and fossil material) and (iii) arising from this, penetrating critiques of site-interpretations which have been accepted as orthodoxy for many years. In some ways *clear* pictures are harder to paint now than say fifteen years ago. This is of course natural in phases when new data and methods have to be integrated.

Readers requiring fuller general accounts of the European Palaeolithic, particularly the Neanderthals, cave art and how our ancestors subsisted, might look at Pfeiffer (1982) or Hadingham (1980).

A *Homo erectus* and the transition to *Homo sapiens*

What was *Homo erectus* – the most successful of our forebears, who arrives rather hazily on the scene in East Africa sometime between 2 and 1½ million years ago and then finally disappears by 100,000 BP? Latest evidence suggests that males at least could reach 1.8m in height, but had smaller brains, on average, than ours and thicker elongated skulls with a sagittal ridge and a nuchal 'torus' or ridge of bone at the occiput. They had not, though, evolved the distinctive brow-ridges and prognathous profile of the later Neanderthals, and may have resembled us more physiognomically than did the Neanderthals. They range from East Africa to China and Java and probably southern Europe. But already we oversimplify. The fossils are few and widely scattered in time as well as place. These sample quite different populations, whose morphological features seem to vary over a considerable range, in which climatic and local ecological factors probably played a major part. There is great variety in the palatal (mouth) region (Bilsborough and Wood, 1984) and in skull thickness, and the status of some early specimens is especially obscure (KNM-ER 3733, KNM-ER 3883). While *H.erectus* was undoubtedly bipedal, there are differences in femur morphology from *H.sapiens* in location of maximum thickness (Kennedy, 1983) suggesting the possibility of associated differences in style of gait. The principal *H.erectus* fossils are listed in Appendix A.

Granted that *H.erectus* represents a single taxon, albeit somewhat polytypic, as befits a widespread species overly nearly 2 million years, how did it live? Had we been asking this question even ten years ago a fairly ready answer would have been forthcoming, a 'gestalt' of its life-style derived primarily from the Peking Man site at Zhoukoudian (Choukoutien) and a few other sparse clues. The FLK site at Olduvai, though associated with *H.habilis* rather than *H.erectus* was also interpreted in such a way as to suggest that a form of home-based life-style had already become established prior to *H.erectus*, thereby somewhat constraining the possibilities for *H.erectus*, who would almost automatically be assumed to have adopted this and further developed it. Our ability to impose a meaningful structure on minimal data where human behaviour is concerned is remarkable, and nowhere more in evidence than here. The earliest writers on Zhoukoudian, Breuil, Weidenreich and Chardin being the foremost, conjured up a fairly detailed picture of life in the cave which was almost ritually recited from then on. In this picture *H.erectus* is depicted as (a) a cave-dweller (since the finds were made in a cave), (b) a hunter of deer (since deer bones were particularly numerous), (c) a fire user (since there appeared to be layers of ash, interpreted as hearth-sites), (d) a cannibal (on the basis of the nature of the damage to the base of the skull and face, combined with lack of post-cranial remains) and (e) a user of bone tools (on the basis of some features of the damage present in the animal bones). Binford (1981) raised some initial doubts about these interpretations and in a more recent paper, co-authored with a Chinese researcher (Binford and Ho, 1985) an attack is mounted on every single one of these conclusions, though fire-use is not entirely ruled out. Without going into their case in detail, some of its main lines should be repeated here as they are rather salutary. Of particular interest is the belief that the Peking Man fossils showed evidence of cannibalism. This is still routinely mentioned in textbooks and popular summaries. Yet the matter rests on a dubious interpretation of two features of the fossils: firstly, the lack of a cranial base and facial area: secondly, the absence of other parts of the anatomy than the skull. The former was thought to indicate that the skulls had been broken in such a fashion that the brains could be extracted, while the latter suggested that they had been brought into the cave separately from the rest of the skeleton. Ritual cerebrophagy by the camp-fire seemed the logical conclusion. But these kinds of damage are now known to be commonplace and 'understandable in terms of a number of taphonomic alternatives' (Binford and Ho, 1985, p. 8), 'hominid skulls recovered from secondary deposits . . . typically lack faces and parts of the skull base' (ibid., p. 9).

> Having reviewed the 'evidence' *we are convinced that there is no support for the cannibalism interpretation*. All such interpretations of the facts appear dependent upon a poor understanding of taphonomy and the modifications that bones can suffer after an animal's death and during the inclusion of bones in geological deposits. (ibid., p. 11, italics in original)

The bias in favour of cranial remains, thought unusual by the original researchers and construed by Breuil as evidence of symbolically motivated behaviour, turns out to be commonplace too; the Zhoukoudian figure of 94% cranial remains comparing with 91% and 95% for Sterkfontein and Swartkrans respectively. The upshot is that the evidence for *H.erectus* cannibalism in China seems, for the time being at any rate, to have evaporated. Imaginative early interpretations made in ignorance of what was taphonomically normal became enshrined in the literature and persisted long after other finds, with similar features, had shown the supposedly 'unnatural' features of the Peking Man fossil assemblage to be unremarkable.

But Binford and Ho go further than this, and argue in detail that the assumptions regarding 'association' between the different kinds of find made in the cave, which enabled earlier workers to create such a coherent image of *H. erectus*'s life-style therein, are generally ill-founded. Unravelling with consummate patience the stratigraphic picture as variously reported at different times during the excavations, a number of factors come to light, for example the accumulation of deer bones – evidence supposedly of the *H.erectus* diet – was introduced 'to the site *when no convincing evidence of hominid use of the site exists*', though good evidence that it was being used by wolves and hyaenas. The hominid remains and deer remains are from different strata. The ash layers (though there are doubts even about whether *all* these are in fact ash), sites it was thought of the fires by which venison-munching forebears warmed themselves are similarly problematical, 'the hominid remains are not covarying with "ash-layer" development since the hominid remains are not generally found within the "ash-layers" or the breccia yielding the hominid remains in the western excavations' (ibid., p. 46, all emphasised in the original).

By the time Binford and Ho have finished, the hominid fossils were as likely to have been brought into the cave by scavenging carnivores as on their own two feet, some of the ash-layers are largely owl-droppings, and there is no discernible patterning in the distribution of tools and faunal remains to suggest any 'cultural' level of life-style being carried on at all. They claim that all we are seeing here are 'the archaeological remains of a non-cultural form of adaptation which is strongly tool-assisted' (p. 56). These negative conclusions, allowing us to go no further really than conceding *H.erectus*'s occasional presence in the cave, possible fire-use and possession of stone tools, does, though, leave us unhampered by certain presuppositions which have entered our image of these people as a result of the earlier analyses. Firstly, the question of whether even by this time (the fossils dating principally from around ½–¼ myr, though dating is still a problem here) hunting had become a major part of the hominid behavioural repertoire is left open; secondly, it reminds us that at this period hominids were but a part of a vaster ecological system and that mere co-occurrence of finds at a site with hominid remains does not mean that there is any necessary connection between them in terms of organised hominid activity. It is premature even to assume that such sites represent 'living-floors' or 'home-bases', which brings us to Potts's (1984) re-evaluation of the Olduvai Bed 1 FLK and FLK N sites. The fact that these *were* for so long interpreted in these ways – as living floors and probable home-bases – meant that later *H.erectus* sites were almost automatically granted such status, as they would not have been likely to have been *less* modern in life-style than *H.habilis*.

It will be recalled that some of the influential behavioural-evolution models referred to in Chapter 4 had as a central feature the notion of a 'home-base' from which males in particular roamed scavenging and gathering (and possibly hunting) while females and young stayed more or less put. The Olduvai sites, also discussed in Chapter 4 when reviewing Isaac's analysis of early meat-eating, have been prime candidates for this, lending such models genuine empirical support. Potts, though, is in this paper claiming a rather weaker form of home-base usage – the first moves towards a home-base rather a home-base in the full sense. Again the case is argued from more sophisticated understanding of taphonomy, wear-analysis and ecological processes than the original investigators possessed. As Potts says, 'the sharing of food at home bases is considered by some to be a crucial expression and perhaps the earliest one, of human social reciprocity' (p. 340). It provided an image of continuity between *Homo habilis* and ourselves, 'the hunter-gatherer adaptation would seem to represent a natural condition with deep roots in our evolutionary past' (ibid.). Taphonomic analysis of the six Olduvai Bed 1 sites in question does not, though, bear this picture out. There are major differences from the kind of traces we would expect hunter-gatherers of anything like the modern kind to have left. Among these are the level of carnivore activity evidenced by tooth-marks on the bones, the relatively incomplete processing of the bones compared to current hunter-gatherer levels of thoroughness (e.g. no splitting of bones to extract marrow) and the nature of site occupation. This last difference emerged from studies of the degrees of weathering to which bones had been exposed. These indicated that sites were used regularly for periods up to ten years, in contrast with modern hunter-gatherer sites which are used for relatively short periods (measured in months) and then abandoned for a long time (due primarily to flies infesting accumulated refuse). 'The available evidence suggests that hominids would have minimized the time spent at these sites, rather than used them as the primary focus of social activity' (pp. 344–5).

How then did *H.habilis* use these sites? Potts introduces a new model which presents a life-style now nowhere evident. Using computer analogs he shows how

> In almost every simulation the production and simultaneous use of multiple caches of stone tools – rather than a single home base – was an energetically efficient way to use both stone and food resources at the same time, for example in the processing of meat with stone tools. (p. 345)

Dismembered limbs of prey or, more likely, of scavenged carcasses, could be brought rapidly to the nearest cache for speedy processing before larger carnivores were attracted, the hominids then leaving the site having either consumed the meat there or taking the now-filleted flesh with them. Their actual base areas need not have differed from sleeping areas typical of most other primates. On the famous DK site stone circle, generally interpreted as the earliest known shelter or wall Potts is uncertain but points out that the level concerned shows 'the most dramatic taphonomic effects' and that the occurrence of the ring as a 'possible result of radial action of tree roots' cannot be ruled out yet (p. 345). The intensity of competition between the, as yet non-fire-using, hominids and other carnivores emerges clearly from Potts's report and the 'stone-cache' model evokes a plausible picture of hominid life in a world they had as yet failed to dominate. It also integrates the data quite economically. For example the stones and artefacts found at these sites range from utilised tools to raw materials, transported from up to 3 kilometres away. Tool-use and scavenging have thus been co-ordinated, via the stone cache technique, in a unique fashion which would imply levels of planning and territorial 'mapping' well beyond non-human primate capabilities. Such sites represent, in Potts's view, areas to which diverse resources (stone, meat) were brought in a way which makes them *antecedents* of the home-bases which would later evolve, when carnivore competition shifted more in the hominids' favour and fire-use was added to their repertoire. But 'we simply do not know as much as we envisioned when it seemed appropriate to extrapolate a human hunter-gatherer model back 2 million years' (p. 347).

Though *H.erectus* may indeed have taken that step towards a 'home-base' centred life-style, even the evidence from their final phase at Zhoukoudien around 1/4 million years ago discloses very little either in confirmation or negation of the matter. We are left with only the dimly discernible picture of a slow improvement in artefact manufacture coupled with improving cognitive competence of the sort described by Wynn, and Parker and Gibson, as demonstrated in the development of the Acheulean handaxe industries and some Far Eastern flake industries (e.g. at Zhoukoudien). We have glimpses only of this probably quite large ancestor (if the new WLT 15,000 is representative) living some kind of semi-nomadic, small-group life-style and spreading across the Eurasian landmass. Of course a single generation might move very little, routinely circulating round a home range with which it is familiar and the resources of which it has learned to exploit, possibly in a seasonal fashion in higher latitudes. Although early fire-use evidence is equivocal (see below, Part D), we can be reasonably sure that it was established by about a million years ago. Whether meat derived from hunting or scavenging or in what proportions from each is in dispute, but the tendency over the last ten years or so has been to bring the advent of hunting ever later, at least for large mammals, possibly even after *H.erectus* (Pilbeam, 1986). The relative contribution of meat to the diet is also unknown (see the discussion in Chapter 4).

As far as social organisation, mating patterns and sex-roles go, we are virtually in the dark. Given its eventual geographical range, it is likely that *H.erectus* evolved a degree of behavioural as well as morphological variation, though this should not be confused with cultural differences. If mating involved some sort of exogamous exchange between neighbouring groups, a genetic continuum across the species' range could have been maintained, preventing complete severance of populations and consequent speciation. There is, we must stress, no evidence as yet for *H.erectus* being a cultural animal, other than tool-use itself. However the very homogeneity and conservatism of tool-making techniques during this period suggests that it as yet retained the character of a general species-specific behaviour, not subject to cultural level processes of stylistic differentiation, formal classification and fairly rapid change. Terra Amata, at 0.3 myr, does contain more convincing evidence and falls within the *H.erectus* time-span, but already 'Archaic *Homo sapiens*' is present in Europe (at Arago for instance, currently estimated at around 0.35–0.4 myr (Gowlett, 1984)), and there is no reason to ascribe Terra Amata to *H.erectus*.

The big enigma we now confront is 'what happened to *H.erectus*?' A number of difficulties arise in unravelling the *H.erectus* to *H.sapiens* transition, the upshot being that the following broad positions each have adherents at the present time:

(a) Asian *H.erectus* split from African *H.erectus* and eventually died out (Andrews, 1984). The transition would then have occurred within the African lineage or its closer European and Middle Eastern branches.
(b) Asian *H.erectus* evolved into Asian *H.sapiens sapiens in situ* and possibly even gave rise to the *H.sapiens* who, moving into Europe, ousted the Neanderthals, who had evolved from *H.erectus* quite separately. This view goes back to the pre-World War II US anthropologist Osborne, a great advocate of 'Asian origins'. It has more recently been powerfully argued by Aigner (1978) and Chinese workers (e.g. Wu and Xingren, 1983). They base this on the fact that Asian *H.erectus* and modern Asian *H.s.sapiens* share what appear to be certain derived characteristics, notably 'shovel incisors' and cheek-bone morphology. W. W. Howells (1983) maintains, contrary to the notion that the *Homo* populations had already split at this time, that ' . . . all modern men have a common ancestor who was later than Peking Man' (p. 299).
(c) *H.erectus* evolved into *H.sapiens* on a broad front. This is hard to visualise as it runs counter to current models of the speciation process and resembles the now-discredited 'orthogenesis' phenomenon. It also suggests extremely early timing of ethnic splits – too far back in the view of most palaeoanthropologists to be consistent with current levels of similarity among humans in regard to language, interfertility, and morphological variability. On the other hand the fossil evidence is of the presence of *Homo* forms apparently intermediate between *H.erectus* and *H.sapiens* in widely different parts of the *H.erectus* range, from Africa to Europe and China (the 'Archaic *H.sapiens*' forms).
(d) *H.sapiens* evolved from *H.erectus* in Africa, or Near East (south west Asia) and moved into Europe in at least two waves, the first as 'Neanderthal', the second as 'Modern' varieties of *H.sapiens*. These waves would follow, progressively ousting the previous *H.erectus* populations. (See also Appendix C.)

The course of events is indeed difficult to disentangle within the models of the nature of evolutionary change currently accepted. Neither a strange synchronous parallel evolution of *Erectus* into *Sapiens* populations across its vast range nor a two (or more)-fold radiative 'pulse' from Africa of first *Erectus*, then Neanderthals and then Moderns, is entirely adequate. The latter would imply that the common derived characteristics of Chinese *H.erectus* and modern Far Eastern *H.sapiens* arose from convergent evolution rather than common descent, while the 'Asian origin' model would seem to entail a 'reversal' – a loss of these derived characteristics and return to a less specialised condition in the populations which spread westward. This is not perhaps impossible, but it would require very firm evidence to convince most evolutionary theorists.

The underlying difficulty is comprehensively integrating the Far Eastern and Asian course of events with the African, Near Eastern and European one. This is not rendered easier by the fact that the actual fossil material is disproportionately European, an arena possibly fairly peripheral to the main action. Africa has yielded several skulls from around the 100,000–200,000-year-old mark, which are difficult to interrelate – some of which seem to resemble skulls of later African ethnic groups, while others are Neanderthaloid or Modern to greater or lesser degrees though retaining *H.erectus* features too (the Bodo, Florisbad, Broken Hill and Border Cave skulls are among these). Dating problems abound. The Border Cave specimen being ascribed dates anywhere between 40,000 and 100,000 years old, while the Greek Petralona skull, considered as a late *H.erectus* sub-species immediately preceding *H.sapiens* by Poulianos (Poulianos, 1982), ranges wildly from up to 0.8 myr (Poulianos, 1984) down to as recent as 0.16–0.24 (Hennig et al., 1981), (see Wintle and Jacobs (1982) and Stringer (1983) for full reviews of the chaos regarding Petralona dating, reviews of which Poulianos (1984) seems to be oblivious). Most current authorities (e.g. Pilbeam, 1986) are reluctant to push *H.s.sapiens* back much beyond 40,000 years and favour an African origin. The various 'archaic' and 'Neanderthal' finds outside Africa would then be products of earlier radiations from Africa or evolved *in situ* from *H.erectus*, only to be ousted and replaced by *H.sapiens* of the modern type. The continuing difficulty in classifying the Mt Carmel fossils, especially the Skhūl ones (see Plate 7), is the only indication that some degree of continuity between Neanderthals and modern *sapiens* forms might have occurred somewhere, possibly due to genetic mixing with the incoming 'modern' wave.

It is, frankly, fruitless at the present time to devote much more space to the precise nature of the transition from *H.erectus* to *H.sapiens*, given the genuine confusion surrounding the topic and the constant flux of the debate (Pilbeam, 1986). What we *do* know is that between the demise of *H.erectus* and the arrival of modern *H.sapiens* the first clearly cultural phase of human evolution is discernible, associated in Europe and south west Asia with the Neanderthals. How and why modern *H.sapiens* replaced Neanderthals is a matter for pure speculation, though we have touched elsewhere on possible factors (Chapter 3) and Lieberman's view (Chapter 5) is as psychologically plausible as any. If there is a clear physiological hiatus between the two forms, the cultural discontinuity is less clear. The first Modern *H.sapiens* stone tools develop from the techniques used by Neanderthals; there is some overlapping of industries and other 'modern' cultural developments are adumbrated in the Neanderthal record. How then did Neanderthals live?

B The Neanderthals

The amount of data – and speculation – on the Neanderthals is vast and precludes anything approaching a complete survey here. My main concern is to sketch the behavioural and cultural level of Neanderthal life as far as this is knowable.

At the outset it is worth emphasising that the gulf between *H.erectus* and Neanderthal life-styles appears to be immense. During the 'Archaic *H.sapiens*' phase, the tenuous fossil chain which links *H.erectus* to both Neanderthals and *H.sapiens sapiens*, dramatic changes in the complexity of hominid life had evidently occurred, including the first glimmerings of something like 'culture'. By the time, around 70,000 BP, that Neanderthals are well established in Europe this cultural dimension is, in some respects, clearly apparent. The stone-tool evidence is striking; after the prolonged Acheulean phase during which progress had been at best lethargic, the tool-kit expands rapidly and new manufacturing techniques are adopted, producing a range of flake-tools which achieves baffling complexity. Although there is no unambiguous evidence that Neanderthals alone 'invented' the 'Mousterian' – on the contrary, the correlation between styles of tool-manufacture and ethnic or sub-species identity is known to be imprecise – there is no doubt that they were its principal masters in Europe.

The beginnings of Neanderthal dominance in Europe coincide with the last interglacial, the milder climatic phase between the last two glaciations, but their real heyday came in the last glaciation itself, the last 'Ice Age'. Contrary to what might be naively assumed, the glaciation facilitated rather than hampered expansion, for not only did it drastically lower sea-levels, it also reduced the pre-existing dense coniferous forest cover in northern Europe. The open grassland steppe which replaced it supported far more animal life than the impenetrable pines had done, providing the conditions in which mammoth herds and other 'megafauna' as well as deer and cattle species, and their predators, could multiply. Among the predators were the Neanderthals, spreading across the north European plain into Siberia. Between the Alpine and Scandinavian glaciers an ice-free belt linked the south of England to central Asia, and excepting rivers, there was land from the Downs to the (albeit shrunken) Mediterranean.

This continuous landmass provided a wide range of habitats and in the reign of the Neanderthals from before 70,000 BP to around 35,000 BP high degrees of variation both in life-style, artefact assemblages and, indeed, morphology, are apparent. With *H.erectus* we were still involved in trying to identify a general species-specific mode of living, with Neanderthals we feel we have moved to a stage where such a mode of living has become overlain with major regional cultural variations. 'Hominisation' was perhaps not yet complete (though some would make the gap between them and us very small indeed in this respect). Neanderthal life was not directly comparable to that of any recorded human group of recent times. Numerous features of their life have been identified or may be reasonably inferred from the abundant archaeological evidence garnered in the last few decades from Shanidar, in Iraq, to Moldova in Russia, and above all, in France. The following summary is gleaned from both primary sources such as the various books and papers by Solecki, Trinkaus and Stringer and secondary reviews such as Pfeiffer (1982) and Hadingham (1980). For psychologists the principal features of Neanderthal life to be noted are these:

1 An awareness of death. This is evidenced by (a) various kinds of special treatment of the dead (e.g. the 'flower burial' at Shanidar referred to in Chapter 3) and (b) some evidence of cannibalism (rather securer than the Peking Man evidence attacked by Binford and Ho). Awareness of death intense enough to generate this kind of behaviour must be assumed to be accompanied by, and express, an equally intense emotional life and consciousness of time. Other species, notably higher primates, elephants and cetaceans, also show grief and some degree of concern for the recently dead, but ritual burials are unique to humans as far as we know. Ritual cannibalism is exclusively human and is generally done in order to acquire the psychological qualities of the deceased, such as bravery or wisdom.

2 A new kind of psychological involvement with other species is also in evidence. This now requires some 'symbolic' or 'non-utilitarian' (not related to basic biological organic needs) expression. Although, characteristically, Binford (1981) has raised doubts about some of the long-cited evidence here, many archaeologists have claimed that collections of cave-bear skulls, and even a cave-bear burial (at Rigordeau, in the Dordogne) have been identified. Aside from these there are various mammoth-bone and mammoth-tooth artefacts of an apparently 'non-functional' kind known. It is perhaps pertinent to point out here that, as we all know, wearing 'fancy dress' can have a profound effect on behaviour. The impact on a human consciousness of wearing animal skins – mammoth, cave-bear, stag, or whatever – could have been equally profound, being experienced perhaps as a 'possession' by the identity of the animal concerned. This is commonplace enough in modern anthropology (e.g. the 'leopard men' cults in parts of West Africa). Retrospectively ascribing this to Neanderthals is going beyond the evidence, strictly speaking, but there is just

about enough evidence to suggest that a degree of psychological involvement with other species at this level had at any rate begun to develop.

3 The existence within the Middle Palaeolithic of tribal or cultural differences was espoused very strongly by Bordes (see Chapter 3), whose knowledge of the French Mousterian industries is unsurpassed. Although Binford's scepticism and the work of British archaeologist P. A. Mellars have somewhat weakened this hypothesis, the current balance of opinion seems to be that in some areas at least, e.g. southern France, distinct groupings, perhaps territorial in character, did occur. In any case, the extraordinary mammoth-bone dwellings of Moldova and other Russian sites prove that Neanderthals could respond to different environments by adapting their life-style and behaviour in novel directions. The difficulty is differentiating the functional from the cultural, and temporal 'evolutionary' changes from geographical or territorial cultural differences. Again, though not proven beyond doubt, the evidence suggests that rudimentary levels of cultural, tribal organisation operated among the Neanderthals.

4 From the 'archaic' sites of Terra Amata and Peche de l'Aze onwards there is evidence of the use of pigmentation and bone engraving. Ochre and manganese dioxide pigments are found at Neanderthal sites but there is no indication that these were used for painting walls. Jewellery of any sort is also absent until the very last days of the Neanderthals (when they seem to have been assimilating new modes of tool-manufacture as well from those who would replace them). The obvious conclusion is that they engaged in some form of body-art or body-painting. Although confirmation of this is unlikely ever to be found, they do appear on occasion to have engaged in artificial skull deformation – something common among moderns – which is strongly suggestive of treatment of the body as art-object. The earliest 'modern' 'art' objects from Vogelherd (30,000 BP+) show fully developed abilities to carve and draw, and it is at least possible that the roots of such behaviour lie in the mysterious non-figurative bone-engravings of which a handful have survived from the preceding Neanderthal era (see below pp. 306–9).

5 Given the environments in which they lived, and the archaeological evidence itself, we may reasonably assume that Neanderthals used fire regularly, had some form of clothing (although no needles indicative of stitching have come to light), and were capable of erecting structures using a wide range of materials. In short they had achieved a fairly high level of domesticity.

6 They also developed seasonal cycles of behaviour. This is especially evident from the L'Hortus cave in Languedoc in which 20,000 years of occupation can be tracked. This was explored by de Lumley (1972). Here

it is possible to identify the seasons at which Neanderthals visited the cave and, furthermore, to see how the pattern of usage changed during the course of its occupation. Between 55,000 BC and 35,000 BC, Hadingham writes, the cave moved from being an occasional butchery site, used between January and March, to a summer site where a broad range of domestic activities were undertaken. Yet the flint tools remain unchanged in style, all being 'Typical Mousterian' (Hadingham, 1980, pp. 50–3). In other parts of their range, Neanderthals would undoubtedly have been nomadically following the herds of the animals on which they preyed during the annual migrations of these species, particularly across the great North European plains. Some researchers have managed to tease out a surprisingly fine-grained picture of Middle Palaeolithic hunting practices for certain areas, for example Gamble (1979).

7 The picture of a highly 'domestic' life-style is further enhanced by the nature of some of the fossil finds themselves. At La Ferrasie in particular we have what is frequently described as a 'cemetery' of a family-sized domestic group, with age-group and sex distributions much as would be expected. The image is further heightened by the fact that many of the Neanderthal fossils show signs of injuries, often severe in nature, from which the sufferer subsequently recovered, as well as long-standing disabilities (e.g. an atrophied arm in one case from Shanidar). Not only does this indicate that Neanderthal life was dangerous, it also implies high levels of caretaking by other group members of the sick and disabled. From this one is moved to conclude, if cautiously (a) that the social value of individuals rested on more than simply physical prowess, but involved affectional bonds and appreciation of the value of experience too, and (b) that, if Neanderthal social organisation was around relatively small domestic family-sized units, loss of individual members could have catastrophic consequences and had to be averted at all costs. (Of course conflict may also have occurred, but there is no evidence of this. One individual from Skhūl had incurred a spear wound in the leg, but this is not usually considered to be a Neanderthal.)

8 Returning to the stone tools, Bordes identified 60 basic types of Mousterian tool. The Mousterian emerges from the Acheulean, and the earliest phase is the 'Mousterian of Acheulean Tradition' (see Chapter 3). It must be stressed again, though, that there is no absolute dividing line at either end of the time-span, between the Mousterian and either its predecessors or successors. At the upper end the Neanderthals, in some areas at least, appear to have moved to a more leptolithic style, creating the 'Chatelperronian' industry intermediate between the Mousterian and Aurignacian, e.g. in the Reindeer Cave at Arcy-sur-Cure. This shift, once thought to demarcate Neanderthal occupation from modern, now appears to be more gradual than is compatible with such a simple 'replacement'

model. The find at St-Césaire (Levèque and Vandermeersch, 1981) has also raised considerable difficulties for the 'sudden replacement' theory, for here a Neanderthal skeleton was discovered with a Chatelperronian industry as late as 31,000 BP. Conversely there is evidence from the Near East of more anatomically modern people of this period using Mousterian tools. Clearly techniques of tool-use and manufacture could spread widely and quickly in the Middle Palaeolithic and Upper Palaeolithic worlds, even though some groups devised variations which are their 'hallmark'.

The stone-tool evidence is in some respects quite confusing because innovation and conservatism seem to co-exist alongside one another for millennia, not only in different regions but even at the same site. If one adopts a quasi-evolutionary model of tool development one might expect that in the absence of any clear *functional* superiority of one style over another, there would be no reason why either should die out. So far as I know no one has invoked *individual* differences to account for the variety, but from a psychological point of view this ought not to be ruled out; people maybe simply varied in the kinds of tool they preferred making or using, and selected from the wide cultural repertoire accordingly. Such preferences might have operated at a family level, too. In all areas of human productivity there is a range of individual differences pertaining to such factors as levels of skill, output, and taste, and Mousterian stone tool-making would not be an exception. The task is to differentiate between this and broader cultural and functional factors as determinants of the total variability such objects display.

9 The hunting of megafauna such as mammoth clearly involves co-ordination among a fairly large number of people, however it was done; whether driving the animal into a swamp or pit where it could be killed with spears, or chasing a herd off a cliff. Analysis of the quarry of Mousterian hunters from the bones found at various sites shows considerable variation from Western Europe to the Caucasus. Sites vary both in the predominant quarry (bison, mammoth, horse, reindeer, for example) and in the number of different species represented. Some finds comprise small numbers of many different animals, others contain only one or two. These differences presumably reflect a complex interaction of seasonal, regional and cultural-preference factors. Though meat was an important dietary factor, and access to it apparently a major determinant of social organisation and life-style, the relative weights of meat and non-meat foods in the total nutrition intake cannot be determined, and in any case would have varied from the northern plains, meat-rich but fairly low in fruits, nuts, and vegetables, to the southern French and Italian regions.

All writers make the point that the situation during the last Ice Age has no current equivalent. For most of the European area we must envisage a combination of colder, drier climate with an annual seasonal cycle of the present kind as far as changing lengths of daylight is concerned. This renders it very different from the Arctic and sub-Arctic conditions in northern Canada, Lapland and Siberia which once tended to be used as a frame of reference.

Having said all this, the impression could easily be given that the Neanderthal cultural repertoire and level of technical competence was comparable to that of some modern or recent tribal societies such as the Tasaday or Tasmanians. Indeed there are writers who would endorse this. There seem to be opposing underlying tendencies to either maximise or minimise the Neanderthal–Modern difference. 'Maximisers' would tend to view Neanderthals as an evolutionary side-branch which became extinct, being replaced with little or no interbreeding by modern *H.sapiens*. 'Minimisers' would, at their most extreme, countenance direct evolution of Neanderthals into Moderns, or at least see the replacement as gradual with some degree of genetic mixing, especially in the Near East. The psychological roots of this polarisation are obscure. Minimisers perhaps fear pandering to already inflated notions of modern human uniqueness, while maximisers wish to avoid the pitfalls of anthropomorphic projection – both imperatives with which one can sympathise. The fairest procedure would be to enumerate those areas in which there does seem to be a likelihood of a major difference without being too dogmatic about the matter, and bearing in mind that, weak though their case now appears to be, there are palaeoanthropologists, such as Wolpoff and Brace, who continue to espouse a 'direct descent' position.

1 If Lieberman is right, the most important single difference would be the lack, among the Neanderthals, of a fully developed phonetic and grammatically structured language. The evidence for this, to briefly recapitulate, is two-fold: the impossibility of fitting a modern larynx into a Neanderthal anatomy (plus the inability of a Neanderthal naso-pharyngeal cavity to produce the full range of sounds now used), and secondly, the case made by both Lieberman and Jaynes that the consequences of possession of modern language are so dramatic that they introduce a dynamic into human affairs of a kind absent from Neanderthal evidence (see Chapter 5). The notion that language facilitated, in modern humans, a level of 'collective processing' beyond the level available to Neanderthals (thereby giving them the edge in the competition between the two) also carries considerable *prima facie* weight and requires further exploration. But how much 'collective processing' did Neanderthal mammoth-hunting entail?

2 Notwithstanding the riddles of the transition period itself, the difference between Mousterian tools and the leptolithic industries which followed them is considerable. More significantly though, after the

Mousterian we find a much higher level of integration between stone- and wood-working (and other raw materials). New techniques such as hafting and manufacture of composite tools become apparent. Bone and ivory occur more regularly. The post-Mousterian industries display a rapid exploration, in short, of the potentials of raw materials other than stone both for tool-making and for non-functional objects such as jewellery. Writers such as Conkey (1980) would see this as closely related to symbolising capacities, and thus one might consider it as further, indirect, evidence of the advent of language. On this reading of the evidence, modern language would permit the regular labelling, and referring to, of a vastly greater number of environmental phenomena than hitherto, a massively enlarged *classification* system. This would be a necessary accompaniment for technologies requiring the bringing together of different raw-materials and their processing in complex ways. How great a gap there is between Neanderthals and the first moderns in this respect is, nonetheless, unclear. The lack of non-lithic material could be largely a function of time and, in the case of wood, if not bone, this is almost certainly a major factor.

Perhaps this is an appropriate point to reiterate something said earlier in the book, that what is crucial in innovation is not the average level of ability in the population but its range. The appearance within a hominid community of a relatively few individuals of higher ability in a given area of activity than any known hitherto *could* lever the entire technological level up beyond that of their neighbours. Such an effect could come about as a result of simple population size – even between genetically identical populations, the 'one-in-a-million' genius will most likely occur sooner in the larger than the smaller. Of course other necessary conditions must be met at the sociological level for such developments to occur, but my point is only that apparent differences between the technological capabilities of two human populations, in this case Neanderthals and Moderns, do not necessarily reflect correspondingly great differences between the average capacities of the members of these populations.

Thus although the post-Neanderthals do fairly rapidly show evidence of novel cultural behaviour and technological achievements, the degree to which this entails an interpretation of the Neanderthals as somehow *essentially* (i.e. genetically or biologically) less advanced at present resists analysis. There is certainly no evidence that the gross morphology of Neanderthal brains differed from that of Moderns in any significant respects, and Maclean's aforecited claim about their frontal lobes being undeveloped is not supported by Holloway (per. comm.), although his public statement of this, in a talk given in 1984, has not yet been published.

3 More speculatively, it is likely that the social organisation of the

Neanderthals was less coherent than that of their successors. Most of the signs point to a small family unit as the basic pattern, which would be appropriate in conditions where resources are relatively scarce. Though such families would have probably belonged to some superordinate 'tribal' population, it is possible that this lacked the sense of tribal identity typical for modern humans. Such hypotheses are not entirely speculative. Strong tribal identity requires the proximity of similar tribes vis-à-vis whom such an identity can be established.[1] The population levels of the Middle Palaeolithic in Europe and Asia are likely to have been extremely low by modern standards. Psychologically, identity may have been centered in the immediate kinship group, and the broader 'tribe' may have been seen simply as 'other people', its tribal character, in terms of stone-tool style, etc., being only apparent from the distant Olympian perspective of the modern archaeologist. While we might be in the realms of speculation as to the nature of the difference, clearly the social organisation of their successors rapidly diverged from that of the Neanderthals. In some regions of France, for example, there is a switch from hills to valleys in the location of sites. Indeed the image given by some recent writers (e.g. Pfeiffer, 1982) is of Neanderthals as laconic loners self-sufficiently wandering the snowy heights, doomed to be ousted by the physically feebler but more socially organised newcomers in the valley bottoms.

Reading Hadingham's summary of the L'Hortus Cave, I was struck by the 20,000 years of Neanderthal occupation; every year for a few months they would tread daily the pathways worn by their forebears, through the accumulated detritus of centuries. The unanswerable question automatically arises, did they ever have a concept of 'time past'? Did they simply take it for granted that their people had always used this cave or were they, less than that, indifferent to or even unrecognising of the nature of the place? Modern Australian natives have woven complex mythologies around the regularly visited sites in their environment, ritually marking them very often with symbols and arrangements of stone (see Gould, R. A., 1969). There appears to be no sign of this at L'Hortus Cave.

Summary

So, Neanderthals were technically skilled, culture-bearing, and innovative (if at a slower pace than us). Physically they were robust and superior to us in muscular strength, while they were capable of adventurous pion-

[1] Tribal naming is often far less similar to proper-naming than one might imagine; frequently what becomes established in western literature as a highly specific tribal name is little more than a translation of the word for 'us', 'humans', 'the people' in the language of the tribe concerned, or that for 'them', 'the people over there', 'the barbarians' or 'the other lot' in the language of one of their neighbours.

eering forays into the unknown territories opened up in the early phases of the last Ice Age. And yet, by just before 30,000 BP they have disappeared altogether, a process which begins in eastern Europe around 40,000 years ago, after which a sort of 'front' of moderns moves westward. Why this demise? The answer is, in spite of much thought over the last eighty years or so, as elusive as ever. If the ability to oust the natives is a sign of superiority, in what did *Homo sapiens sapiens*'s superiority consist? At this juncture it is more reasonable to look for an answer in terms of social organisation, or behavioural, psychological, factors rather than in gross morphology and physiology. The possibility of interbreeding between the two human subspecies is often canvassed, and one writer (Gooch, 1977) makes this the centre of his theory, arguing that modern humans stemmed from the hybridisation of Neanderthal and non-Neanderthal stock. This view receives no support from experts in the field. The majority opinion is that if interbreeding occurred it was to a very limited extent, and unlikely to have happened in Europe. The physical differences between late Neanderthals and Moderns are widely construed as ruling out direct descent, it being thought more likely that modern humans descended from a different branch of 'Archaic *sapiens*' in Africa or, if Aigner and Wu are correct, the Far East.

Although unanswerable, it is worth raising to consciousness such points as whether or not the arriving Moderns themselves actually 'saw' the Neanderthals as different, and to what extent. At one time dramatic genocidal massacres were envisaged, but no evidence exists for this, no caches of dozens of bones of Neanderthals slain in battle, and on balance the study of sites like Arcy-sur-Cure goes against this. The demise was likely to have been a more long-drawn out affair, piecemeal, mosaic, in character, with attitudes towards the Neanderthals among Moderns varying across the spectrum as widely as they do on most subjects today, and perhaps vice versa. Whatever the truth, the Neanderthal 'Achilles heel' in confronting the Moderns cannot be simply diagnosed in the ways once customary.

C The spread of the Moderns

In a sense the evolution of our behavioural *capacities* is virtually completed by the time we actually appear on the scene; it is the exploitation and exploration of these that occupies the succeeding centuries. In Part C, then, I am going to deal with a relatively restricted range of topics only. Firstly, expanding on what was said in Chapter 3, we return to the spread of modern *H.sapiens* to America and Australia in order to get the situation in focus. Secondly, I will consider some of the views put forward on the nature of Upper Palaeolithic European cave art, as the most extensively researched of the achievements of early modern people, moving on in the third section to Marshack's studies of notational aspects of Upper Palaeolithic artefacts. Section four contains a sketch of the character of the phase between the Upper Palaeolithic and the advent of 'civilisation' in the Near East, during which so many of the basic human achievements were accomplished.

As emerged in Chapter 5, there is a genuine difficulty, a paradox even, in knowing what attitude to adopt to this material. This stems from the fact that on present showing our species is a planetary disaster of the first order, a fact which previous writers on the 'rise of civilisation' theme have been able to repress, downplay or turn a blind eye to. This becomes annually more difficult. If we adopt a negative position, the initial manifestations of our disastrous character occurred quite early on, with the extinction of the European and North American megafauna, in which human hunters almost certainly played a major, if not exclusive role. The uninhibited use of fire for forest clearance and hunting also began early (though its use on a small scale is not necessarily damaging to the ecology, as Mellars (1976) has shown). Appreciation of this has led Bargatzky (1984) to pull together a number of criticisms of the earlier notion of culture as an 'adaptation' to the environment in some broad ecological sense. On the contrary, human 'culture' rather defines the environment in terms of its potential for human exploitation as a resource, always oversimplifying its true complexity, proceeding then, with a blithe disregard for consequences, to 'assimilate' it. This is quite consistent with the author's 'physiomorphic' hypothesis proposed in Chapter 5, though couched in anthropological rather than psychological terms. And yet, one can hardly be immune to the fascination of the tale, and perhaps what is going on is a shift in the tenor of the 'myth'-making after a century of optimistic heroic 'ascent' to a revised version of 'the fall'.

1 *Radiation to America and Australia*

We may begin by recalling that, as mentioned in Chapter 3, the ethnic diversification of *H.sapiens* is not yet understood. The genetic evidence on the timing of the various splits of *H.s.sapiens* into its current ethnic groups is too equivocal to justify further discussion here. The species is in any case geographically highly uniform; according to J. S. Jones (1986), 85% of the genetic diversity that exists is *within* group difference, about 6% due to differences between countries and 10% apparently 'racial'. The 'most average' country genetically turns out to be China, though if mitochondrial rather than structural DNA is used as a basis for comparison the average lies in Africa. However useful for interspecies comparisons, as yet molecular biology can do little to illuminate the ethnic diversification of the human race.

One feature of the modern radiation which is particularly interesting, and researchable, is the nature and timing of the spread of our species from its homelands in the Afro-Asian landmass to the distant regions of America and Australia.

(a) America

The move into America almost undoubtedly occurred during the last Ice Age, when the lowered sea-level provided a land-bridge, at times hundreds of kilometres wide, between Siberia and Alaska across the Bering Straits (this lost landmass is referred to in the literature as Beringia). Ethnic affinities between indigenous American peoples and Asian groups have long been recognised by physical anthropologists. The last Ice Age lasted, in human terms, for a very long time, from around 70,000 to 10,000 years ago, and precisely when during this stretch of time humans first ventured into America – and how often they did so subsequently – is a riddle which many have tried to crack. The background picture is of a largely ice-free land-bridge and a further narrow ice-free corridor down the inland side of the Canadian Rockies (with another possibly along the Pacific coast). These enabled the migrating newcomers to move south from Alaska into the more hospitable heartlands of North America, and thence into South America.

Details of this process are controversial, since the evidence is nearly all highly problematical. A popular article by Canby in the *National Geographic* (1979) spread the notion that there was strong evidence for occupation as long ago as 50,000 BP (a date then being offered for a site called Taber in Canada), and pictured a child's jaw associated with artefacts from Old Crow Basin on the Yukon, dated to 27,000 BP. By 1984, more reliable dating tests and Binford's perspicacious criticisms had taken their toll, leaving the 'early origins' case in something of a shambles (Binford, 1981). Gowlett's review (in Gowlett, 1984) is tentative in suggesting anything much prior to 20,000 years ago, though two Mexican hearth sites may be 22,000 and 30,000 years old. The great Alaskan bone-tool industry of Old Crow Basin which figured so prominently in Canby's article seems set to evaporate under Binford's taphonomic critique – and it is in any case odd why, given the amount of stone available, the immigrants would have suddenly abandoned that medium altogether. The oldest Alaskan sites now turn out to be around 14–15,000 BP, and until the Mexican datings are confirmed the most plausible entry date for humans into America on current evidence cannot be much earlier. There is some similarity between the style of making bifacial projectile points characteristic of the earliest north-east Siberian culture (the Dyuktia who go back to 30,000 BP) and that commonly found later in America. If this affinity is genuine it makes the Alaskan bone-tool culture even more anomalous.

The best known early site in North America is Meadowcroft in Pennsylvania. Here a culture older than the dominant 'Clovis' culture (found across North America and once thought to be its earliest industry) has been found, and securely dated to 16,000 BP. The Peruvian site of Flea Cave has a sequence going back at least 15,000 years and possibly 19,000. But early fossil hominids are rare and most likely candidates have fallen by the wayside under closer examination (e.g. the Sunnyvale skeleton). The jaw illustrated in Canby's article was still undated at the time Gowlett was writing, while the Laguna specimen for which he was then prepared to accept 17,000 BP has since been shown, in a paper co-authored by Gowlett (Bada et al., 1984) to be 5,100 + 500 years old. Only 'Los Angeles Man' – with a suggested date of 23,000 (Gowlett, 1984) is holding out as possibly 'early'. At the time of writing, then, there is no hard evidence of humans in America prior to around 23–25,000 years ago and certainly not before 30,000, while some archaeologists would consider even this nearly 10,000 years too early.

The numbers coming over may in any case have been small, to judge by blood-group evidence, for all indigenous Americans south of the Canadian border are blood-group O, while above it, closer to the entry point, some A group is found, but B is entirely absent. Since the earliest South American dates are virtually contemporaneous with the oldest northern ones we may plausibly assume a fairly rapid movement south of populations heading for the warmer tropical climes, with the Panamanian isthmus acting as a kind of valve, preventing backward movement. Darwin's benighted Fuegians may just have been the last remnants of the first wave. Though out of date, Canby's article remains a good introduction to the personalities and positions involved in the US academic debate, and readers may expect new developments at any time.

(b) Australia

Turning to Australia, the picture described by Mulvaney (1975) and Flood (1985) is intriguing. Archaeological sites as old as 30,000 BP are known (e.g. Keilor, Victoria) and modern *H.sapiens* fossils as early as 25,000 BP have been found at Lake Mungo, New South Wales, in the Wilandra Lakes system. A particularly striking find from here was the opalised skull W.L.H.50. From Kow Swamp, also in Victoria, have come 40 fossils which though only dating from about 15,000 years ago are so archaic in morphology, verging on *H.erectus* in character, that their existence is generally accepted as implying an initial entry into the Australia-New Guinea-Tasmania landmass of the Ice Age ('Sahul', as it is known) as early as 50,000 BP. Since these fossils were accompanied by grave-goods, their archaic physiology would not seem to have stopped them acquiring some degree of modern behaviour. The morphological diversity of the numerous Australian and Tasmanian fossil hominids is considerable, and the theory endorsed by Flood is that modern native Australians result from hybridisation between the gracile, anatomically modern Wilandra Lakes people and the robust people represented at Kow Swamp. Thorne and Wolpoff (1981) have reported and analysed in more detail the morphological affinities between Kow Swamp skulls and the Indonesian *H.erectus* specimen Sangiran 17. While there is no question of the fact that they had achieved full *H.sapiens* status, they share certain regional morphological characteristics of facial and cranial structure apparent in Sangiran 17 and other Far Eastern specimens but absent from western skulls such as those from Petralona and Broken Hill. One kind of explanation for this continuity, favoured by earlier authorities such as Coon, would have been to postulate a 'semi-isolated subspecies'. Kow Swampers would, on this view, be the last representatives of a separate lineage descended from *H.erectus* which had evolved in the 'sapiens' direction in parallel, but isolated from, true *H.sapiens*. Thorne and Wolpoff reject this model, offering a second, more sophisticated account based on Thorne's 'centre and edge' hypothesis. Though too technical for full elaboration here, the principal thrust of this is that peripheral, or 'edge', populations show far less within-group 'polytypism', far more homogeneity, than those at the centre of the species' distribution. Edge populations are smaller, relatively isolated, sometimes subject to different selection pressures, and may also retain, in their scattered, separate groups, random genetic features of the different founder populations. Low gene-flow and differences in selection pressure, may maintain what the authors term a 'morphological clade' over a long period, but this

> need not imply any special lineage relation for a specific region over the timespan involved (just as the water level may remain constant in an overflowing bathtub without the water at the surface of the tub remaining the same). The resulting appearance of morphological continuity in a region *simulates the appearance* of a longlasting subspecific taxon without necessarily being one in any meaningful genetic sense (p. 347). We hypothesize that initial differences between Australasian and other populations, as well as a degree of relative homogeneity within the populations of this area were established during the first habitation of the region by *Homo erectus*. Subsequent to this, a dynamic balance of geneflow and opposing selection, in both cases probably of very low magnitude, provided the basis for a long-standing clinical equilibrium involving a number of morphological features. The geneflow that was required to maintain this dynamic clinical equilibrium was a critical factor in the dissemination of evolutionary changes that characterized all populations of *Homo* during this timespan. (p. 348)

Thus Kow Swampers were not genetically isolated from the rest of the evolving human population, but in a situation where selection forces were such as to sustain a degree of robustness lost elsewhere. Nevertheless, this is still *relative* robustness – they too had become more gracile by comparison with *H.erectus*, but they had not moved so far in this direction as regards the features in question as other populations had. The 'centre-edge' hypothesis might also be invoked to account for the apparent continuity between Chinese *H.erectus* and *H.sapiens* referred to earlier.

Clearly Kow Swamp *was* a very peripheral point in the distribution of the hominid population world-wide. The presence there of a population retaining such robust features as late as ten millennia ago clearly suggests that they *arrived* in the general region at a time prior to the emergence of more fully morphologically modernised peoples in the Indonesian parent population. The gene-flow level between the main hominid centres and this outpost was thereafter likely to be low, though sufficient for the major adaptive changes to spread. Those aspects of robust morphology which remained adaptive in the conditions in which this population lived could, though, be retained, even while they vanished or decreased more rapidly elsewhere, where other conditions prevailed.

Kow Swamp peoples therefore represent the earliest wave of settlement, and while some researchers have argued for a three-wave scenario, two certainly – at around 50,000 BP (early in the last glaciation) and another much later, around 25,000 BP – are strongly on the cards.

Unlike the entry into America, such a settlement could not have been accomplished entirely over land. Even at its lowest levels during the last Ice Age, a channel at least 100 km wide would have come between Java and Sahul. The first wave, if not the second, is unlikely to have ventured forth on this expanse of open sea on purpose, especially since it seems probable that at this time their craft could not have amounted to much more than rafts of lashed tree-trunks. The first settlers then could have been very few in number, their arrivals in Sahul intermittent.

In terms of popular stereotyping, current archaeological evidence is definitely redeeming the indigenous peoples of Australia from the 'world's most primitive culture' status so long ascribed to them. For example, they are known to have begun using grinding and polishing techniques for working stone as early as 15,000 years ago, earlier than anywhere else in the world as far as we know. Such methods have often been taken as diagnostic of the Neolithic in Europe, around 10,000 BP. Rock paintings over 20,000 BP are also to be found. At present the prehistoric archaeology of Australia is proceeding rapidly under conditions far more favourable than those obtaining elsewhere, with relatively little urbanisation and destruction of sites and a history mercifully free of two millennia of grave-robbers, treasure hunters, or even methodologically naive Victorian gentlemen archaeologists. Flood's book is the most recent comprehensive account, Mulvaney's is the first major text and Gowlett also has a synopsis of the situation.

Obscure though the details are, our species had managed to extend its range across all the Earth's major land-masses, except Antarctica and New Zealand, by at least 20,000 BP and possibly before. Some no doubt retreated before more advanced invaders to yet further afield, some got lost or blown astray accidentally, some migrated more purposefully forward, and most from time to time were driven by climatic or ecological changes to uproot themselves. Even if we cannot disentangle the sequence of splits in any detail a handful of uncontroversial propositions may be made. The African and Asian populations, for example, must have parted fairly early on with the Europeans following closely, if not coincidentally (though they probably were not in Europe at the time). Americans split later from a small section of the 'mongoloid' or oriental population who had ventured into north-east Siberia. In the southern part of Asia a sort of 'fraying' rather than 'splitting' of the population was probably going on at the same time, resulting in the occupation of Australia and the western Pacific and South East Asian region generally. The rest of the picture remains chaotic. The successive prehistoric populations of Europe, from Cro-Magnons onwards, displayed a variety of physical anthropological features suggesting considerable ethnic range. The older, happily 'racial' (if not 'racialist') works of Coon (1963), Keane (1899) and Hooton (1946) revelled in human diversity, and identified great numbers of 'races' and 'subraces' in true Linnean fashion. Hooton identified 10 'White' subraces alone. Given the rapid tempo with which change can sometimes occur anyway (e.g. the increase in average height among North Americans of European descent within a century), it is hardly likely that the streams, eddies, cross-currents and tides of prehistoric human diversification will ever be mapped without some major breakthroughs in the analysis of genetic material at the molecular level, way beyond those already achieved.

2 *European cave art in the Upper Palaeolithic*

In Chapter 3 I summarised the general character of European cave art. The descriptive literature on this is both extensive and far from complete, in that some important sites have yet to be fully explored, let alone written up. While we have images galore, classifications of the animals and symbols represented, frequency counts and studies of distribution patterns within caves, the interpretation of the meaning of this corpus is rudimentary. The behavioural and psychological significance of European cave art, which is our primary interest in the context of this book, is inaccessible, and in the opinion of most authorities likely to remain so.

In a situation not unakin to that outlined above regarding *H.erectus*, older theories, such as they were, have recently fallen from favour, while equally encompassing alternatives have not appeared. I will nevertheless summarise these theories, identified with l'Abbé Breuil and A. Leroi-Gourhan, since they dominated the scene until the last fifteen years and left important methodological legacies. If any future progress is possible on this front, it seems to rest on the use of sophisticated statistical analysis to test clearly formulated psychological hypotheses. Even so, the Popperian view that truth can only be approached by progressive falsification of erroneous theories rather than directly demonstrated is nowhere better illustrated than here. Any theory regarding the meaning of European Cave Art must somehow account for what currently appear to be a number of contradictory facts. (Although facts cannot logically be severed from the theoretical structures or frameworks which define them, in this case the 'facts' in question are those defineable in general archaeological terms, such as datings, not 'theories of cave art' as such.) Five such salient facts, of a broad character, may be enumerated straight away, along with some of their implications, in order to demonstrate the difficulties this material presents for psychological or behavioural analysis.

1 The variability of the work is primarily chronological, not regional, and even so is in the main a matter of style rather than content. This is acknowledged by almost everyone concerned, and the boundary with the adjacent 'Mediterranean' art is clear in all respects – the Mediterranean art is later, stylistically different, *and* different in content.

This can, it seems, only be interpreted to mean that the people responsible for European cave art constituted a single coherent culture with high levels of internal communication and inter-regional movement. They were in fact almost certainly nomadic. Although in terms of quantity the majority of the work dates from the Magdalenian end of the sequence, the evidence from earlier periods back to the Aurignacian bears out the picture of an extremely durable and longstanding cultural tradition extending over a progressively wider area as the Palaeolithic drew to an end. (Though Conkey, 1983 is cautious on this.)

2 The inner cave sites rarely show evidence of having been frequently used or visited and certainly were not occupation sites (unlike the daylit shelter sites).

This is a most striking, if easily overlooked, piece of information, since it rules out any explanations for the impressive surviving masterpieces of cave art which involve regular assemblies of people or mass visits. It is an extremely enigmatic feature and one difficult to incorporate into an overall model of the function of the work.

3 With the exception of the rarer human and bird figures, the standard of drawing is high, and though conventions for depicting different species are adopted, the figures are individually unique, not routinised repetitions of a standard 'ideogram'. This is true to a large extent of the non-figurative 'tectiforms' and 'claviforms' as well.

This implies several things: firstly, that the artists were experts working within a continuous tradition; secondly, that the function of the art was not purely 'symbolic' or 'notational'; thirdly, that a system for transmitting the requisite skills must have existed. What this last would have been we can only guess, and suggestions range from apprenticeship to clan-specialist to family tradition.

4 The content of the paintings and engravings is not only unchanging but also restricted. Horses and bovines, for instance, account for 60% of all animals depicted. The repertoire consists basically of little over a dozen species, represented, though, in a wide variety of postures. The individual images appear to bear no relationship to one another in a design sense, though pairs and even rows do occasionally occur (the conditions under which the work was done may well have precluded such overall patterning and organisation). As we will see, there are some, albeit elusive, cross-site regularities in the within-cave patterns of distribution and relative frequencies of the various animals (though the latter also shifts chronologically). Non-representational elements are similarly universal (though whether they were originally related to the representative elements, whether they are even really non-representational, is at present unknowable). There are chronological, and to a lesser extent regional, differences in the frequency of all these elements, in the manner, speed and skill of their execution too, but these are all variations on a common content within narrow parameters.

This conservatism, over as long as twenty centuries, points to an enduring role and significance for Upper Palaeolithic cave art. The only phenomenon we know which exhibits comparable long-term stability is religious symbolism. What is meant by describing something as 'religious' or 'a religion' is not perhaps as clear as we often imagine, but this cave art certainly seems to lie in the general direction of being a perennial expression of fundamental cosmological beliefs, attitudes or feelings. Current anthropological evidence, particularly from Australia (Gould, 1969, 1980) regarding the role of mural markings and rock art might be interpreted as lending credence to such a view. At least such long-term coherence convincingly eliminates theories which would see the work as purely decorative (for this is highly subject to fashion, in constant change even in traditional societies, even though more slowly than in modern western countries), or personal in inspiration.

5 Many figures are incomplete, they also overlie one another in often chaotic disarray. This holds for both paintings and engravings. In short there is an apparent disregard for existing images among successive generations of artists. This is *not* due to pressure of space, for large untouched expanses of cave-wall are often found at the same sites. Nor are the overdrawn figures attempts to rejuvenate or restore existing ones – though this may happen, as may incorporation of parts of earlier figures into later ones.

This feature is especially mysterious, and hints at a psychological gap between the artists and ourselves, offsetting the pyschological affinity we feel in viewing the naturalistic and painstaking animal images as such. It certainly seems to imply the absence of a *monumental* motive, if I can call it that. It almost suggests that the value lay only in the act of creation, and not in its subsequent appreciation by others. This of course is very *unlike* most religious art.

The reader will appreciate then, in the light of just these five rather elementary observations, how difficult it is to come up with an explanation which is both plausible and satisfactory – how do we integrate in one model the apparent contradictions involved? The combinations of care with negligence, creative fervour with abandonment of the product, conservative devotion to tradition with spontaneity of expression, endow this art with a tantalising fascination. We now turn to Breuil and Leroi-Gourhan's theories.

(a) Breuil

Breuil's principal focus of concern was on the development of the genre stylistically, and the identification of its main phases. He proposed a two-cycle model for this, based on an unrivalled knowledge of both style and technique, and a lengthy unravelling of the sequences of these by close study of superimpositions. The first cycle as far as painting was concerned he called the 'Aurignacian', which he believed began with hand-prints, and moved through yellow and red drawings to black linear ones. A feature of this was '*perspective tordue*' – representation of horns and antlers in front perspective on side-view profiles. The second, 'Magdalenian', cycle began he argued with simple line drawings, went on to polychromes and finished with black outlined polychromes and some red linear drawings. Antlers and horns were now shown correctly. This scheme has not been entirely supported by later research. Breuil proposed it first in the 1930s and reiterated it in Breuil and Windels (1952). Modifications in his chronological attributions of particular paintings have had to be made and the trends are not so clear as he thought them to be, hand-prints for instance, occur in abundance in some Magdalenian caves. Of more interest to us here though is his view of this art's function.

Breuil of course began his career in the first years of the century and anthropologists such as Sir James Frazer were at that time extensively interested in the nature of 'magical thought' and what they believed to be primitive, pre-rational thought processes. It was perhaps natural then for Breuil to view cave art in terms of magic, in particular that form of it termed 'sympathetic magic', whereby the magician strives to bring about a desired outcome by imitating it – the most widely known example, I suppose, being sticking pins in an effigy of one's enemy. This was combined with an assumption (almost certainly erroneous on modern anthropological evidence) that preoccupation with food supplies was a dominant feature of Upper Palaeolithic life. Put together, this indicated 'hunting magic' and 'reproductive magic' – figures represented desired prey for the Stone Age hunters, who would try to ensure success by depicting animals with arrows in them, or by stabbing the image with spears. Pregnant female species members would have been the focus of fertility rites, ensuring plentiful supply of the animal in question Non-representative symbols were 'spirit houses' by analogy with the beliefs of Siberian shamanistic tribes and American Indian folk-lore.

It would be wrong to say even now that Breuil was entirely mistaken in all this. There is enough circumstantial evidence to indicate that some sites at least were used for ritualistic purposes, though on a very small scale, suggestive to some of 'initiation rites'. Extant footprints in these caves include a high proportion made by children (though these are not necessarily contemporary with the art), though the very fact that the footprints *do* remain clearly visible on otherwise undisturbed floors testifies to the infrequent use made of these sites after completion. The 'magic' and 'shaman' theories were bolstered by the presence, e.g. at Les Trois Frères, of the famous 'sorcerer' figures – semi-human, semi-animal figures adorned with horns, and placed in apparently dominating positions. One such appears to be playing a flute, another a musical bow. The reasons for scepticism about Breuil's all-embracing 'sympathetic magic' account are nevertheless compelling. Only 11% of the figures show any signs of the kind of damage or imagery which would be expected of 'hunting magic' rituals, and the correlation between representation of a species in cave art and presence of its bones among food-remains is poor. Indeed for long periods the reindeer is very rarely depicted, although it was the animal most frequently hunted. Modern hunter-gatherers, in circumstances of great food-scarcity compared with the abundance enjoyed in the Upper Palaeolithic in Europe, display no such over-riding concern with food-acquisition. Current anthropological thought has in any case moved on from the way of conceptualising 'primitive' thought that Breuil knew, and the term 'magic' has lost its clarity of meaning. (Indeed it is sometimes used [in popular accounts] as a sort of dustbin category for phenomena not understood, somewhat similarly to the use of the phrase 'cult object'.) The influence of the later 'structuralist' approach can be seen in André Leroi-Gourhan's approach to which we go next.

(b) Leroi-Gourhan

Breuil had tended to treat images individually, accepting that their distribution was, as it initially seemed, largely random. Leroi-Gourhan initiated a rather different approach by studying the distribution within caves of the various kinds of animal and symbol. Did a pattern emerge from this? And if so, how was it to be interpreted?

In a key work published in 1965, Leroi-Gourhan argued that spatial arrangement of figures within caves was based on a principle of masculine-feminine opposition, with each species or sign being classified as one or the other. Thus horses were masculine and bison feminine. A figure of one gender will always occur with at least one figure of the opposite gender (though the numbers may be very unequal). The caves themselves, Leroi-Gourhan believes, are divisible into seven types of zone and particular animals and genders are associated typically (though not exclusively) with particular zones. There is thus an underlying formula or plan which, albeit imperfect and modified to suit the idiosyncrasies of each actual cave-site, is imposed on all cave art. (See Fig. 6.1.) Leroi-Gourhan's

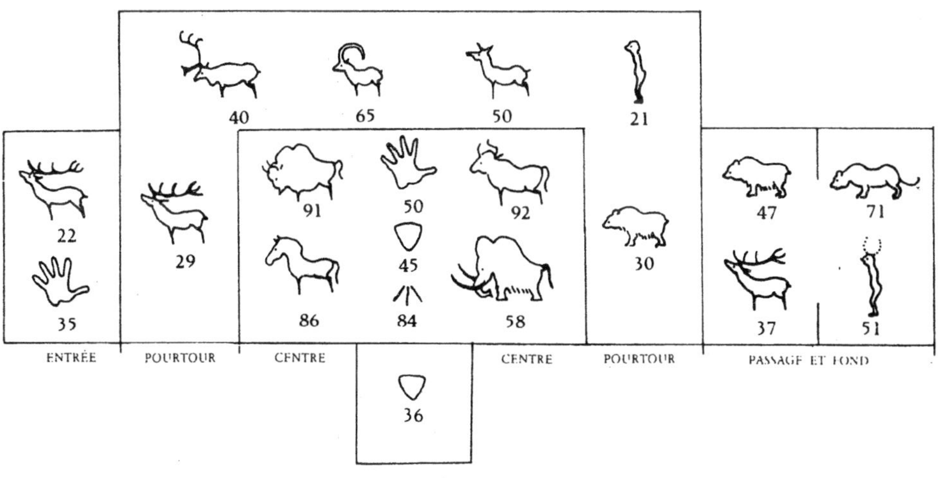

Figure 6.1 Ideal arrangement of animal figures in a Palaeolithic sanctuary after Leroi-Gourhan. The numbers below each drawing are percentages based on 865 subjects in 62 caves. The subject is here counted once each time it is represented regardless of the *number* of times it may be repeated in the same panel.

analysis of stylistic development also broke from Breuil's two-cycle pattern; in particular he contends that *deep* cave art occurs in a much more restricted time-range, flourishing in the 5,000 years from the mid-Solutrean to mid-Magdalenian, and not, as Breuil held, going back to the Aurignacian. A four-stage model is identified, from two primitive through archaic and classic phases, referred to as styles I–IV (see Table 6.1). This analysis seems to be replacing the older Breuil one in all current literature.

TABLE 6.1 *Upper Palaeolithic cultural and art periods of Leroi-Gourhan*[1]

Years before present	*Cultural stages*	*Art periods*
	Azilian	
10,000		
	Late Magdalenian VI	
	V	Style IV
	Middle Magdalenian IV	Classical
15,000	III	
	Early Magdalenian II	Style III
	I	Archaic
	Late Solutrean	
20,000	Middle Solutrean	
	Early Solutrean	Style II
	Late Gravettian[2]	2nd Primitive phase
25,000	Middle Gravettian[2]	
	Early Gravettian[2]	
30,000		Style I
		1st Primitive phase
	Aurignacian	
35,000	Chatelperronian[3]	Prefigurative period

[1] After Sieveking (1979).

[2] Sometimes referred to as Upper Perigordian.

[3] Sometimes referred to as Lower Perigordian.

The most controversial aspect of Leroi-Gourhan's work was the masculine-feminine classification, which he himself has since downplayed somewhat. Sieveking (1979), on whom I am primarily drawing here, sees the underlying importance of the theory as being that it shifted away from narrative interpretations of the work to 'an explanation at a more probably correct intellectual level than any advanced before . . .' (p. 64).

Regardless of the specific nature of the meaning of the work, the important move was to recognise that it expressed some 'metaphysical' or 'cosmological' concepts, was 'mythographic', rather than depicting particular incidents, events or scenes. The impact of this was dramatic, triggering numerous analyses of the spatial distribution of figures within caves in order to identify possible regularities or formulae regarding their placement, not least to test and challenge Leroi-Gourhan's own hypotheses on these matters, and whether a 'preconceived structural relationship' could, as he claimed, be identified from the north of Spain to Perigord. This, Sieveking claims, revolutionised the study of cave art, while she holds his identification of patterns to have been substantially correct.

Nevertheless there are those challenging both Leroi-Gourhan's statistics and his conclusions. Stevens (1975) has severely criticised Leroi-Gourhan's hypothesis using frequency counts for 10 species across 40 caves, while work by Tuohy (in preparation) has reached somewhat mixed conclusions. Tuohy (per. comm.) has adopted a statistical approach derived from Berlyne (1971), the psychologist, which enables him to express the *amount* of organisation (i.e. departure from random distribution) in each cave as a single number. This has led him to a number of novel conclusions. Firstly, he finds an increase in the level of organisation from Style II to Style III. Secondly, by examining the effect of the number of figures present on level of organisation, he concludes that 'a change in style meant a change of plan, but . . . there was some attempt to incorporate old figures in the new plan (per. comm.) – the reverse of Leroi-Gourhan's view that new figures were fitted into the old plan. Thirdly, regional differences occur suggesting an increase in organisation 'as a function of . . . proximity to the south east section of the distribution' (per. comm). Of course these are tentative findings at this stage, but are interesting as an example of the application of ideas drawn from the psychological literature, by a psychologist, to this material. They also show how Leroi-Gourhan's work has opened up the topic, providing testable hypotheses and methodologies which may leave his own specific conclusions behind.

But Upper Palaeolithic art is not confined to caves, indeed it extends across northern Europe into Siberia and probably down into Africa too. Engraved plaques, decorated ivory tools and statuettes, and engraved fragments of ivory and bone form the principal media for this broader body of work. Although they often share certain features with the cave art, such objects differ considerably in content, and, we may assume, function. This material often takes the form, especially in the Eastern European and Russian settings, of non-figurative, apparently iconic, but also visually chaotic, engraving of lines, zigzags, 'meanders' and criss-cross patterns which can overlay one another to add to the impression of utter confusion. Elsewhere, the best cases being the rock shelter of La Marche and Feldkirchen-Gönnersdorf on the Rhine, figurative engraving different in character from the cave art occurs. At La Marche this takes the form of engravings on 1500 Limestone blocks, where the tangled lines include what Hadingham calls an 'intriguing Ice Age portrait gallery of fifty-seven isolated heads, with an additional fifty-one more or less complete human heads and bodies', this in addition to animals such as bears and felines rare on the cave walls. Many of these are superficially reminiscent of modern cartoons. The La Marche work has been described by L. Pales and his associates (Pales, 1976; Pales and St Pereuse, 1964, 1968). For the psychologist, however, some of the most interesting investigations of this whole *oeuvre* have been those undertaken by Harvard's Alexander Marshack since the mid-1960s, which have placed its study on a whole new footing.

3 *Alexander Marshack's research*

Initially the dominating theme of Marshack's studies of many of the long-famous French ivory artefacts was a conviction that much which had hitherto been seen as mere ornamental decoration in these objects was actually notational. What he brought to the study of these objects was a microscope, patience, and a flair for code-breaking and pattern-recognition. These soon yielded results. He learned that it was possible to identify where changes of engraving tool had occurred, how lines were superimposed on one another, and degrees of wearing. Often he could show that pieces long thought to be broken were in fact engraved *after* the damage had taken place, when the object was in more or less its current form. Different engraving points leave different groove profiles as distinctive as fingerprints, some V-sectioned, some U-sectioned and so forth. He discovered that, to judge from the number of tool-changes, many of the markings on these objects were made over a relatively long period of time, rather than at a single sitting. Apparently minor differences in line-length, spacing, and other variations also rewarded his scrutiny. In his first major work, *The Roots of Civilisation* (1972a) Marshack proposed that such features were often notational and that mathematical analysis strongly pointed to them as calendrical in function, probably related to the lunar cycle.

Following this, Marshack further explored the psychological implications of such artefacts, investigating how far they could disclose the cognitive capacities and thought-modes of their makers (though not, like others we have discussed, within a specific theoretical framework). Following a key statement in 1972 (Marshack, 1972b) came a series of papers in 1976, 1977 and 1979 though more recent work is unavailable as yet in the United Kingdom. Marshack's work must now be considered central to any understanding of the later phases of psychological evolution. Among his numerous conclusions, drawn from detailed examination of thousands of prehistoric engraved artefacts and paintings, the most important are the following:

1 The practice of deliberate, controlled, carving or engraving of an apparently symbolic character extends back to the Peche de l'Aze rib (300,000 BP) and is far commoner among Neanderthal (or at any rate Mousterian) artefacts than once thought. A tradition of personal adornment also seems to have been present prior to the advent of the Moderns. This reinforces the growing feeling that the Neanderthal Modern differences are primarily a matter of degree, and that there is continuity between the two at the level of cultural development, if not genealogically. Similar elements, notably 'meanders' and 'zigzags', occur from the earliest objects through to the end of the Palaeolithic and into some modern tribal cultures (Marshack, 1976).

2 The engravings on these objects were invariably accumulated over a period of time during which the artefact was regularly handled and curated. This is so even though they possess no 'utilitarian' function visible to us. Indeed, engravings are often found on the broken remains of functional tools, microscopic examination showing the engraving to postdate the damage, in the fashion already referred to. We are not, though, dealing with casual doodles, or point-sharpening blocks, but with a non-representative iconography or symbol-system of some now obscure kind. One of Marshack's characteristic skills is to be able to reconstruct the precise behavioural movements involved in engraving, unveiling thereby the complexity of the actions involved in the manufacture of a particular image, which are often belied by its, to us, casual or meaningless aspect.

3 From this Marshack argues we must posit the operation of a developed cultural context involving at least some level of linguistic communication. It is only within such a setting that activities like symbol-engraving can acquire any meaning:

> Just as human infant babbling does not lead to language outside a cultural context, human or chimpanzee doodling does not lead to symbol outside a symboling cultural context. It is not the capacity for, or even the expression of, such pre-language or pre-art processes that is important for ontogeny, phylogeny and for the development of cultural forms, but the utilization of these capacities and expressions within established symboling processes. The ontogenetic and phylogenetic development of the hominid symboling capacity requires feedback between the individuals in a group and the ongoing cultural structures. (1979, p. 274)

4 In his 1979 paper he makes an important advance in his theoretical interpretation of the significance of the early material in the light of a superb study of a number of Russian objects, mostly ivory plaques and engraved bone fragments (Plate 13). These, with a characteristic collection

of symbols – schematic fishes, 'ladders', zigzags, meanders, parellel lines, 'trees' and hatches – he had already come to consider as somehow being a water-related symbolism, and some speculations are offered as to the way this would have fitted into what is known of the late Ice Age life-style, with judicious invocation of ethnological evidence of an analogical type. But Marshack goes further than this, to argue for a proto-human capacity for symbolic representation intimately related to the emergence of language. This capacity forms the base-line from which specific cultural traditions subsequently derive, and can be taken *either* in the direction of naturalistic representation *or* in the direction of abstraction:

> From the viewpoint of both the evolution of the hominid cognitive-and-symbolic capacity and the development of human cultural forms and structures, this mode of motif use may have been one of the early, pre-art symbol-making processes out of which art evolved. (1979b, p. 274)
>
> . . . the use of near-iconic and non-representational images in the Upper Paleolithic developed its own internally consistent strategies and cultural forms without recourse to or development from any naturalistic or representational traditions. Once instituted as a functional mode, these early symbolic strategies and forms could develop either towards naturalism or towards an increasing elaboration of the near-iconic and the schematic. With the dispersal of the individual *H.sapiens sapiens* cultures, regional and temporal styles appear and individual strategies and skills enter into image and motif production. (ibid., p. 274)
>
> The analytic data and the theoretical questions presented here have relevance for almost all the widely dispersed symbolic traditions of the post-Würm and, more particularly, for the rock art of all known pre-literate, pre-writing cultures, from Eurasia to Africa to the Americas and even to Australia. The rock art evidences the same modes of image use, reuse, association and accumulation as are found in the earlier Eastern and Western European traditions. In this sense, the cognitive mode, involving the periodic use of image and symbol, approaches a universal *H.sapiens sapiens* capacity comparable to, but not equivalent to, the capacity for language use.
>
> Different specialized areas of the brain and different functional anatomy are involved in the use of language and the use of icon and image. Nevertheless, the level of evolved capacity and the cortical architectonics involved in the two modalities, including their associative functions, suggest that they are aspects of one development and evolved together. It would probably be impossible to develop complex artificial iconographic traditions without language and

> impossible to have language without reference to the near-iconographic conceptual models implied in linguistic semantics. (ibid., pp. 293–4).

5 This conclusion, that the making of such engravings is part and parcel of a universal proto-human behavioural complex in which language and community life as a whole are involved, makes the issue of 'diffusion' of particular symbol traditions very difficult to unravel:

> The question whether language was dispersed from a single origin or was repeatedly invented may be asked of these iconographic systems as well. If we assume the presence of certain regularities in the phenomenological and relational worlds of early man, similarities in the symbolic products referring to these may not be surprising, but this does not solve the problem of possible dispersal of cultural strategies and modes.
>
> Within the Eurasian heartland and contiguous worlds (perhaps then, the human world), the periodic mode of symbol use and the presence of a water related iconography are abundantly documented. One also finds a broad distribution of the animal image, the female image, the human or anthropomorphic image, and even the more specialized vulvar image. It is difficult to make any choice between diffusion and independent invention without more and better archaeological evidence . . . We may . . . have evidence not of a diffusion of the motif, but of a fundamental human recognition of the seasonal, periodic, variable and life-supporting nature of water-related phenomena. (ibid., p. 294)

Since Marshack's expositions are necessarily interwoven with detailed illustrations and analyses of specific objects, it is hard to do justice to him here. His views on language evolution were briefly touched on in the previous chapter, and the quotations just given amplify them a little further. Marshack (1975) is an excellent popular account of his work and introduction to his ideas. It might be apposite here to remark that Marshack's work on the calendrical nature of much of the later Upper Palaeolithic marking relates in some ways to the work of Thom (e.g. Thom, 1971) on the astronomical functions of the later megalithic constructions and stone circles of Britain and France (but see also Ellegard, 1981, for a critical perspective). This earlier material perhaps represents temporal tracking and coding in a more rudimentary form, from which later more sophisticated or ambitious methods eventually stemmed.

Another writer influenced by the psychological perspective is Pfeiffer (1982). In this work he offers what is probably the most comprehensive account of the Upper Palaeolithic period and draws eclectically on numerous disciplines for possible insights. His underlying thesis is nevertheless broadly psychological in character, as he argues that the 'creative explosion' can best be explained in terms of the need to process, store, and transmit an ever-growing information load (a trend which continues from then on, eventually giving rise to writing, printing and computers). The connections between this perspective and those of Lieberman and Marshack, for example, are fairly self-evident; however, Pfeiffer's eclecticism is perhaps a little too uncritical for most psychologists to be entirely happy with.

4 *Between the Stone Age and civilisation*

The ice retreats. The caves are deserted. Mammoth and cave-bear disappear. The human condition, in Eurasia particularly, but to some extent universally, enters a new phase of more rapid and continuous change. Never again will a tradition last twenty millennia. The dynamic is powered from several different directions. The environment itself has altered, of course, requiring a change in survival and subsistence strategies, but this alone would not have brought about *continuous* change, only a shift to a new equilibrium. There is little doubt that population was increasing also, if not dramatically, as a result of more effective methods of exploiting natural resources. At the heart of the situation is the momentum of human culture itself, set on a cumulative, irreversible course. It is this which makes the more effective methods of resource exploitation possible, this which, as Pfeiffer sees it, both facilitates and reinforces the steady growth in 'information processing', and 'information storage' capacity. In another four thousand years or so the first cities will appear in the Middle and Near East. But even where city-based civilisations do not arise and literacy is never achieved, things do not stand still – intermittent discoveries and inventions, cross-cultural contact and centuries of ritualised oral storage and transmission of knowledge and beliefs move even the most 'backward' groups some way away fom their Upper Palaeolithic level – even if not in directions which appear 'progressive' to modern western eyes.

For the purposes of the present work, the post-Palaeolithic period is something of a coda. Although there are dissenting voices, such as Jaynes, the general consensus is that by the end of the Palaeolithic humans had acquired all the resources both psychological and biological which they now possess. Between the end of the Palaeolithic, coinciding in Europe with the retreating ice, and the beginning of 'history' falls what Grahame Clark has called the 'Mesolithic Prelude'. The ensuing Neolithic, once seen as itself revolutionary by Childe and others (e.g. S. Cole, 1965), now appears to emerge more gradually from this, fading into the earliest phases of 'history', still roughly classed as the Bronze and Iron ages, though such terms fit less well on American and African developments. The list of Mesolithic 'firsts' is extensive (see Part D), ranging from the domestication of the dog to the invention of the dug-out canoe. Though not yet farmers, as were the Neolithic peoples who replaced them in Europe, much of Asia and the Nile Valley, if not elsewhere, the inhabitants of Mesolithic Europe could deploy a great range of food-gathering techniques, including animal management, fishing, hunting with bow and arrow, and harvesting of wild cereals using sickles. Rudimentary trading networks were probably in existence, and had been so since the late Upper Palaeolithic, but now a wider range of goods seems to be involved than hitherto. Seasonal migrations in pursuit of food resources continue. Pottery only comes after agriculture in the Neolithic. Most of the apparently 'primitive' contemporary societies would be better described as Mesolithic in level than Palaeolithic, and even so have by now acquired some post-Mesolithic features. The fact that certain natural resources, such as metal, are simply not available to a people, does not mean that their cultural level as a whole is irredeemably 'stuck' in a form comparable to the late European Stone Age, but rather that it will evolve along a different trajectory. Thus these labels ought not to be applied too strictly when comparing regions widely different in ecology, physical geography and resources.

Such boundaries as that between Mesolithic and Neolithic are retrospectively drawn for the convenience of modern scholars. Even if the Mesolithic can be characterised as a period of transition from hunter-gatherer foraging to pottery-using settled farming, there was no point when somewhere in 6,500 BC the day dawned over the Jordan Valley and everyone awoke to find themselves in the Neolithic at last! Already in the Upper Palaeolithic, regional variations are coming to dominate the universal species level; the vagaries and varieties of 'cultural evolution' obscuring general species-wide psychobiological evolutionary trends. After the Ice Age this becomes even more acute, and the evolutionary perspective yields to the cultural historical one. What befalls populations from then on will depend primarily on the luck of the environmental draw; it is this which will determine the primary phenomenological background to cultural behaviour – whether the seasons are regular, as in Egypt, or erratic, as in Mesopotamia, or hardly perceptible, as in Equatorial Africa, whether life is dominated by forests, mountains, desert, or the sea, whether life is harsh, or easy, whether the resources require skill and ingenuity to obtain or whether they are easy of access – these and a hundred other factors will determine the 'informational input', the raw material for 'physiomorphic assimilation' that will determine, in its turn, the unique course of each culture's development, and the unique modes of consciousness, the 'psychologies', of those who comprise it. But even this process eventually reaches its culmination at the point, now presumably reached, where regional cultures begin to yield to a global one.

D First known cases

First known cases (Table 6.2) do not of course necessarily reflect the first actual occurrences, whether of fire-use, flint-tool or basketry. Different activities leave different traces, some of which are more ephemeral than others, while raw materials vary from the transient (who sang the first song?) to the nigh-imperishable (some rocks used in tool-making). Even so, though the chronological sequence itself is unreliable in terms of giving priorities, it can at least provide *ante quem* dates, dates before which an innovation must have occurred. Furthermore, the sudden appearance of a new class of object or behaviour trace in the archaeological record, such as pottery shards, might give a relatively clear indication of when it became culturally significant. We can never know whether or not an idle australopithecine ever made a rude clay figure while playing in the mud by an African river, but it does not matter, for even if they did, it never caught on, and pottery had to wait three million years more before it entered hominid life. Anticipations such as this are of no significance if they are not followed up, and perhaps some things have to be rediscovered many times before they really 'take'. The table then should not be over-interpreted. First-time dates in the post-Palaeolithic are somewhat more likely to be realistic than the earlier ones as far as those items likely to leave enduring traces are concerned – it is thus now highly improbable that our picture of the beginnings of pottery or metal use is seriously in error. But other products like netting are a different matter entirely, and a freak preservation can still push first-time dating back by millennia. Nor are all developments clearcut – what, for example, *is* 'domestication'? With animals this covers a continuum from some form of herding or even mere provision of fodder to attract wild herds, to controlled breeding and use as work animals. With plants it ranges from taking some degree of rudimentary care of wild-plant sites (keeping them weed-free, leaving enough unharvested to ensure seeding for the next season) to selective breeding and planting which gives rise to recognisably new strains. The picture of high levels of domestication going on between ten and five thousand years ago is nonetheless undoubtedly correct. I could find nothing, incidentally, on the domestication of cats.

Fire-use data is especially problematical, and has been discussed elsewhere. Again this covers everything from opportunistic use of natural fires, keeping them going perhaps for a short period of time, to efficient fire-making. It is nigh on impossible to tell from the earliest sites whether the fire was naturally started, 'curated', or artificially created. Chemical processes and other natural phenomena can also leave traces deceptively like crude hearth sites, which adds to the difficulties of identification.

TABLE 6.2 *Earliest dates of various behavioural categories, artefacts, etc*

Category	*Site*	*Date BP*[1]	*Notes*
Artificial structure	Olduvai, Tanzania	1.8 myr	Doubts raised by Potts (1984)
	Terra Amata, France	0.3 myr	
Basketry and weaving	Danger Cave, Utah, USA	>10,000	Extremely favourable preservation conditions
Bow and arrow	Ahrensberg, Germany	10,000	Arrows found
Burial	La Ferrassie, France	0.1 myr	
	Shanidar, Iran	0.05 myr	The famous 'Flower Burial'
Butchery shown by cut marks	Olduvai Gorge, Tanzania	1.8 myr	
'City'	Jericho, Israel	>9,000	Definition of 'city' problematical
Domestication:			
Animals –			
Cattle	Lukenya Hill, Kenya	13,000?	Both dates are probably substantially later than the actual origin of domesticated cattle
	Uan Muhuggiag, Tassili, Algeria	7,500	
Dog	?	'late Ice Age'	Gradual differentiation in anatomy from wild canidae
Horse	La Quina, France	35,000!	Strong circumstantial evidence for these very early dates. See Hadingham (1980) for discussion of the mystery
Reindeer	Abri Pataud, France	22,000!	
Sheep	Zawi Chemi Shanidar, Iran	10,000	Shift from hunting to herding
Plants –			
Maize	Tehuacén, Mexico	7,000	
	Guitarrero, Peru	6,000	
Rice	Homutu, China	6,000	
Wheat	Abu Hureya, Syria	9,000	Not undisputed
Engraving	Peche de l'Aze, France	0.2 myr	On an ox-rib
Fire-use	Chesowanja, Kenya	1.5 myr	Controversial
	Karari, Kenya	1.4 myr	Some uncertainty
	Gadeb, Ethiopia	1.2 myr	Some uncertainty
	Escale, France	0.7 myr	Earliest in Europe
	Yuanmou, China	1.3 myr	Dating very uncertain
	Olorgesailie, Kenya	0.45 myr	
	Terra Amata, France	0.3 myr	
	(Petralona, Greece	0.8 myr	Dating very dubious)

Category	*Site*	*Date BP*[1]	*Notes*
	Choukoutien, China	0.5 myr	Interpretation now in doubt
Grindstones	Wadi Kubbaniya, Egypt	17,000	
Homo arrival in			
America	?	>30,000	Highly contentious
Australia	–	50,000	Inference from morphology of Kow Swamp fossils
China	Hsi Hontu, Shangsi Province	1.8 myr	Tools only
Europe	Isernia, Italy	>0.7 myr	
	(Petralona, Greece	0.8 myr	Dating very dubious)
Irrigation	Catal Hüyük, Turkey	>7,000	
	Deh Luran, Iran	>7,000	
'Jewellery'	La Quina, France	35,000	Bone and tooth pendants in Mousterian context
Laws	Sumer (Iraq)	4,100	Ur Nammu Code
Metal use	Cayönü Tapesi, Turkey	9,000	Hammered copper, not smelted
Musical instrument	Haua Fteah, Libya	60,000	Bone whistle
Needles	Found in Gravettian industries, France	26,000	
Pigment use	Terra Amata, France	0.3 myr	Utilised ochre lump. This date is likely to grossly under-estimate ochre-use origin
Pottery (fired)	'Two Japanese Sites'	10–12,000	Not firmly established
	Spirit Cave, Thailand	8,800	
	Jericho, Israel & Jarmo, Iran	8,000	
	(America	4,000)	
but	Pavlov, USSR	27,000!	Baked clay *figures*
Scalping	Bodo, Ethiopia	1.0 myr?	This skull is undated but late *H.erectus* from Acheulean Site 0.5–1.5 myr
Script	Mesopotamia	5,000	Definition is a problem here, *notation* of some kind existed in Upper Palaeolithic
Script (alphabetic)	Ugarit, Syria	3,500	Cuneiform

Category	*Site*	*Date BP*[1]	*Notes*
Skull deformation	Shanidar, Iran	>50,000	
Statuettes (Ivory)	Vogelherd, Germany	>32,000	Quality suggests well-established tradition
Stone tools	Hadar, Ethiopia	2.6 myr	
Levalloisian	Kapthurin, Lake Baringo, Kenya	0.23 myr	
Thread, rope	Lascaux Caves, France Danger Cave, Utah, USA	15,000 >10,000	But much earlier existence of needles proves these dates to be deceptively late
Walking bipedally	Laetoli, Tanzania	3.5 myr	Footprint trail. Anatomical evidence shows *A.afarensis* had been walking well before this
Wheel	Caucasus area?	8,000	Very inferential. The wheel was unknown in Pre-Columbian America
Wooden artefact	Clacton-on-Sea UK Lehringen, Germany	>0.3 myr 0.12 myr	A thrusting spear A thrusting spear

This table was drawn up primarily from Gowlett (1984), with assistance from Hadingham (1980), and Wymer (1982).

[1] All these dates are approximations. The Petralona Cave dating by Poulianos of as much as 800,000 years is given here since it is cited in some of the literature, but as explained elsewhere it is most doubtful, likely to be less than half this age.

E Conclusion

This chapter has merely scratched the surface of current knowledge regarding the final phase of our evolution. Its intention has been to draw attention to some of the work being done, and the kinds of problem which are arising, to which the behavioural sciences might fruitfully contribute. New methodologies, such as microwear studies and advanced forms of site analysis, must, it seems to the author, draw Psychology into the picture if the behavioural and motivational implications of the data are to be understood. But beyond this, it has been the underlying intention of much of this book to show how the fundamental issue of the nature of 'hominisation' cannot be understood without addressing the psychological dimension. From the perspective I adopted in Chapter 4, the period which we have just been looking at, the Upper Palaeolithic and its aftermath, appears to be an undamming of the 'physiomorphic' process, caution is thrown to the wind, the human race – or at least a section of it – embarks on 'conquering nature' by taking into itself the very properties which were once hers and remixing them in ever more effective permutations. Against this strategy, no ecological check could work in the long run. The process continues to accelerate. As Gowlett points out, 10,000 years ago 100% of the race were hunter-gatherers; by AD 1500 this had reduced to 1%, and is now less than 0.001%. World population in the meantime has increased from an estimated 10 million to 4 billion, an increase of 40,000%. Rarely can the planet have seen such an unstoppable and sustained explosion in a single species, now deploying abilities with a potential for affecting events at the climatic and geophysical levels. We swarm like flies in the wake of the retreating ice. Only by our own understanding of our situation from this perspective will we come to be able to do what no other force can now do: get ourselves under control.

APPENDIX A

Principal Pre-Modern Hominid Fossils

The following table lists most of the pre-modern hominid fossil material referred to in the main text. It does not include Miocene hominoids, since their inclusion would have gone beyond the scope of the book, as well as the author's current range of expertise. Well-known material has not been annotated, but sources for some more recent finds and data have been given.

The normal practice in cataloguing fossils is to number them consecutively as found, with a letter prefix indicating the site. In cases where a number of bones of a single individual are found they may be classified with a single number but suffix letters added for each specific bone, thus KNM-ER 1481a refers to the femur of an individual whose tibia and distal fibula were also found (KNM incidently stands for Kenya National Museum).

This listing makes no pretence of being comprehensive. For a fuller introduction to the hominid fossil material see the revised edition of M. H. Day's *Guide to Fossil Man* (1986, replacing the 1977 edition). The massive British Museum (Natural History) Catalogues only cover the material up to the early–mid 1970s.

Taxon[1]	*Site*	*Fossils*[2]	*Dating*[3]
Australopithecus afarensis and earlier hominids	Baringo, Ethiopia	'Ngorora' tooth	12–9 myr
	Bodo, Ethiopia	MAK-VP-1/1 femur BEL-VP–1/1 cranial fragment	>4 myr[4]
	Hadar, Ethiopia	*A. afarensis* type site. 200 + bones inc. AL 288-1 'Lucy'. (Are robust represented too?)	c.3 myr[5]
	Kanapoi, Kenya	part left humerus	4 myr

Taxon[1]	*Site*	*Fossils*[2]	*Dating*[3]
	Laetoli, Tanzania	20+ 'dental & gnathic fragments', footprint trail	3.5–3.8 myr
	Lothagam, Kenya	KNM-LT 329 mandible	5–5.5 myr
	Lukeino, Kenya	KNM–LN–335 molar	c. 6.3 myr

The taxon name '*Australopithecus afarensis*' should perhaps be reserved strictly speaking for the Hadar fossils at present, with the remainder as indeterminate early hominids which may or may not be identical with Hadar species.

Australopithecus africanus (i.e. gracile australopiths)	Makapansgat, S.A.	26 early hominid finds. Sample may inc. robust specimens too or even be at gracile-robust speciation point	>2 myr, perhaps as early as 3.7 myr (when cave is estimated to have opened)
	Omo, Ethiopia	fragments	2–3 myr
	Sterkfontein, S.A.	Numerous, inc. Sts-14, near complete skeleton	2–3 myr
	Taungs, S.A.	'Taungs skull' Dart's type specimen. Child's skull + endocast	? as late as 0.9 estimated
Australopithecus robustus varieties	Chesowanja, Kenya	KNM-CH-1 (part face)	c. 1.1 myr
	Kromdraai, S.A.	Numerous frags, '*Paranthropus*'	1.5–2 myr?
	Olduvai, Tanzania	O.H.5 'Zinjanthropus' *A.r.boisei*	c. 1.7 myr
	Omo, Ethiopia	Numerous fragments (most abundant species at Omo)	From 2 to 1 myr
	Peninj, Tanzania	A robust jaw	c. 1.35 myr
	Sterkfontein, S.A.	Inc. Sts-5, 'Mrs Ples', 'Plesianthropus'	2–3 myr
	Swartkrans, S.A.	Over 200 fossils inc. '*A. crassidens*', finds inc. SK 1585 endocast	c. 2 myr (younger than Makapansgat and Sterkfontein)

Dating of the South African sites has been very problematical. Classification of the Australopithecines is again in flux (see Chapter 3).

Taxon[1]	*Site*	*Fossils*[2]	*Dating*[3]
Homo habilis	East Lake Turkana, Kenya.[6]	KNM-ER 1470, 1481, 1580, 1802, 3228, 3722. KNM-ER 1805, 1813 ?[7]	c. 2 myr (1470)
Homo habilis	Modjokerto, Java	skull, '*H. modjokertensis*' morphologically similar to both *A. robustus* & *H. habilis*.	?
	Olduvai, Tanzania	O.H. 4, 6, 7, 8, 13, 25. O.H. 14 & 16?[7] (O.H. 16 – 'Olduvai George')	1.7 myr
	Omo, Ethiopia	A few teeth	1.85 myr
	Sterkfontein, S.A.	Some fragments akin to *H. habilis*	1.5 myr

Although this problematical taxon still seems legitimate, there is a continuous shuffling of some specimens which possess intermediate features between *H.habilis*, *A.robustus* and/or *H. erectus*. These are better described at present as 'indeterminate'.

Homo erectus	Choukoutien, China (now 'Zhoukoudian')	Remains of over 40 individuals of the original 'Peking Man' collection lost in WWII but more found since inc. Skull V (1966) described as 'more progressive'	From 0.46 myr to 0.23 myr[9]
	Lake Turkana, Kenya	KNM-ER 1813, 3733, WLT 15,000	c. 1.6 myr (WLT 15,000)
	Hexian, China	Cranium, teeth 'evolved form'	'Middle Pleistocene.'[9]
	Lantian, China	Mandible and cranium. Older than 'Peking Man'	'Early Pleistocene.'[9]
	Olduvai Gorge, Tanzania	Numerous fragments	c. 1.5 myr
	Omo, Ethiopia	Fragments	1.1 myr
	Sangiran, Java	Sangiran II–VIII (VIII = Sangiran 17)	? c. 0.75 myr
	Ternifine, Algeria	Ternifine skull	>1.0 myr

Taxon[1]	*Site*	*Fossils*[2]	*Dating*[3]
	Trinil, Java	Original 'Pithecanthropus' skull and post-cranial fragments, the femur is now considered very doubtful	0.5 myr ?
	Yuanmou, China	2 teeth	>0.73 myr (revised down from 1.7 myr)

Javanese dating is very problematic, and the stratigraphy of Dubois' original Trinil find far more complex than he realised, raising questions about the coherence of the finds. The femur in particular is now clearly aberrant compared with other *H. erectus* femurs.

Homo erectus – *Homo sapiens* transition	Arago, Tautuval, France	Numerous fragments inc. Arago 21 cranium	> 0.3 myr
('archaic *H. sapiens*' included here)	Bodo, Ethiopia	Cranium with signs of having been scalped	? undated as yet, but early
	Castel di Guido, Italy	Post-cranial fragments	c. 0.3 myr[8]
	Dali, Shangsi, China	skull	'late middle[9] Pleistocene.'
	Fontéchevade, France	skull	c. 140 kyr
	Heidelberg, Germany	'Mauer jaw'	0.5 myr
	Maba, Guandong, China	Skull	n.a.[9]
	Omo, Ethiopia	Skulls Omo I and II in Kibish formation	c. 1.0 myr
	Petralona, Greece	Skull	c. .3–.4 myr (disputed)[10]
	Steinheim, Germany	Skull	c. 0.3 myr
	Swanscombe, U.K.	Part skull	0.25–0.3 myr
	Vertesszöllös, Hungary	Rear skull	0.2 myr
Homo sapiens neanderthalensis	La Chapelle-aux-Saints, France	Skeleton 'Old Man' reconstructed by Boule	40,000 BP +
	Djebel Irhoud, Morocco	Skulls I and II	40,000 BP
	La Ferassie, France	Several burials inc. ritual burial of child	40,000 BP +

Taxon[1]	*Site*	*Fossils*[2]	*Dating*[3]
	Gibraltar	Skulls I and II	35–70,000 BP, c. 50,000 BP
	Haua Fteah, Libya	Fragments I and II (part jaw)	47,000 BP
	Krapina, Yugoslavia	800 fragments = c. 80 individuals	60,000 BP?
	Monte Circeo, Italy	I–IV	35–70,000 BP
	Solo River, Java (Ngandgong)	Solo I–XI. Class. *H. soloensis.* Probably a Neanderthaloid subspecies	c. 0.1 myr?
	La Quina, France	Fossils I–XXVII, various	Range from 33–55,000 BP
	Saccopastore, Italy	Skull fragments I, II	70,000 BP +
	Saint-Césaire, France	Skull fragments in Chattelperronian context	32,000 BP[11]
	Sala, Czechoslovakia	Front part of skull	125,000 BP
	Shanidar, Iraq	9 partial skeletons, inc. Shanidar 4 – 'flower burial'	60,000 BP + 40,000 BP +
	Tabūn, Israel (Mt. Carmel)	Remains of 6 individuals	38–39,000 BP
	Vindija, Yugoslavia	35 Neanderthal fragments inc. Vindija 206 mandible	'Riss-Würm' Interglacial
Homo sapiens, early non-Neanderthals	Border Cave, S.A.	Skull	? 39–120,000!
	Boskop, S.A.	Skull	c. 15,000 BP
	Broken Hill, Zimbabwe. (Rhodesia Man)	Skull, class. *H.s. rhodesiensis* ('Archaic' classification prob. now preferable)	c. 125,000, rev. up from 40,000
	Florisbad, S.A.	Skull	c. 35,000 BP
	Kow Swamp, Australia	40 skeletons, anatomically archaic	9–10,000 BP
	Laetoli, Tanzania	'Ngaloba skull' LH–18	c. 0.11 myr[12]
	Nazlet Khater, Egypt	2 skeletons	30–35,000 BP[13]
	Qafzeh, Israel	I–VIII	40–60,000 BP
	Saldhana, S.A.	Skull, similar *H.s. rhodesiensis*	57,000 BP
	Skhūl, Israel (Mt Carmel)	Skhūl I–X inc. skulls	40,000 BP +

Taxon[1]	*Site*	*Fossils*[2]	*Dating*[3]
	Wadjak, Java	Skulls I and II 'H. *wadjakensis*'	?
H.sapiens neand. – *H.s. sapiens* intermediate?	Hahnofersand, Germany	Front of cranium	c. 36,000 BP[14]

Notes

[1] I have used the term 'taxon' in an effort to avoid contentious issues about the status of these terms as referring to clades, grades, genera, etc. These are the broad groupings in which hominid fossils are currently classified.

[2] I have only included enough here to indicate the nature of the find(s), though specimen numbers are given for some sites of fossils referred to frequently in the literature in this way.

[3] Dating figures are those in the most recently available sources, these may sometimes differ from those given elsewhere and should be preferred to dates given in pre-1980 sources if these differ significantly.

[4] J. D. Clark et al. (1984) dating.

[5] Hadar dating from R. C. Walter and J. L. Aronson (1982); J. T. Schmitt and A. E. M. Nairn (1984).

[6] The general 'East Lake Turkana' area includes Koobi Fora formation finds, specimen numbers prefixed KNM-ER, the ER standing for 'East Rudolph', the colonial name for the lake.

[7] Status of these currently being reviewed.

[8] F. Mallegni et al. (1983).

[9] Wu Rukang (1982), Wu Rukang and Xingren Dong (1983).

[10] C. B. Stringer at al. (1979); A. G. Wintle and J. A. Jacobs (1982); A. N. Poulianos (1984).

[11] A. M. ApSimon (1980).

[12] M. H. Day, M. D. Leakey and C. Magori (1980).

[13] P. M. Vermeersch et al. (1984); P. M. Vermeersch, G. Gijselings and E. Paulissen (1984).

[14] G. Brauer (1981).

APPENDIX B

Development of the hominid jaw

Since jaws and teeth are the most frequently occurring hominid fossils, they have assumed a central place in the study of our evolution. The following drawings, from the British Museum of Natural History leaflet *Our Fossil Relatives* illustrate some of the most important of these, enabling us to track, if somewhat hazily, the emergence of the modern *H.sapiens* mandible.

A Ramapithecine jaw from Turkey
B Sivapithecine jaw from Potwar Plateua, Pakistan
C A chimpanzee
D AL-288-1, 'Lucy', *Australopithecus afarensis*, from Hadar, Ethiopia
E A larger australopithecine (AL-333w60) also from Hadar
F Gracile *Australopithecus africanus* (Sts 52b) from Sterkfontein, South Africa
G Robust australopithecine from Koobi Fora, Kenya (KNM-ER 729)
H OH 13, a small *Homo habilis* from Olduvai
I KNM-ER 1802, a large *Homo habilis* from Koobi Fora
J Modern *Homo sapiens*

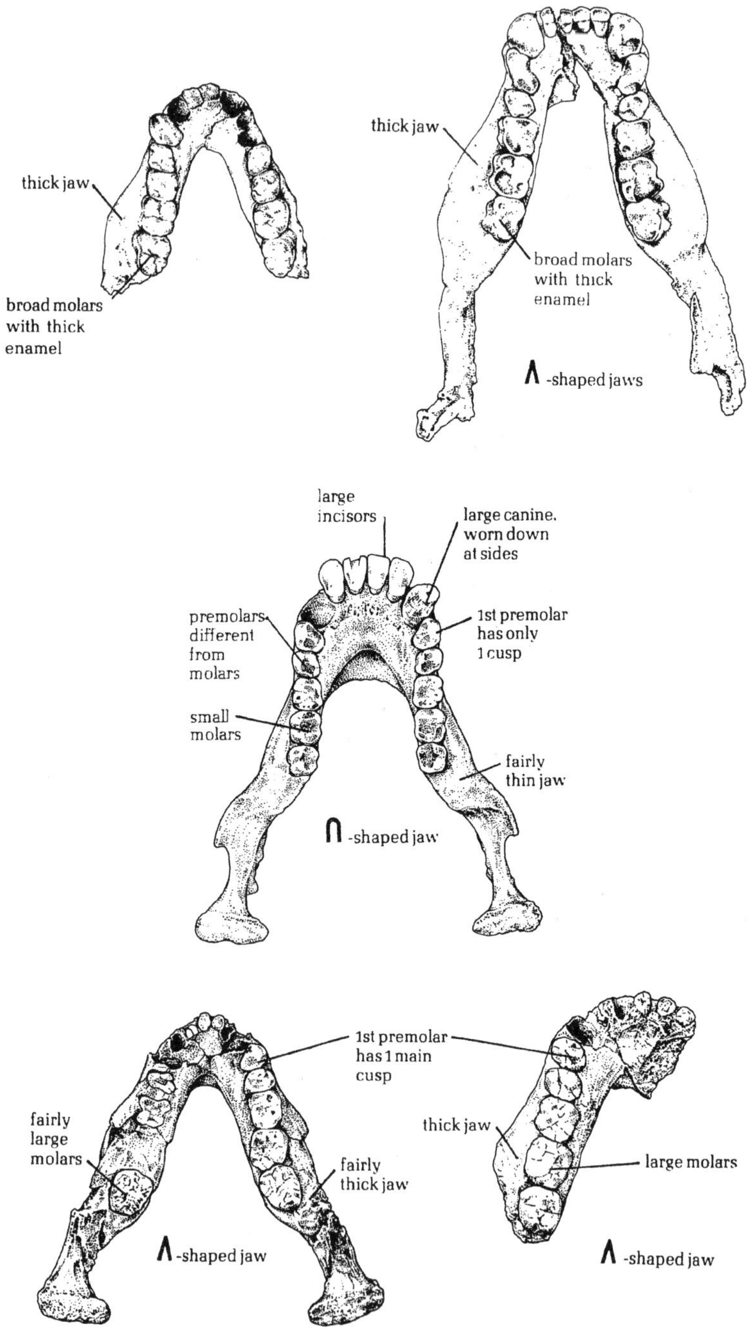
thick jaw
broad molars
with thick
enamel
thick jaw
broad molars
with thick
enamel
Λ -shaped jaws
large
incisors
large canine.
worn down
at sides
premolars
different
from
molars
1st premolar
has only
1 cusp
small
molars
fairly
thin jaw
∩ -shaped jaw
1st premolar
has 1 main
cusp
fairly
large
molars
fairly
thick jaw
Λ -shaped jaw
thick jaw
large molars
Λ -shaped jaw

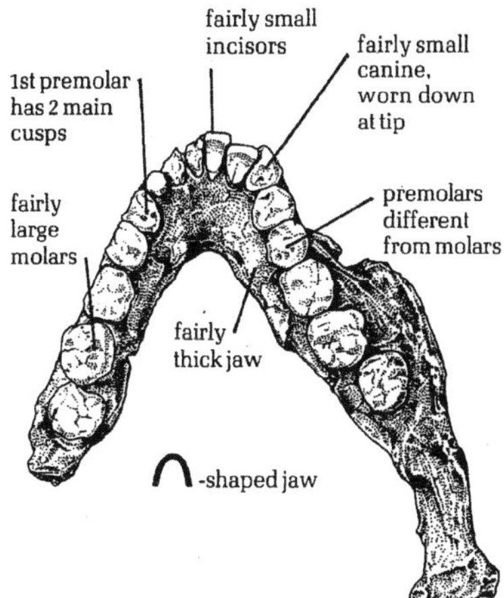
fairly small incisors
fairly small canine, worn down at tip
1st premolar has 2 main cusps
premolars different from molars
fairly large molars
fairly thick jaw
-shaped jaw

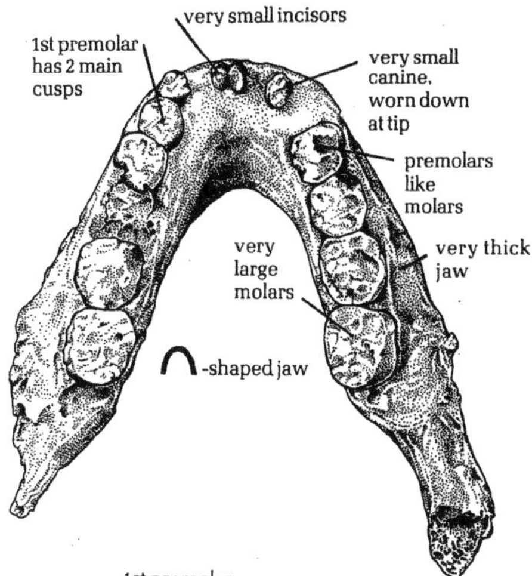
very small incisors
1st premolar has 2 main cusps
very small canine, worn down at tip
premolars like molars
very large molars
very thick jaw
-shaped jaw

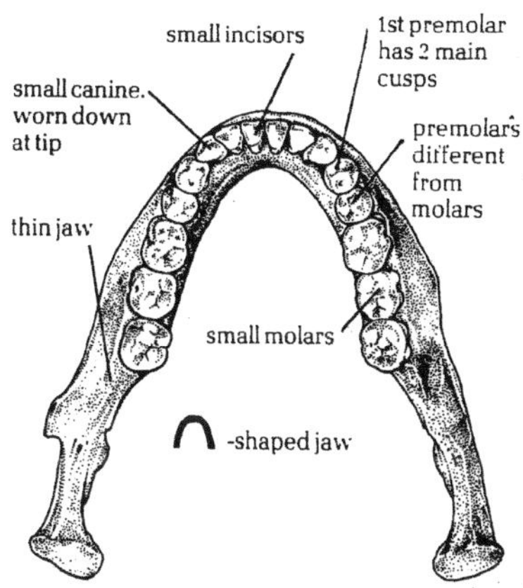
small incisors
1st premolar has 2 main cusps
small canine, worn down at tip
premolars different from molars
thin jaw
small molars
-shaped jaw

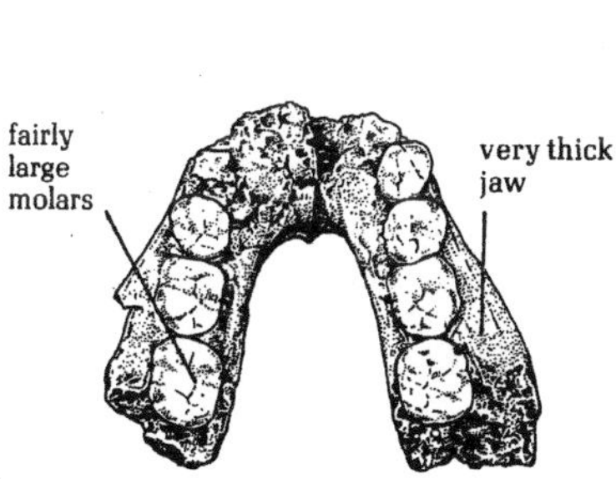
fairly large molars
very thick jaw

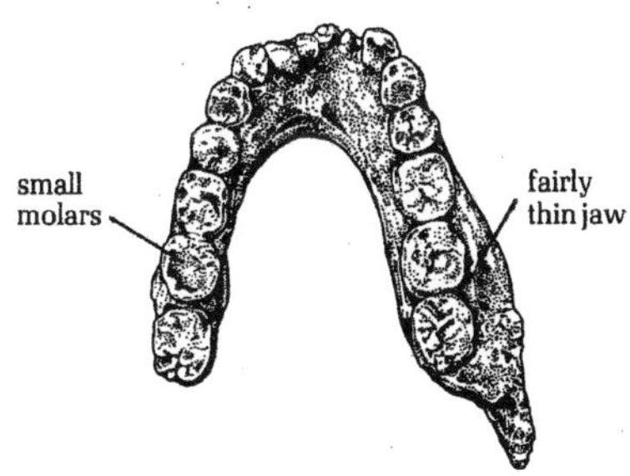
small molars
fairly thin jaw

APPENDIX C

Hominid genealogy – seventy-two possibilities

Stage 1

Splitting from the apes. Kluges (1983) identifies four logically possible relationships.

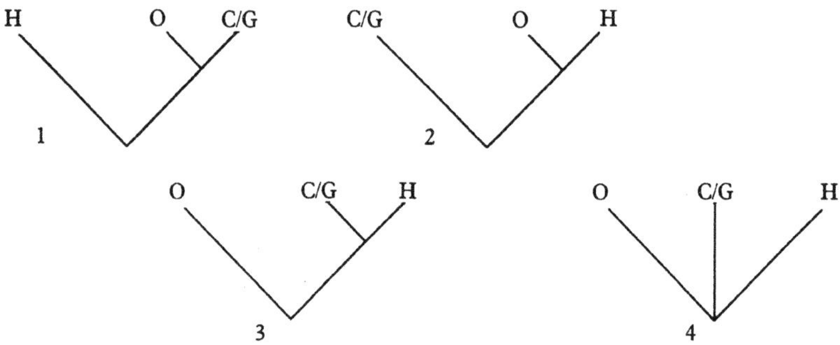

1 The hominid split from the great apes precedes the orang utan (O) separation from the Chimpanzee/Gorilla (C/G) lineages. The common ancestor is thus shared by all three extant pongid species and hominids, but no pongid species is any closer to us than the other two.
2 Hominids are closer to orang utans than to gorillas and chimpanzees.
3 Hominids are closer to gorillas and chimpanzees than to orang utans (the most favoured current position).
4 All three groups diverged from the common ancestor at the same point, thus being equally distant from one another.

Stage 2

The relationships between the australopithecines and *Homo habilis*. This yields six major possibilities. A.af = *Australopithecus afarensis*, A.afric = *Australopithecus africanus*, A.r. = *Australopithecus robustus*, H.h = *Homo habilis*. This is, even so, a simplification in view of the revival of '*Paranthropus*' for robust australopithecines (see main text) and the possibility that more than one species is represented in the Hadar '*A.afarensis*' material.

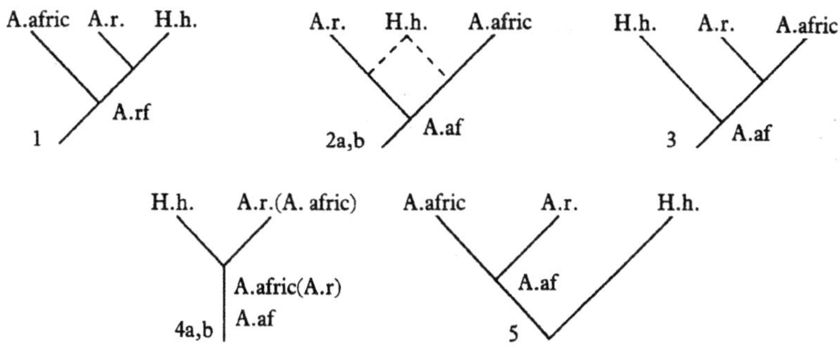

1 *H.habilis* evolved from *A.afarensis*, other australopithecines splitting off from this main lineage (either sequentially as depicted here, or in reverse order – very unlikely – or via a single intermediate species which then divided).
2 The *A.afarensis* lineage divided into the two australopithecine families from one or other of which *H.habilis* later evolved or split (giving two sub-versions according to whether *A.robustus* or *A.africanus* is the ancestor – a & b respectively).
3 The lineage leading to *H.habilis* split from *A.afarensis* prior to that dividing into the later australopithecine species.
4 The *A.afarensis* lineage evolved into *A.africanus* which subsequently split into *H.habilis* and *A.robustus*. A possible alternative would be to have *A.robustus* and *A.africanus* reversed in this scheme of things, but this is actually virtually impossible since the robust fossils generally postdate the gracile ones.
5 *H.habilis* split off from a common ancestor with *A.afarensis* prior to the emergence of *A.afarensis* itself. *A.afarensis* then gives rise to the australopithecine family, but is not a direct ancestor of the hominids.

The favoured positions at present tend to be versions of 3 or 5 (which are similar insofar as they do not place known australopithecines directly in the *H.habilis* ancestry except perhaps via an early version of *A.afarensis*). On the other hand the morphological similarities between some robust australopithecines and *H.habilis* are such that positions 1 and 2a remain very much in the picture.

Stage 3

From *H.habilis* to *H.sapiens sapiens*. Three general kinds of position can be adopted here, none devoid of problems, and all yielding a variety of detailed possibilities regarding the inter-relationships of specific fossils. H.h. = *Homo habilis*, H.e. = *Homo erectus*, N. = Neanderthals, A. & I. = Archaic & Intermediate forms, S = Solo Man (Asian Neanderthaloids), P = Peking Man, = = extinction

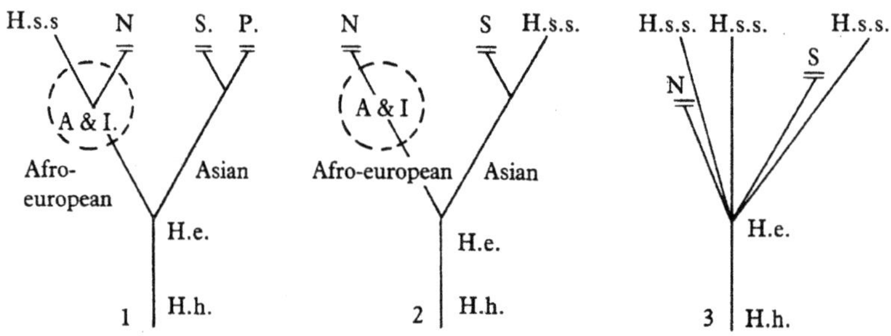

1 The position favoured by Andrews (1984). *H.erectus* splits into Asian and Afro-European populations. The former becomes extinct after Peking Man. The latter gives rise both to the Neanderthals and to *H.s.sapiens* as well as the various Archaic *H.sapiens* species and others apparently intermediate morphologically between *H.erectus*, Neanderthals and Moderns. *H.sapiens* is considered to have evolved in Africa as a phylogenetically distinct population 'lacking the many specialisations of *H.erectus*' (p. 25). *H.erectus* thus represents the first wave or waves of radiation from Africa, and is strictly speaking a non-African species; it is then followed by a further *sapiens* wave which displaces or absorbs it.

2 A second possibility is that the Afro-European population, culminating in the Neanderthals although incorporating various 'Archaic' populations, is the one which becomes extinct, while the Asian population is the one which gives rise to *H.s.sapiens*, as well

as to the Neanderthal-like Solo Man of Java. This is compatible with the evidence that Moderns entered Europe from the East and the archaic elements in the Kow Swamp specimens, as well as accounting for the shared derived characteristics of both *H.erectus* and Modern Asian populations. How, though, were these lost again in the European populations? Is such evolutionary reversal possible or likely?

3 Thirdly, it may be possible to conceive of *H.erectus* as a polytypic species in which no genuine split in the population occurs, although regional differences in morphology arise. There would be continuous gene-flow through the population, perhaps radiating from a central African/Near Eastern region to the peripheries. Some types (Neanderthals) become extinct, some (Asian populations for example) 'sapienise' *in situ*, retaining some regional characteristics. On this model 'sapienisation' could occur across a broad front as a result of gene-flow. This is roughly the position of Thorne and Wolpoff (1981). Such an evolutionary process would, though, be unparalleled elsewhere in the animal kingdom and does present some difficult technical and theoretical problems.

Given 4 possibilities at the first stage, 6 at the second and 7 at the third we clearly have no fewer than 72 genealogical possibilities for the course of hominid ancestry, even in the rather simple terms adopted here.

Bibliography

The abbreviations used are the following:

AJPA	*American Journal of Physical Anthropology*
Am. Psy.	*American Psychologist*
Anth.	*Anthropology*
BBS	*The Behavioural and Brain Sciences*
CA	*Current Anthropology*
EMM	D. S. Bendall (ed), (1983), *Evolution from Molecules to Men*, Cambridge, UK, Cambridge University Press
JHE	*Journal of Human Evolution*
L'Anth	*L'Anthropologie* (Paris)
LDC	H. C. Plotkin (ed.) (1982), *Learning, Development and Culture: Essays in Evolutionary Epistemology*, Chichester, John Wiley.
MTP	B. A. Wood, L. B. Martin, and P. Andrews (eds) (1986), *Major Topics in Primate and Human Evolution*, Cambridge UK, Cambridge University Press
Nat.	*Nature*
NI	R. L. Ciochon and R. S. Corruccini (eds) (1983), *New Interpretations of Ape and Human Ancestry*, New York, Plenum Press
N.Sc.	*New Scientist*
OEL	S. R. Harnad, H. D. Steklis and J. Lancaster (eds) (1976), *Origins and Evolution of Language*, New York, New York Academy of Sciences
PBE	E. Armstrong and D. Falk (eds) (1982), *Primate Brain Evolution; Methods and Concepts*, New York and London, Plenum Press
PPS	*Proceedings of the Prehistoric Society*
Sci.	*Science*
Sci.Am.	*Scientific American*

Aarsleff, H. (1976), 'An outline of language origins theory since the Renaissance', *OEL*, pp. 4–13.

Aigner, J. S. (1978), 'Important archaeological finds from North China' in F. Ikawa-Smith (ed.), *Early Paleolithic in South and East Asia*, The Hague, Mouton, pp. 163–233.

Albritton, C. C. (1980), *The Abyss of Time: Changing Conceptions of the Earth's Antiquity after the Sixteenth Century*, San Francisco, Freeman, Cooper.

Allen, G. E. (1983), 'The several faces of Darwin: materialism in nineteenth and twentieth century evolutionary theory', *EMM*, pp. 81–102.
Altmann, S. A. and Altmann, J. (1970), *Baboon Ecology: African Field Research*, Bibliotheca Primatologica, No. 12, Basel, Karger.
Andrews, P. (1984), 'The descent of man', *N.Sc.* 3rd May, pp. 24–5.
Andrews, P. and Cronin, J. E. (1982), 'The relationships of *Sivapithecus* and *Ramapithecus* and the evolution of the orang-utan', *Nat.*, vol. 297, pp. 541–6.
Andy, O. J. and Stephan, H. (1976), 'Septum development in primates', in J. F. De France (ed.), *The Septal Nuclei*; *Advances in Behavioural Biology*, vol. 20, New York & London, Plenum Press, pp. 3–36.
Apel, K. O. (1977), 'Types of social science in the light of human interests of knowledge', *Social Research*, vol. 44, part 3, pp. 425–70.
ApSimon, A. M. (1980), 'The last neanderthal in France?' *Nat.*, vol. 287, pp. 271–2.
Arambourg, C. (1955), 'Sur l'attitude, en station verticale, des Néanderthaliens', *Compte Rendu de l'Académie des Sciences, Paris*, vol. 240, pp. 804–6.
Ardrey, R. (1967), *The Territorial Imperative*, London, Collins.
Ardrey, R. (1970), *The Social Contract*, London, Collins.
Ardrey, R. (1976), *The Hunting Hypothesis*, New York, Atheneum.
Armstrong, E. (1981), 'A quantitative comparison of the hominoid thalamus: IV. Posterior association nuclei – the pulvinar and lateral posterior nucleus', *AJPA*, vol. 55, pp. 369–83.
Armstrong, E. (1982), 'Mosaic evolution in the primate brain: differences and similarities in the hominid thalamus', *PBE*, pp. 131–62.
Armstrong, E. and Falk, D. (eds) (1982), see p. 330.
Armstrong, E. and Onge, M. St. (1981), 'Number of neurons in the anterior thalamic complex of the primate limbic system', *AJPA*, vol. 54, p. 197.
Ayala, F. J. (1983), 'Microevolution and macroevolution', *EMM*, pp. 387–402.
Bada, J. L., Gillespie, R., Gowlett, J. A. and Hedges, R. E. M. (1984), 'Accelerator mass spectrometry radiocarbon ages of amino acid extracts from Californian palaeo-indian skeletons', *Nat.*, vol. 312, pp. 442–4.
Baer, C. E. von (1828), *Über Entwickelungsgeschichte der Thiere*, Königsberg.
Bahn, P. G. (1978), 'Water mythology and the distribution of Palaeolithic parietal art', *PPS*, vol. 44, pp. 125–34.
Bahn, P. G. (1984), 'How to spot a fake Azilian pebble', *Nat.*, vol. 308, p. 229.
Bain, A. (1859), *The Emotions and The Will*, London, Longmans, Green.
Barbizet, J., Duizabo, Ph. and Poirier, J. (1978), 'Étude anatomo-clinique d'un cas encéphalite amnésiante d'origine herpétique', *Revue Neurologique*, vol. 134, pp. 241–53.
Bargatzky, T. (1984), 'Culture, environment, and the ills of adaptionism', *CA*, vol. 25, no. 4, pp. 399–415.
Bartstra, G–J. (1983), 'Some remarks upon: Fossil Man from Java, His Age and His Tools', *Bijdragen Tot de Taal-, Land- en Volkenkunde*, vol. 139, no. 4, pp. 421–34.
Bateson, W. (1915), 'Heredity', *Annual Report of the Board of Regents of the Smithsonian Institution*, Washington, Government Printing House, pp. 359–94.
Bauchot, R. and Stephan, H. (1969), 'Encéphalisation et niveau évolutif chez les simiens', *Mammalia*, vol. 33, pp. 235–75.
Beard, R. (1969), *An Outline of Piaget's Developmental Psychology*, London, Routledge & Kegan Paul.

Behrensmeyer, A. K. (1978), 'The habitat of Plio-Pleistocene hominids in East Africa: taphonomic and micro-stratigraphic evidence' in C. Jolly (ed.), *Early Hominids in Africa*, London, Duckworth, pp. 165–90.
Bendall, D. S. (1983), see p. 330.
Bennett, P. G. and Dando, M. R. (1983), 'The arms race: is it just a mistake?', *N.Sc.*, 17th February, pp. 432–5.
Berckhemer, F. (1933), 'Ein Menschen-Schädel aus den diluvialen Schottern von Steinheim a.d. Murr', *Anthropologie Anz.*, vol. 10, pp. 318–21.
Berlyne, D. E. (1971), *Aesthetics and Psychobiology*, New York, Appleton-Century-Crofts.
Berndt, C. H. and Berndt, R. M. (1965), *The World of the First Australians*, Chicago, Chicago University Press.
Bernor, R. L. (1983), 'Geochronology and Zoogeographic relationships in Miocene Hominoidea', *NI*, pp. 21–64.
Bicchieri, M. G. (ed.) (1972), *Hunters and Gatherers Today*, New York, Holt, Rinehart & Winston.
Bilsborough, A. and Wood, B. A. (1984), 'The origin and fate of *Homo erectus*', *MTP*.
Binford, L. R. (1972), 'Contemporary model building: Paradigms and the current state of Palaeolithic research', in D. L. Clarke (ed.), *Models in Archaeology*, London, Methuen, pp. 109–66.
Binford, L. R. (1973), 'Interassemblage variability – the Mousterian and the 'Functional' argument' in C. Renfrew (ed.), *The Explanation of Culture Change and Models in Prehistory*, London, Duckworth, pp. 227–54.
Binford, L. R. (1981), *Bones: Ancient Men and Modern Myths*, New York, Academic Press.
Binford, L. R. and Binford, S. R. (1966), 'A preliminary analysis of the functional variability of the Mousterian of Levallois facies', *American Anthropologist*, vol. 68, pp. 238–95.
Binford, L. R. and Ho, C. K. (1985), 'Taphonomy at a distance: Zhoukoudian – The Cave Home of Beijing Man', *CA*. (page refs to pre-circulated draft).
Binford, S. R. and Binford, L. R. (1969), 'Stone tools and human behavior', *Sci. Am.*, vol. 220, no. 4, pp. 70–84.
Bingham, H. C. (1932), *Gorillas in a Native Habitat*, Washington, Carnegie Institute.
Black, D. (1926), 'Tertiary man in Asia: The Chou Kou Tien Discovery', *Nat.*, vol. 118, pp. 733–4.
Bleek, D. F. (1930), *Rock Paintings of South Africa from Parts of the Eastern Province and Orange Free State*, London, Methuen.
Bloor, D. (1978), 'Polyhedra and the Abominations of Leviticus', *British Journal of the History of Science*, vol. 11, pp. 245–72.
Blumenbach, J. F. (1865), *The Anthropological Treatises of Johann Friedrich Blumenbach*, London, Anthropological Society of London.
Blumenberg, B. (1983), 'The evolution of the advanced hominid brain', *CA*, vol. 24, no. 5, pp. 589–623.
Boakes, R. (1984), *From Darwinism to Behaviourism: Psychology and the Minds of Animals*, Cambridge, UK, Cambridge University Press.
Boas, Franz (1911), *The Mind of Primitive Man*, New York, Macmillan.
Bodmer, W. F. and Cavalli-Sforza, L. L. (1976), *Genetics, Evolution, and Man*, San Francisco, W. H. Freeman
Boesch, C. and Boesch, H. (1981), 'Sex differences in the use of natural hammers by wild chimpanzees. A preliminary report', *JHE*, vol. 10, pp. 585–93.

Boesch, C. and Boesch, H. (1983), 'Optimisation of nut-cracking with natural hammers by wild chimpanzees', *Behaviour*, vol. 83, pp. 265–86.
Bolk, L. (1926), *Das Problem der Menschwerdung*, Jena, Gustav Fischer.
Bordes, F. (1968), *The Old Stone Age*, London, Weidenfeld & Nicolson.
Boswell, P. G. H. (1932), 'The Oldoway Human Skeleton', *Nat.*, vol. 130, pp. 237–8.
Boule, M. (1911–1913), 'L'Homme fossile de La Chapelle-aux-Saints', *Annales de Paleontologie*, vols 6, 7, 8.
Boule, M. (1923), *Les Hommes Fossiles*, Paris, Masson.
Bowlby, J. (1969), *Attachment and Loss, Vol. 1: Attachment*, London, Hogarth Press.
Bowler, P. J. (1984), *Evolution: the History of an Idea*, University of California Press.
Bradshaw, J. L. and Nettleton, N. (1983), *Human Cerebral Asymmetry*, Englewood Cliffs, New Jersey, Prentice-Hall.
Brain, C. K. (1976), 'Some principles in the interpretation of bone accumulations associated with man', in G. Ll. Isaac and E. R. McCown (eds), *Human Origins: Louis Leakey and the East African Evidence*, Menlo Park, California, W. A. Benjamin Inc., pp. 97–116.
Brainerd, C. J. (1978), 'The stage question in cognitive-developmental theory', *BBS*, vol. 1, no. 2, pp. 173–213.
Brannigan, A. (1979), 'The reification of Mendel', *Social Studies of Science*, vol. 9, pp. 423–54.
Braudel, F. (1972), *Mediterranean and the Mediterranean World in the Age of Philip II*, London, Collins.
Brauer, G. (1981), 'New evidence on the transitional period between Neanderthal and Modern Man', *JHE*, vol. 10, pp. 467–74.
Breuil, H. (1934), 'Presidential Address', *Proceedings of the Prehistoric Society of East Anglia*, vol. VII, part ii, pp. 289–322.
Breuil, H. and Berger-Kirchner, L. (1961), 'Franco-Cantabrian Rock Art' in H-G. Bandi et al., *The Art of the Stone Age: Forty Thousand Years of Rock Art*, London, Methuen, pp. 11–65.
Breuil, H. and Windels, F. (1952), *Four Hundred Centuries of Cave Art*, Montignac.
Broca, P. (1869), 'Crâne et ossements humains des cavernes de Gibraltar', *Bulletin de la Société d'Anthropologie de Paris* (2me série), vol. 4, p. 154.
Broom, R. and Schepers, G. W. H. (1946), 'The South African fossil ape-men. The Australopithecinae. Part I. The occurrence and general structure of the South African ape-men', *Transvaal Museum Memoir*, vol. 2, pp. 7–144.
Brown, A. (1893), 'On the continuity of the neolithic and palaeolithic periods', *Journal of the Royal Anthropological Institute*, vol. XXII, pp. 66–98.
Brown, C. H. (1979), 'Folk zoological life-forms: their universality and growth', *American Anthropologist*, vol. 81, no. 4, pp. 791–816.
Buckland, W. (1823), *Reliquiae Diluvianae*, London, John Murray.
Buckland, W. (1836), *Geology and Mineralogy considered with reference to Natural Theology*, London, William Pickering (Bridgewater Treatise No. 6).
Bunn, H. T. (1981), 'Archaeological evidence for meat-eating by Plio-Pleistocene Hominids from Koobi Fora and Olduvai Gorge', *Nat.*, vol. 291, pp. 574–7.
Burkitt, M. (1963), *The Old Stone Age, A Study of Palaeolithic Times* (4th edition), London, Bowes & Bowes.

Busk, G. (1864), 'Pithecoid Priscan Man from Gibraltar', *The Reader*, 23rd July.
Bygott, J. D. (1972), 'Cannibalism among wild chimpanzees', *Nat.*, vol. 238, pp. 410–1.
Cahen, D., Keeley, L. H. and Van Noten, F. L. (1979), 'Stone tools, toolkits and human behavior in prehistory', *CA*, vol. 20, no. 4, pp. 661–83.
Campbell, B. G. (1974), *Human Evolution: An Introduction to Man's Adaptations* (2nd edition), Chicago, Aldine.
Campbell, B. G. (1978), 'Some problems in hominid classification and nomenclature', in C. J. Jolly (ed.), *Early Hominids in Africa*, London, Duckworth, pp. 567–81.
Campbell, D. T. (1975), 'On the conflicts between biological and social evolution and between psychology and moral tradition', *Am. Psy.*, December, pp. 1103–26.
Canby, T. Y. (1979), 'The search for the first Americans', *National Geographic*, vol. 156, no. 3, pp. 330–63.
Carey, J. W. and Steegman, A. T. (1981), 'Human nasal protrusion, latitude and climate', *AJPA*, vol. 56, pp. 313–9.
Carpenter, C. R. (1940), 'A field study in Siam of the behavior and social relations of the gibbon, *Hylobates lar.*', *Comparative Psychology Monographs*, vol. 16, pp. 112–41.
Cartailhac, E. (1902), 'Les cavernes ornées de dessins: La grotte d'Altamira. Mea culpa d'un sceptique', *L'Anth.* vol. xiii, p. 348.
Casson, S. (1940), *The discovery of Man*, London, Readers Union and Hamish Hamilton.
Chagnon, N. A. (1968), *Yanomamö: The Fierce People*, New York, Holt, Rinehart & Winston.
Chambers, R. (1844), *Vestiges of the Natural History of Creation*, London, Churchill.
Chapman, D. (1837), *A Comprehensive Theological and Philosophical Dissertation on Man*, London, Hamilton, Adams.
Childe, V. G. (1925), *The Dawn of European Civilization*, London, Kegan Paul, Trench, Trubner.
Childe, V. G. (1942), *What Happened in History*, Harmondsworth and New York, Penguin Books.
Childe, V. G. (1958), *The Prehistory of European Society*, Harmondsworth, Penguin Books.
Childe, V. G. (1963 [1951]), *Social Evolution*, London, Watts.
Ciochon, R. L. and Corrucini, R. S. (1983), see p. 330.
Clark, G. (1980), *Mesolithic Prelude*, Edinburgh, University Press.
Clark, J. D. (1950), *The Stone Age Cultures of Northern Rhodesia*, Claremont (Cape Province), South African Archaeological Society.
Clark, J. D. (1959), *The Prehistory of Southern Africa*, Harmondsworth, Penguin Books.
Clark, J. D. (1982), 'The cultures of the Middle Palaeolithic/Middle Stone Age', in J. D. Clark (ed.), *Cambridge History of Africa*, Vol. 1, Cambridge, UK Cambridge University Press, pp. 248–341.
Clark, J. D. et al. (1984), 'Palaeoanthropological discoveries in the Middle Awash Valley, Ethiopia', *Nat.*, vol. 307, pp. 423–8.
Clark, W. E. LeGros (1959), *The Antecedents of Man*, Edinburgh, University Press.
Clutton-Brock, T. H. (1983), 'Selection in relation to sex', *EMM*, pp. 457–82.

Cochrane, L. (1984), 'On the categorization of traits', *Journal of the Theory of Social Behavior*, vol. 14, no. 2, pp. 183–210.
Cohen, D. (1979), *J. B. Watson, the Founder of Behaviorism: A Biography*, London, Routledge & Kegan Paul.
Cohen, R. (1981), 'Evolutionary epistemology and human values', *CA*, vol. 22, no. 3, pp. 201–18.
Cole, S. (1954), *The Prehistory of East Africa*, Harmondsworth, Penguin Books.
Cole, S. (1965), *The Neolithic Revolution*, London, British Museum (Natural History).
Cole, S. (1975), *Leakey's Luck: The Life of Louis Seymour Bazett Leakey 1903–1972*, London, Collins.
Coles, J. M., Heal, S. V. E. and Orme, B. J. (1978), 'The use and character of wood in prehistoric Britain and Ireland', *PPS*, vol. 44, pp. 1–46.
Coles, J. M. and Higgs, E. S. (1969), *The Archaeology of Early Man*, London, Faber & Faber.
Combe, G. (1860, 1st ed. 1828), *The Constitution of Man*, Edinburgh, MacLachlan & Stewart.
Conkey, M. W. (1980), 'Context, structure, and efficiency in Palaeolithic art and design' in M. LeC. Foster and S. H. Brandes (eds), *Symbol as Sense*, New York, Academic Press, pp. 225–48.
Conkey, M. W. (1983), 'Palaeolithic art and accessing symbolic domains', Paper presented to the 48th Annual Meeting, *Society of American Archaeologists*, Pittsburg.
Conroy, G. C. (1982), 'A study of cerebral vascular evolution in primates', in *PBE*, pp. 247–61.
Coon, C. S. (1962), *The History of Man*, Harmondsworth, Penguin Books.
Coon, C. S. (1963), *The Origin of Races*, London, Jonathan Cape.
Corruccini, R. S. and Ciochon, R. L. (1983), 'Overview of Ape and Human Ancestry. Phyletic Relationships of Miocene and Later Hominoidea', *NI*, pp. 3–19.
Corruccini, R. S. and McHenry, H. M. (1980), 'Cladometric analysis of Pliocene hominids', *JHE*, vol. 9, pp. 209–21.
Crawford, O. G. S. (1923), 'Air photos show Celtic fields on palimpsest of English soil', *Christian Science Monitor*, 14th December, p. 11.
Creelan, P. G. (1974), 'Watsonian behaviorism and the Calvinist conscience', *Journal of the History of the Behavioral Sciences*, vol. 10, pp. 95–118.
Cronin, J. E. (1983), 'Apes, humans and molecular clocks: a reappraisal', *NI*, pp. 115–35.
Crookshank, F. G. (1931), *The Mongol in Our Midst*, London, Kegan Paul, Trench, Trubner.
Daniel, G. E. (1950), *A Hundred Years of Archaeology*, London, Duckworth.
Daniel, G. E. (1967), *The Origins and Growth of Archaeology*, Harmondsworth, Penguin Books.
Darlington, C. D. (1959), *Darwin's Place in History*, Oxford, Basil Blackwell.
Darlington, C. D. (1969), *The Evolution of Man and Society*, London, Allen & Unwin.
Darlington, C. D. (1978), *The Little Universe of Man*, London, Allen & Unwin.
Darlington, C. D. (1979), 'Cock on evolution', *Nat.*, vol. 281, p. 326.
Dart, R. (1925), '*Australopithecines africanus*, the man-ape of South Africa', *Nat.*, vol. 115, p. 195–9.
Dart, R. (1957), 'The Makapansgat Australopithecine Osteodontokeratic

culture', in J. D. Clark and S. Cole (eds), *Proceedings of the Third Pan-African Congress on Prehistory, Livingstone*, London, Chatto & Windus, pp. 161–71.
Darwin, C. (1859), *The Origin of Species by Means of Natural Selection*, London, John Murray.
Darwin, C. (1868), *Variation of Animals and Plants under Domestication* (2 vols), London, Murray.
Darwin, C. (1874 – 2nd edn.), *The Descent of Man and Selection in Relation to Sex*, London, John Murray. (All text references are to the 1913 reprint.)
Darwin, C. (1877), 'Biographical Sketch of an Infant', *Mind*, vol. II, pp. 285–94.
Darwin, C. (ed. F. Darwin), (1901 [1872]), *The Expression of the Emotions in Man and Animals*, London, John Murray.
Dawkins, R. (1976), *The Selfish Gene*, London, Oxford University Press.
Dawkins, W. B. (1874), *Cave Hunting*, London, Macmillan.
Dawson, J. W. (1883 [1889]), *Fossil Men and Their Modern Representatives*, London, Hodder & Stoughton.
Day, M. H. (1977), *Guide to Fossil Man*, London, Cassell (4th edition in press, due for publication 1986).
Day, M. H., Leakey, M. D. and Magori, C. (1980), 'A new hominid fossil skull (L.H. 18) from the Ngaloba Beds, Laetoli, northern Tanzania', *Nat.*, vol. 284, pp. 55–6.
Day, M. H., Leakey, R. E. F., Walker, A. C. and Wood, B. A. (1975), 'New hominids from East Rudolf, Kenya, I.', *AJPA*, vol. 42, pp. 461–76.
Day, M. H. & Wickens, E. H. (1980), 'Laetoli Pliocene hominid footprints and bipedalism', *Nat.*, vol. 286, pp. 385–7.
Della Porta, G. B. (1623), *Della Celeste Fisonomia di . . .*, Padua.
Desmond A. (1979), *The Ape's Reflexion*, London, Blond & Briggs.
Desmond, A. (1982), *Archetypes and Ancestors: Palaeontology in Victorian London 1850–1875*, London, Blond & Briggs.
DeVore, I. and Washburn, S. L. (1964), 'Baboon ecology and human evolution' in F. Clark Howell and F. Boulière (eds), *African Ecology and Human Evolution*, London, Methuen, pp. 335–67.
De Winter, K. W. (1984), 'Biological and cultural evolution: different manifestations of the same principle: A systems-theoretical approach', *JHE*, vol. 13, pp. 61–70.
Dickemann, M. (1981), 'Comments' on R. Cohen (1981), pp. 209–10.
Dimond, S. J. and Blizard, D. A. (eds), (1977), *Evolution and lateralization of the Brain*, New York, New York Academy of Sciences.
Dobzhansky, T. (1944), 'On the species and races of living and fossil man', *AJPA*, vol. 2, pp. 251–65.
Dobzhansky, T. (1955), *Evolution, Genetics, and Man*, New York, John Wiley.
Dobzhansky, T. (1962), *Mankind Evolving: The Evolution of the Human Species*, New Haven and London, Yale University Press.
Dobzhansky, T. and Montagu, M. F. A. (1947), 'Natural selection and the mental capacities of mankind', *Sci.*, vol. 105, pp. 587–90. Reprinted in M. F. A. Montagu (ed.), (1962), *Culture and the Evolution of Man*, New York, Oxford University Press.
Douglas, M. (1970), *Natural Symbols*, London, Barrie & Rockliff (reprinted 1973, Harmondsworth, Penguin Books).
Dover, G. (1982), 'Molecular drive: a cohesive mode of species evolution', *Nat.*, vol. 299, pp. 111–7.

Downie, R. A. (1940), *James George Frazer. The Portrait of a Scholar*, London, Watts.
DuBrul, E. L. (1958), *Evolution of the Speech Apparatus*, Springfield, Ill., Charles C. Thomas.
DuBrul, E. L. (1977), 'Origin of the speech apparatus and its reconstruction in fossils', *Brain and Language*, vol. 4, pp. 365–81.
Dupont, E. (1872), *Les Temps antéhistorique en Belgique: L'Homme pendant les Ages de la Pierre dans les environs de Dinant-sur-Meuse*, Brussels, C. Muquardt.
Durant, John R. (1981), 'The myth of human evolution', *New Universities Quarterly*, Autumn, pp. 425–38.
Eldredge, N. and Gould, S. J. (1972), 'Punctuated equilibria: an alternative to phyletic gradualism', in T. J. M. Schopf (ed.), *Models in Paleobiology*, San Francisco, Freeman, Cooper, pp. 82–115.
Eldredge, N. and Tattersall, I. (1982), *The Myths of Human Evolution*, New York, Columbia University Press.
Ellegard, A. (1981), 'Stone Age Science in Britain', *CA*, vol. 22, no. 2, pp. 99–125.
Elliot Smith, G. (1924), *Essays on the Evolution of Man*, Oxford, Oxford University Press and London, Milford.
Elliot Smith, G. (1934), *Human History*, London, Jonathan Cape.
Engels, F. (1884), *Der Ursprung der Familie des Privateigenthums und des Staats*, Zurich, Huttingen.
Evans, Sir J. (1872, 1897 2nd ed.), *Ancient Stone Implements of Great Britain*, London, Longmans, Green.
Falk, D. (1975), 'Comparative anatomy of the larynx in man and the chimpanzee: implications for language in Neanderthal', *AJPA*, vol. 43, pp. 123–32.
Falk, D. (1980a), 'Hominid brain evolution: the approach from paleoneurology', *Yearbook of Physical Anthropology*, vol. 23, pp. 93–107.
Falk, D. (1980b), 'A Reanalysis of the South African Australopithecine natural endocasts', *AJPA*, vol. 53, pp. 525–39.
Falk, D. (1980c), 'Language, handedness, and primate brains: did the Australopithecines sign?', *American Anthropologist*, vol. 82, no. 1, pp. 72–8.
Falk, D. (1982), 'Mapping fossil endocasts', *PBE*, pp. 217–26.
Falk, D. (1983), 'The Taung endocast: a reply to Holloway', *AJPA*, vol. 60, pp. 17–45.
Falk, D. and Conroy, G. C. (1983), 'The cranial venous system in *Australopithecus afarensis*', *Nat.*, vol. 306, pp. 779–81.
Feyerabend, P. (1976), *Against Method*, London, New Left Books.
Figuier, L. (1882), *L'Homme Primitif*, Paris, Hachette et Cie.
Fishbein, H. D. (1979), 'Peer Commentary', to S. T. Parker and K. R. Gibson (1979), pp. 384–5.
Fisher, R. A. (1930), *The Genetical Theory of Natural Selection*, Oxford, University Press.
Fleagle, J. G. (1983), 'Locomotor adaptations of Oligocene and Miocene hominoids and their phyletic implications', *NI*, pp. 301–24.
Fleming, S. (1976), *Dating in archaeology: a guide to scientific techniques*, London, Dent.
Flood, J. (1983), *Archaeology of the Dreamtime*, Sydney and London, Collins.
Flynn, L. J. and Guo-Qin, Q. (1982), 'Age of Lufeng, China, hominoid locality', *Nat.*, vol. 298, pp. 746–7.

Fossey, D. (1984), *Gorillas in the Mist*, New York, Houghton-Mifflin.
Fox, R. (1979), 'The evolution of mind: an anthropological approach', *Journal of Anthropological Research*, vol. 35, no. 2, pp. 138–56.
Frazer, J. G. (1890–1915), *The Golden Bough, A Study of Comparative Religion*, London, Macmillan.
Freeden, M. (1979), 'Eugenics and progressive thought: a study in ideological affinity', *History Journal*, vol. 22, no. 3, pp. 645–71.
Freud, S. (1913), *Totem und Tabu*, Vienna, Hugo Heller.
Fristrup, K. (1983), 'Comments' on B. Blumenberg (1983), pp. 602–3.
Fuller, J. L. (1978), 'Genes, brains and behaviour' in M. S. Gregory, A. Silvers and D. Sutch (eds), *Sociobiology and Human Nature*, San Francisco, Jossey-Bass, pp. 98–115.
Gamble, C. (1979), 'Hunting Strategies in the Central European Palaeolithic', *PPS*, vol. 45, pp. 35–52.
Gantt, D. G. (1983), 'The enamel of Neogene hominoids: structural and phyletic implications', *NI*, pp. 249–98.
Garrod, D. A. E. and Bate, D. M. A. (1937), *The Stone Age of Mount Carmel: Excavations at the Wady el-Mughara (Vol. 1)*, Oxford, Clarendon Press.
Gazzaniga, M. S., Smylie, C. S. and Baynes, K. (1984), 'Profiles of right hemisphere language and speech following brain bisection', *Brain and Language*, vol. 22, pp. 206–20.
Geist, V. (1978), *Life Strategies, Human Evolution, Environmental Design*, New York, Springer-Verlag.
Gleadow, A. J. W. (1980), 'Fission track age of the KBS Tuff and associated hominid remains in northern Kenya', *Nat.*, vol. 284, pp. 225–30.
Godfrey, L. and Jacobs, K. E. (1981), 'Gradual, autocatalytic and punctuational models of hominid brain evolution: a cautionary tale', *JHE*, vol. 10, 255–72.
Goldschmidt, W. (1976), in L. G. Wispé and J. N. Thomson Jnr (eds) (1976), pp. 355–7.
Gombrich, E. H. (1950), *The Story of Art*, London, Phaidon Press.
Gooch, S. (1977), *The Neanderthal Question*, London, Wildwood House.
Goodall, A. (1979), *Wandering Gorillas*, London, Collins.
Goodall, J. (1971), *In the Shadow of Man*, London, Collins.
Goodall, J. (1977), 'Infant killing and cannibalism in free-living chimpanzees', *Folia Primatologica*, vol. 28, pp. 262–4, 279.
Goodman, M., Baba, M. L. and Darga, L. L. (1983), 'The bearing of molecular data on the cladogenesis and times of divergence of hominoid lineages', *NI*, pp. 67–86.
Gould, R. A. (1968), 'Living Archaeology: the Ngatatjara of Western Australia', *South Western Journal of Anthropology*, vol. 24, no. 2, pp. 101–22.
Gould, R. A. (1969), *Yiwara; Foragers of the Australian Desert*, London and Sydney, Collins.
Gould, R. A., (1980), *Living Archaeology*, Cambridge UK, Cambridge University Press.
Gould, S. J. (1973), *Ever Since Darwin*, Harmondsworth, Penguin Books.
Gould, S. J. (1977), *Ontogeny and Phylogeny*, Cambridge, Mass., Harvard University Press.
Gould, S. J. (1983a), *The Panda's Thumb*, Harmondsworth, Penguin Books.
Gould, S. J. (1983b), 'Irrelevance, submission and partnership: the changing role of palaeontology in Darwin's three centennials and a modest proposal for macroevolution', *EMM*, pp. 347–66.

Gould, S. J. and Eldredge, N. (1977), 'Punctuated equilibria: the tempo and mode of evolution reconsidered', *Paleobiology*, vol. 3, pp. 115–51.
Gowlett, J. (1984), *Ascent to Civilization: The Archaeology of Early Man*, London, Collins.
Green, S. (1981), *Prehistorian: A Biography of V. Gordon Childe*, Bradford-on-Avon, Moonraker Press.
Greenfield, L. O. (1979), 'On the adaptive pattern of "*Ramapithecus*" ', *AJPA*, vol. 50, pp. 527–48.
Gribbin, J. and Cherfas, J. (1982), *The Monkey Puzzle: Reshaping the Evolutionary Tree*, New York, McGraw-Hill.
Gruber, H. E. and Barrett, P. H. (1974), *Darwin on Man: A Psychological Study of Scientific Creativity*, London, Wildwood House. (Reprinted 1981 by University of Chicago Press with Gruber as sole author.)
Hadingham, E. (1980), *Secrets of the Ice Age: The World of the Cave Artists*, London, Heinemann.
Haeckel, Ernst (1866), *Generelle Morphologie* (2 vols), Berlin, G. Reimer.
Haeckel, Ernst (1883), *The Evolution of Man* (2 vols), London, Kegan Paul, Trench.
Haeckel, Ernst (1892, 4th ed. [1868]) *The History of the Creation* (2 vols.) London, Paul, Trench, Trübner & Co.
Hall, G. S. (1904, 1905 in UK), *Adolescence* (2 vols), London, Sidney Appleton.
Halstead, L. B. (1978), 'New Light on the Piltdown hoax?', *Nat.*, vol. 276, pp. 11–13.
Hamilton, W. D. (1963), 'The evolution of altruistic behavior', *American Naturalist*, vol. 97, pp. 354–6.
Hamilton, W. D. (1964), 'The genetical evolution of social behaviour, I and II', *Journal of Theoretical Biology*, vol. 7, pp. 1–16, 17–62.
Hammond, M. (1979), 'A framework of plausibility for an anthropological forgery: the Piltdown Man', *Anth.*, vol. 3, pp. 47–58.
Hammond, M. (1982), 'The expulsion of the Neanderthals from human ancestry: Marcellin Boule and the social context of scientific research', *Social Studies of Science*, vol. 12, pp. 1–36.
Hardy, A. C. (1960), 'Was man more aquatic in the past?' *N.Sc.*, vol. 7, pp. 642–5.
Hardy, A. C. (1977), 'Was there a *Homo aquaticus*?', *Zenith*, vol. 15, no. 1, pp. 4–6.
Harlow, H. F. and Harlow M. K. (1965), 'The affectional systems' in A. M. Schrier, H. F. Harlow and F. Stollnitz (eds), *Behavior of Non-Human Primates*, New York, Academic Press.
Harlow, H. F. and Suomi, S. J. (1970), 'Nature of love-simplified', *Am. Psy.* vol. 25, pp. 161–8.
Harlow, H. F. and Zimmerman, R. R. (1959), 'Affectional responses in the infant monkey', *Sci.* vol. 130, pp. 421–32.
Harnad, S. R., Steklis, H. D. and Lancaster, J. (eds) (1976), *Origins and Evolution of Language and Speech*, New York, New York Academy of Sciences, issued as *Annals of the New York Academy of Sciences*, vol. 280.
Harvey, P. and Clutton-Brock, T. (1983), 'The survival of the theory', *N.Sc.* 5th May, pp. 313–5.
Hayden, B. (1979), *Lithic Use-Wear Analysis*, New York, Academic Press.
Hearnshaw, L. S. (1964), *A Short History of British Psychology*, London, Methuen.

Heim, J-L. (1982a), 'Le dimorphisme sexuel du crâne des hommes de Néandertal', *L'Anth.*, vol. 85, no. 2, pp. 193–218.
Heim, J-L. (1982b), 'Le dimorphisme sexuel du crâne des hommes de Néandertal (suite)', *L'Anth.*, vol. 85/86, no. 3, pp. 451–69.
Heim, J-L. (1983), 'Les variations du squelette post-crânien des hommes de Néandertal suivant le sexe', *L'Anth.*, vol. 87, no. 1, pp. 5–26.
Hennig, G. J., Herr, W., Weber, E., and Xirotiris, N. I. (1981), 'ESR-dating of the fossil hominid cranium from Petralona Cave, Greece', *Nat.*, vol. 292, pp. 533–6.
Herman, L. M., Richards, D. G. and Wolz, J. P. (1984), 'Comprehension of sentences by bottlenosed dolphins', *Cognition*, vol. 16, pp. 129–219.
Hewes, G. W. (1973), 'An explicit formulation of the relationship between tool-using, tool-making, and the emergence of language', *Visible Language*, vol. 7, no. 2, pp. 101–27.
Hewes, G. W. (1976), 'The current status of the gestural theory of language origin', *OEL*, pp. 482–503.
Hewes, G. W. (1979), 'Peer Commentary' on S. T. Parker and K. R. Gibson (1979), pp. 387–8.
Hill, K. (1982), 'Hunting and human evolution', *JHE*, vol. 11, pp. 521–44.
Hill, K. and Hawkes, K. (1983), 'Neotropical hunting among the Ache of Eastern Paraguay' in R. Hames and W. Vickers (eds), *Adaptive Responses of Native Amazonians*, New York, Academic Press.
Hill, K., Hawkes, K., Hurtado, M. and Kaplan, H. (1984), 'Seasonal variance in the diet of Ache hunter-gatherers in Eastern Paraguay', *Human Ecology*, vol. 12, no. 2, pp. 101–35.
Ho, M–W. (1984), Where does biological form come from?', *Rivista di Biologia*, vol. 77, no. 2, pp. 147–79.
Ho, M–W. and Saunders, P. T. (1982), 'The epigenetic approach to the evolution of organisms – with notes on its relevance to social and cultural evolution', *LDC*, pp. 343–62.
Ho, M–W. and Saunders, P. T. (1984), 'Pluralism and convergence in evolutionary theory', in M–W. Ho and P. T. Saunders (eds), *Beyond Neo-Darwinism. An Introduction to the New Evolutionary Paradigm*, London, Academic Press, pp. 3–12.
Hockett, C. F. (1960), 'Logical considerations in the study of animal communication', in W. E. Lanyon and W. N. Tavolga (eds), *Animal sounds and communication*, Washington D.C., American Institute of Biological Sciences.
Holloway, R. L. (1969), 'Culture: a human domain', *CA*, vol. 4, pp. 135–68.
Holloway, R. L. (1976), 'Paleoneurological evidence for language origins', *OEL*, pp. 330–48.
Holloway, R. L. (1981a), 'Culture, symbols and human brain evolution: a synthesis', *Dialectical Anthropology*, vol. 5, pp. 287–303.
Holloway, R. L. (1981b), 'Revisiting the South African Taung Australopithecus endocast: the position of the lunate sulcus as determined by the stereoplotting technique', *AJPA*, vol. 56, no. 1, pp. 43–58.
Holloway, R. L. (1983), 'Human brain evolution: a search for units, models and synthesis', *Canadian Journal of Anthropology*, vol. 3, no. 2, pp. 215–30.
Holloway, R. L. and La Coste-Lareymondie, M. C. de (1982), 'Brain endocast asymmetry in pongids and hominids: some preliminary findings on the paleontology of cerebral dominance', *AJPA*, vol. 58, pp. 101–10.

Holloway, R. L. & Post, D. C. (1982), 'The relativity of relative brain measures and hominid mosaic evolution', *PBE*, pp. 57–76.
Hooton, E. A. (1946), *Up From The Ape*, New York, Macmillan.
Howell, F. C. (1966), *Early Man*, Amsterdam, Time-Life International.
Howells, W. W. (1967), *Mankind in the Making: The Story of Human Evolution*, Harmondsworth, Penguin Books.
Howells, W. W. (1983), 'Origins of the Chinese People: Interpretations of Recent Evidence' in D. N. Keightley (ed.), *The Origin of Chinese Civilization*, Berkeley, University of California Press.
Hrdlička, A. (1913), 'The most ancient skeletal remains of man', *Annual Report of the Board of the Smithsonian Institution*, Washington D.C., Government Printing Office, pp. 491–552.
Hrdlička, A. (1930), 'The skeletal remains of early man', *Smithsonian Miscellaneous Collections*, vol. 83, pp. 1–379.
Hull, D. L. (1982), 'The naked meme', *LDC*, pp. 273–327.
Hull, D. L. (1983), 'Darwin and the nature of science', *EMM*, pp. 63–80.
Huxley, J. (1942), *Evolution, The Modern Synthesis*, London, Allen & Unwin.
Huxley, J. (1953), *Evolution in Action*, London, Chatto & Windus.
Huxley, T. H. (1901), *Man's Place in Nature and other Anthropological Essays*, London, Macmillan.
Ikawa-Smith, F. (ed.) (1978), *Early Palaeolithic in South and East Asia*, The Hague, Mouton.
Irvine, W. (1955), *Apes, Angels, and Victorians*, New York, McGraw-Hill.
Isaac, G. Ll. (1982), 'The earlier archaeological traces' in J. D. Clark (ed.), *Cambridge History of Africa Vol. 1*, Cambridge UK, Cambridge University Press, pp. 157–247.
Isaac, G. Ll. (1983), 'Aspects of human evolution', *EMM*, pp. 509–45.
Isaac, G. Ll. and Crader, D. C. (1981), 'To what extent were early hominids carnivorous? An archaeological perspective' in R. S. O. Harding and G. Teleki (eds), *Omnivorous Primates: Hunting and Gathering in Human Evolution*, New York, Columbia University Press, pp. 37–102.
Isaac, G. Ll. and McCown, E. R. (eds) (1976), *Human Origins: Louis Leakey and the East African Evidence*, Menlo Park, W. A. Benjamin.
Ivanova, I. K. and Chernysh, A. P. (1965), 'The Palaeolithic site of Moldova V on the Middle Dnestr (USSR)', *Quaternaria*, vol. 7, pp. 197–217.
Jaynes, J. (1976a), 'The evolution of language in the late Pleistocene', *OEL*, pp. 312–25.
Jaynes, J. (1976b), *The Origin of Consciousness in the Breakdown of the Bicameral Mind*, New York, Houghton Mifflin.
Jenkins, F. A. (ed.), (1974), *Primate Locomotion*, New York, Academic Press.
Jerison, H. (1973), *Evolution of the Brain and Intelligence*, New York, Academic Press.
Jerison, H. (1982a), 'Allometry, brain size, cortical surface, and convolutedness', *PBE*, pp. 77–84.
Jerison, H. (1982b), 'The evolution of biological intelligence' in R. J. Sternberg (ed.), *Handbook of Human Intelligence*, Cambridge UK, Cambridge University Press, pp. 723–92.
Jerison, H. (1983), 'Comments' on B. Blumenberg (1979), pp. 604–5.
Johanson, D. C. and Edey, M. E. (1981), *Lucy, The Beginnings of Humankind*, London, Granada.
Johnson, J. P. (1907), *The Stone Implements of South Africa*, London, Longmans.

Johnson, S. C. (1981), 'Bonobos: generalized hominid prototypes or specialized insular dwarfs?', *CA*, vol. 22, no. 4, pp. 363–75.
Jones, J. S. (1986), 'The genetic evidence', *MTP*, pp. 317–30.
Kalb, J. E. and 12 others (1982), 'Fossil mammals and artefacts from the Middle Awash Valley, Ethiopia', *Nat.*, vol. 298, pp. 25–9.
Kavanagh, M. (1984), *A Complete Guide to Monkeys, Apes and other Primates*, New York, Viking.
Kay, R. F. and Covert, H. H. (1983), 'True grit: a microwear experiment', *AJPA*, vol. 61, no. 1, pp. 33–8.
Keane, A. H. (1899), *Man Past and Present*, Cambridge, UK, Cambridge University Press.
Keeley, L. H. (1974), 'Technique and methodology in microwear studies', *World Archaeology*, vol. 5, pp. 323–36.
Keeley, L. H. (1980), *Experimental Determination of Stone Tool Uses: A Microwear Analysis*, Chicago and London, University of Chicago Press.
Keeley, L. H. and Toth, N. (1981), 'Microwear polishes on early stone tools from Koobi Fora, Kenya', *Nat.*, vol. 293, pp. 464–5.
Keith, A. (1925), *The Antiquity of Man* (2 vols), London, Williams & Norgate.
Keith, A. (1948), *A New Theory of Human Evolution*, London, Watts.
Kennedy, G. E. (1983), 'Some aspects of femoral morphology in *Homo erectus*', *JHE*, vol. 12, pp. 587–616.
Kidd, B. (1894), *Social Evolution*, London, Macmillan.
Kidd, J. (1837 [1833]), *On the Adaptation of External Nature to the Physical Condition of Man* (5th ed), Second Bridgewater Treatise, London, William Pickering.
Kimbel, W. H. (1984), 'Variation in the pattern of cranial venous sinuses and hominid phylogeny', *AJPA*, vol. 63, pp. 243–63.
Kitahara-Frisch, J. (1980), 'Symbolizing technology as a key to human evolution' in M. LeC. Foster and S. Brandes (eds), *Symbol as Sense*, New York, Academic Press, pp. 211–23.
Klein, R. G. (1973), *Ice-Age Hunters of the Ukraine*, Chicago, University of Chicago Press.
Kluges, A. G. (1983), 'Cladistics and the classification of the Great Apes', *NI*, pp. 151–77.
Kochetkova, V. I. (1978), *Paleoneurology*, New York, Wiley.
Koffka, K. (1935), *Principles of Gestalt Psychology*, London, Kegan Paul.
Köhler, W. (1925), *The Mentality of Apes*, London, Kegan Paul, Trench Trubner.
Kohts, N. N. (1935), 'Infant ape and human children: instincts, emotions, play, habits', *Scientific Memoirs of the Museum Darwinianum in Moscow*, vol. 3, pp. 1–596.
Kokkoros, P. and Kanellis, A. (1960), 'Découverte d'un crâne d'homme paléolithique dans la peninsule Chalcidique', *L'Anth*, vol. 64, no. 5–6, pp. 438–66.
Kolb, B. and Whishaw, I. Q. (1985), *Fundamentals of Human Neuropsychology* (2nd ed.), New York, W. H. Freeman.
Kortlandt, A. (1972), *New Perspectives on Ape and Human Evolution*, Amsterdam, Stichting voor Psychobiologie.
Krantz, G. E. (1980), 'Sapienization and speech', *CA*, vol. 21, no. 6, pp. 773–92.
Krogman, W. M. (1951), 'Scars of human evolution', *Sci.Am.* vol. 185, no. 6, pp. 54–7.

Kropotkin, P. (1939 [1902]), *Mutual Aid, a Factor of Evolution*, Harmondsworth, Penguin Books.

Kummer, H. (1968), *Social Organisation of Hamadryas Baboon; a Field Study*, Chicago and London, Chicago University Press.

Kuper, A. (1973), *Anthropologists and Anthropology: The British School 1922–1972*, London, Allen Lane.

Ladefoged, P., De Clerk, J., Landau, M. and Papcun, G. (1972), 'An auditory-motor theory of speech production', *UCLA Working Papers in Phonetics*, no. 22, pp. 48–76.

Laitman, J. T. and Heimbuch, R. C. (1982), 'The basicranium of Plio-Pleistocene hominids as an indicator of their upper respiratory systems', *AJPA*, vol. 59, pp. 324–43.

La Lumière, L. P. (1981), 'Evolution of human bipedalism: a hypothesis about where it happened', *Philosophical Transactions of the Royal Society of London*, B. vol. 292, pp. 103–8.

Laming, A. (1959), *Lascaux: Paintings and Engravings*, Harmondsworth, Penguin Books.

Lankester, E. R. (1880), *Degeneration. A Chapter in Darwinism*, London, Nature Series.

La Peyrère, Isaac de (1655), *Systema theologicum ex Prae-Adamitorum hypothesi*, Amsterdam, Elzevir.

Lartet, E. (1858), 'Sur les migrations anciennes des Mammiferes de l'époque actuelle', *Compte rend. de l'Académie des Sciences*, vol. 46, p. 409.

Latham, R. G. (1850), *The Natural History of the Varieties of Man*, London, John van Voorst.

Latham, R. G. (1851), *Man and His Migrations*, London, John van Voorst.

Lavater, J. C. (1797), *Essays on Physiognomy calculated to extend the Knowledge and the Love of Mankind*, Vol. II, London, H. D. Symonds.

Lawrence, W. (1819), *Lectures on Physiology, Zoology and the Natural History of Man*, London, Callow. (See C. D. Darlington (1959), pp. 100–101 for bibliographical complexities.)

Leakey, L. S. B. (1934), *Adam's Ancestors*, London, Methuen.

Leakey, L. S. B. (1936), *Stone Age Africa*, Oxford, Oxford University Press.

Leakey, L. S. B. (1937), *White African*, London, Hodder & Stoughton.

Leakey, L. S. B. (ed.), (1951), *Olduvai Gorge: A Report on the Evolution of the Hand-axe Culture in Beds I–IV*, Cambridge, UK, Cambridge University Press.

Leakey, L. S. B. (1974), *By the Evidence. Memoirs 1935–1951*, New York and London, Harcourt, Brace & Jovanovich.

Leakey, L. S. B. and Cole, S. (eds) (1952), *Proceedings of the First Pan-African Congress on Prehistory, Nairobi*, Oxford, Blackwell.

Leakey, L. S. B., Reck, H., Boswell, P. G. H., Hopwood, A. T. and Solomon, J. D. (1933), 'The Oldoway Human Skeleton', *Nat.*, vol. 131, pp. 397–8.

Leakey, M. D. (1979), *My Search for Early Man*, London, Collins.

Leakey, M. D. (1983), *The Vanishing Art of Africa*, London, Hamish Hamilton.

Leakey, R. E. (1981), *The Making of Mankind*, London, Michael Joseph.

Le Bon, G. (1896), *The Crowd, A Study of the Popular Mind*, London, T. Fisher Unwin.

Lee, R. B. (1979), *The !Kung San: Man, Woman and Work in a Foraging Society*, Cambridge, UK, Cambridge University Press.

LeMay, M. (1975), 'The language capability of Neanderthal man', *AJPA*, vol. 42, pp. 9–14.

LeMay, M., Billig, M. S. and Geschwind, N. (1982), 'Asymmetries of the brains and skulls of nonhuman primates', *PBE*, pp. 263–78.
Leroi-Gourhain, A. (1968), *The Art of Prehistoric Man*, London, Thames & Hudson.
Leroi-Gourhain, Arl. and Allain, J. (1979), *Lascaux inconnu*, Paris, Editions de C.N.R.S.
Leutenegger, W. (1984), 'Encephalisation of *Proconsul africanus*', *Nat.*, vol. 309, pp. 287–8.
Lévêque, F. and Vandermeersch, B. (1981), 'Le néandertalien de Saint-Césaire', *Recherche*, vol. 12, pp. 424–44.
Lévi-Strauss, C. (1966), *The Savage Mind*, London, Weidenfeld & Nicolson.
Lewin, R. (1982), 'How did humans evolve big brains?', *Sci.*, vol. 216, pp. 840–1.
Lewin, R. (1983), 'Were Lucy's feet made for walking?', *Sci.*, vol. 220, pp. 700–2.
Lewin, R. (1984), 'Man the scavenger', *Sci.*, vol. 224, pp. 861–2.
Lewontin, R. C. (1982), 'Organism and environment', *LDC*, pp. 151–72.
Lewontin, R. C. (1983), 'Gene, organism and environment', *EMM*, pp. 273–86.
Lhote, H. (1959), *The Search for the Tassili Frescoes*, London, Hutchinson.
Lhote, H. (1961), 'The Rock Art of the Maghreb and Sahara', in H-G. Bandi et al., *The Art of the Stone Age: Forty Thousand Years of Rock Art*, London, Methuen, pp. 123–71.
Libby, W. F. (1955), *Radiocarbon Dating*, Chicago, Chicago University Press.
Lieberman, P. (1975), *On the Origins of Language: An Introduction to the Evolution of Human Speech*, New York, Macmillan.
Lieberman, P. (1984), *The Biology and Evolution of Language*, Cambridge, Mass. and London, Harvard University Press.
Lieberman, P. and Crelin, E. S. (1971), 'On the speech of Neanderthal Man', *Linguistic Inquiry*, vol. 2, pp. 203–22.
Linnaeus, C. (1735), *Systema Naturae*, Lugduni, Theodorum Haak.
Linton, S. (1971), 'Woman the gatherer: male bias in anthropology', in S–E. Jacobs (ed.), *Women in Cross-cultural Perspective: A Preliminary Sourcebook*, Urbana, University of Illinois Press, pp. 9–21. (Reprinted under the name S. Slocum (1975), in P. R. Reiter (ed.), *Toward an Anthropology of Women*, New York, Monthly Review Press, pp. 36–50.)
Lopreato, J. (1984), *Human Nature and Biocultural Evolution*, London, Allen & Unwin.
Lorenz, K. (1966), *On Aggression*, London, Methuen.
Lovejoy, A. O. (1936), *The Great Chain of Being: A Study of the History of an Idea*, Cambridge, Mass., Harvard University Press.
Lovejoy, C. O. (1981), 'The Origin of Man', *Sci.*, vol. 211, pp. 341–50.
Lubbock, J. (1865), *Pre-Historic Times as Illustrated by Ancient Remains and the Manners of Modern Savages*, London, Williams & Norgate.
Lubbock, J. (1882), *Ants, Bees, and Wasps*, London, Kegan Paul, Trench, Trubner.
Lumley, H. de (1969), 'A Palaeolithic camp-site at Nice', *Sci. Am.*, vol. 220, no. 5, pp. 42–50.
Lumley, H. de (1972), *La Grotte Moustérienne de l'Hortus*, Marseille, Université de Provence.
Lumsden, C. J. and Wilson, E. O. (1981), *Genes, Minds, and Culture*, Cambridge, Mass., Harvard University Press.

Lyell, C. (1863), *The Geological Evidence of the Antiquity of Man*, London, John Murray.
McBurney, C. B. M. and Callow, P. (1971), 'The Cambridge Excavations at La Cotte de Saint Brélade, Jersey – a preliminary report', *PPS*, vol. 37, part 2, pp. 167–207.
McCown, T. D. and Keith, A. (1939), *The Stone Age of Mount Carmel. The Fossil Human Remains from the Levalloiso-Mousterian* (Vol. 2), Oxford, Clarendon Press.
MacCurdy, G. G. (1924), *Human Origins: A Manual of Prehistory*, London & New York, D. Appleton.
MacDonald, K. (1984), 'An ethological-social learning theory of the development of altruism: implications for human sociobiology', *Ethology and Sociobiology*, vol. 5, pp. 97–109.
McDougall, W. (1908), *An Introduction to Social Psychology*, London, Methuen.
McHenry, H. M. and Corruccini, R. S. (1980), 'On the status of *Australopithecus afarensis*', *Sci.*, vol. 207, pp. 1103–4.
McIntosh, S. K. and McIntosh, R. J. (1983), 'Current directions in West African archaeology', *Annual Review of Anthropology*, vol. 12, pp. 215–58.
MacKinnon, J. (1974), 'The behavior and ecology of wild orang-utans', *Animal Behavior*, vol. 22, pp. 3–74.
MacLean, P. D. (1982), 'The origin and progressive evolution of the triune brain', *PBE*, pp. 291–316.
McWhirter, N. (1984), *The Guinness Book of Records*, London, Guinness.
Maddi, S. (1976), *Personality Theories: A Comparative Analysis* (3rd ed.), Homewood, Ill., Dorsey Press.
Maerth, O. K. (1973), *The Beginning was the End*, London, Joseph.
Mallegni, F. et al. (1983), 'New European fossil hominid material from an Acheulian site near Rome, (Castel di Guido)', *AJPA*, vol. 62, pp. 263–74.
Malthus, T. R. (1803), *An Essay on the Principle of Population* (2nd ed.), London, J. Johnson.
Marais, E. (1969), *The Soul of the Ape*, New York, Atheneum.
Marler, P. (1976), 'Social organisation, communication and graded signals: the chimpanzee and gorilla', in P. P. G. Bateson and R. A. Hinde (eds), *Growing Points in Ethology*, Cambridge, UK, Cambridge University Press.
Marshack, A. (1972a), *The Roots of Civilization*, New York, McGraw Hill and London, Weidenfeld & Nicolson.
Marshack, A. (1972b), 'Cognitive aspects of Upper Palaeolithic engraving', *CA*, vol. 13, pp. 445–77.
Marshack, A. (1975), 'Exploring the mind of Ice Age man', *National Geographic*, vol. 147, no. 1, pp. 62–89.
Marshack, A. (1976), 'Some implications of the Paleolithic symbolic evidence for the origin of language', *CA*, vol. 17, no. 2, pp. 274–82.
Marshack, A. (1977), 'The meander as a system: the analysis and recognition of iconographic units in Upper Palaeolithic compositions' in P. J. Ucko (ed.), *Form in Indigenous Art*, Canberra, Australian Institute of Aboriginal Studies.
Marshack, A. (1979a), 'Peer Commentary' on S. T. Parker and K. R. Gibson (1979) pp. 394–6.
Marshack, A. (1979b), 'Upper Palaeolithic symbol systems of the Russian Plain: cognitive and comparative analysis', *CA*, vol. 20, no. 2, pp. 271–311.
Martin, R. D. (1981), 'Relative brain size and basal metabolic rate in terrestrial vertebrates', *Nat.*, vol. 293, pp. 57–60.

Martin, R. D. (1982), 'Allometric approaches to the evolution of the primate nervous system', *PBE*, pp. 39–56
Marzke, M. W. (1983), 'Joint functions and grips of the *Australopithecus afarensis* hand, with special reference to the region of the capitate', *JHE*, vol. 12, pp. 197–211.
Medawar, P. (1984), 'Eliciting imperfect assent' (review of J. Lopreato (1984)) *Times Literary Supplement*, 2nd Nov., p. 1263.
Meijer, O. G. (1983), 'The essence of Mendel's discovery' in V. Orel and A. Matalova (eds), *Gregor Mendel and the Foundation of Genetics*, Brno, Mendelianum of the Moravian Museum.
Mellars, P. A. (1969), 'The Chronology of Mousterian Industries in the Perigord Region' *PPS*, vol. 35, pp. 134–71.
Mellars, P. A. (1976), 'Fire, ecology, animal populations and man', *PPS*, vol. 42, pp. 15–46.
Merchant, C. (1983), *Death in Nature: Women, Ecology and the Scientific Revolution*, New York, Harper & Row.
Merker, B. (1984), 'A note on hunting and hominid origins', *American Anthropologist*, vol. 86, no. 1, pp. 112–14.
Midgley, M. (1979a), *Beast and Man: The Roots of Human Nature*, London, Methuen.
Midgley, M. (1979b), 'Gene-juggling', *Philosophy*, vol. 54, pp. 108–34.
Midgley, M. (1984), 'De-dramatizing Darwinism', *The Monist*, vol. 67, no. 2, pp. 200–15.
Millar, R. (1972), *The Piltdown Men*, London, Gollancz.
Mischel, W. (1973), 'Toward a cognitive social learning reconceptualization of personality', *Psychological Review*, vol. 80, pp. 252–83.
Moncalm, M. (1905), *The Origin of Thought and Speech*, London, Kegan Paul, Trench, Trübner.
Monckton, C. A. W. (1920, rep. 1936), *Some Experiences of a New Guinea Resident Magistrate: Second Series*, Harmondsworth, Penguin Books.
Montagu, M. F. A. (1962), 'Time, morphology and neoteny in the evolution of man' in M. F. Ashley Montagu (ed.), *Culture and the Evolution of Man*, London, Oxford University Press, pp. 324–42.
Morel, B. A. (1857), *Traité des dégénérescences physiques, intellectuelles et morales de l'espèce humaine . . .* Paris, J. B. Baillière.
Morgan, C. L. (1900), *Animal Behaviour*, London, Edward Arnold.
Morgan, E. (1972), *The Descent of Woman*, New York, Stein & Day.
Morgan, E. (1982), *The Aquatic Ape: A Theory of Human Evolution*, London, Souvenir Press.
Morgan, E. (1984), 'The Aquatic hypothesis', *N.Sc.* no. 1405, pp. 11–13.
Morgan, L. H. (1877), *Ancient Society*, Chicago, Kerr.
Morris, D. (1967), *The Naked Ape*, London, Cape.
Morrison, A. (1980), *Early Man in Britain and Ireland*, London, Croom Helm.
Mortillet, G. de (1869), 'Essai de Classification des Cavernes et des Stations sous Abrifondée sur les produits de l'industrie humaine', *Comptes Rendus de L'Académie des Sciences*.
Mortillet, G. de (1883), *Le Préhistorique: Antiquité de l'Homme*, Paris.
Mulvaney, D. J. (1971), 'Prehistory from Antipodean perspectives', *PPS*, vol. 38, part 2, pp. 228–52.
Mulvaney, D. J. (1975), *The Prehistory of Australia*, Harmondsworth, Penguin Books.

Mumford, L. (1967), *The Myth of the Machine: Technics and Human Development*, London, Secker & Warburg.
Myers, R. E. (1976), 'Comparative neurology of vocalization and speech: proof of a dichotomy', *OEL*, pp. 745–57.
Nance, J. (1975), *The Gentle Tasaday: Stone Age People in the Philippine Rain Forest*, London, Gollancz.
Neel, J. V. (1983), 'Some base lines for human evolution and the genetic implications of recent cultural development', in D. J. Ortner (ed.), *How Humans Adapt. A Biocultural Odyssey*, Washington D.C., Smithsonian Institution Press, pp. 67–102.
Neumann, E. (1954), *The Origins and History of Consciousness*, New York, Pantheon Books for Bollingen Foundation.
Newcomer, M. H. and Keeley, L. H. (1979), 'Testing a method of microwear analysis with experimental flint tools', in B. Hayden (ed.), *Lithic Use-wear Analysis*, New York, Academic Press.
Nilsson, S. (translation and introduction by Sir J. Lubbock), (1869), *The Primitive Inhabitants of Scandinavia*, London, Longmans, Green.
Nilsson, T. (1983), *The Pleistocene. Geology and Life in the Quaternary Age*, Dordrecht, Boston and London, D. Reidel.
Nissen, H. W. (1931), *A Field Study of the Chimpanzee: Observations of Chimpanzee Behavior and Environment in Western French New Guinea*, Comparative Psychology Monographs No. 8, Baltimore, Johns Hopkins University Press.
Noback, C. R. (1982), 'Neurobiological aspects in the phylogenetic acquisition of speech', *PBE*, pp. 279–90.
Nordau, M. (1895), *Degeneration*, London, Max Heinemann.
Oakley, K. P. (1969), *Frameworks for Dating Fossil Man*, London, Weidenfeld & Nicolson.
Oakley, K. P. (1972), *Man the Toolmaker*, London, British Museum (Natural History).
Oakley, K. P. and Leakey, M. (1937), 'Report on Excavations at Jaywick Sands, Essex (1934), with some observations on the Clactonian industry, and on the Fauna and Geological significance of the Clacton Channel', *PPS*, vol. 3, part 2, pp. 217–60.
Oakley, K. P. and Randall Hoskins, C. (1950), 'New evidence on the Dating of Piltdown Man', *Nat.*, vol. 165, pp. 379–82.
Oakley, K. P. Andrews, P., Keeley, L. H. and Clark, J. D. (1977), 'A reappraisal of the Clacton Spearpoint', *PPS*, vol. 43, pp. 13–30.
Odling-Smee, F. J. (1983), 'Multiple levels in evolution: an approach to the nature-nurture issue via "applied epistemology" ', in G. C. L. Davey (ed.), *Animal Models of Human Behaviour*, Chichester, John Wiley, pp. 135–58.
Oken, L. (trans. A. Hulk), (1847), *Elements of Physiophilosophy*, London, The Ray Society.
Olby, R. C. (1966), *Origins of Mendelism*, London, Constable.
Olby, R. C. (1984), 'Galton's reply to Buckle: an hereditarian defends the data of human heredity', Paper delivered to the *British Psychological Society* Annual Conference, University of Warwick.
Oldroyd, D. R. (1980), *Darwinian Impacts, An Introduction to the Darwinian Revolution*, Milton Keynes, Open University Press.
Oxnard, C. E. (1981), 'The place of man among the primates: anatomical, molecular and morphometric evidence', *Homo*, vol. 32, nos 3+4, pp. 149–76.

Paddayya, K. (1977), 'The Acheulian Culture of the Hungsi Valley', *Proceedings of the American Philosophical Society*, vol. 121, no. 5, pp. 383–406.

Paget, R. (1930), *Human Speech: Some Observations as to the Nature, Origin, Purpose and Possible Improvement of Human Speech*, London, Kegan Paul, Trench, Trübner.

Pales, L. (1976), *Les Gravures de La Marche, II – Les Humains*, Paris, Ophrys.

Pales, L. and St Pereuse, M. de (1964), 'Une Scène Gravée Magdalénienne', *Objets et Mondes*, vol. 4, p. 77.

Pales, L. and St Pereuse, M. de (1968), 'Humains Superposés de La Marche', in *La Préhistoire, Problèmes et Tendances*, Paris, CNRS, p. 327.

Paley, W. (1802), *Natural Theology*, London, R. Faulder.

Parker, S. T. (1982), 'Higher Intelligence as an adaptation for social and technological strategies in early *Homo sapiens*', Paper presented at British Psychological Association (*sic*), *University of Sussex*, March, pp. 1–29.

Parker, S. T. (1985), 'Higher intelligence as an adaptation for social and technological strategies in early *Homo sapiens*', in G. Butterworth, J. Ratkowska and M. Scafe (eds), *Evolution and Development*, Chicago, University of Chicago Press and Brighton, Harvester.

Parker, S. T. and Gibson, K. R. (1979), 'A developmental model for the evolution of language and intelligence in early hominids', *BBS*, vol. 2, pp. 367–408.

Parker, S. T. and Gibson, K. R. (1982), 'The Importance of Theory for Reconstructing the Evolution of Language and Intelligence in Hominids', in B. Chiarelli and R. Corruccini (eds), *Advanced Views in Primatology*, Berlin and New York, Springer-Verlag, pp. 42–64.

Passingham, R. E. (1981), 'Broca's area and the origins of human vocal skill', *Philosophical Transactions of the Royal Society, London*, B. 292, pp. 167–75.

Passingham, R. E. (1982), *The Human Primate*, Oxford and San Francisco, W. H. Freeman.

Passingham, R. E. and Ettlinger, G. (1974), 'A comparison of cortical function in man and other primates', *International Review of Primatology*, vol. 16, pp. 233–99.

Patte, E. (1955), *Les Néanderthaliens: Anatomie, Physiologie, Comparaisons*, Paris, Masson et Cie.

Patterson, F. and Linden, E. (1982), *The Education of Koko*, London, André Deutsch.

Paul, D. (1984), 'Eugenics and the Left', *Journal of the History of Ideas*, vol. 45, no. 4., pp. 567–90.

Penfield, W. and Rasmussen, T. (1950), *The Cerebral Cortex of Man*, New York, Macmillan.

Penfield, W. and Roberts, L. A. (1959), *Speech and Brain Mechanisms*, Princeton, Princeton University Press.

Peringuey, L. A. (1911), 'The stone ages of South Africa as represented in the collection of the South African Museum', *Annals of the South African Museum*, vol. 8, part 1, pp. 1–218.

Perkell, J. S. (1969), *Physiology of Speech Production: Results and Implications of a Quantitative Cineradiographic Study*, Cambridge, Mass., MIT Press.

Perthes, B. de (1847), *Antiquités Celtiques et Antédiluviennes, Tome I*, Abbéville.

Petitto, L. A. and Seidenberg, M. S. (1979), 'On the evidence for linguistic abilities in signing apes', *Brain and Language*, vol. 8, pp. 162–83.

Peyrony, D. (1948), *Eléments de Préhistoire*, Paris, Schleicher.

Pfeiffer, J. (1978), *The Emergence of Man*, New York, Harper.

Pfeiffer, J. (1982), *The Creative Explosion: An Inquiry into the Origins of Art and Religion*, New York, Harper & Row.
Phillipson, D. W. (1982), 'The Later Stone Age in sub-Saharan Africa', in J. D. Clark (1982), pp. 410–77.
Phillipson, D. W. (1985), *African Archaeology*, Cambridge UK, Cambridge University Press.
Piaget, J. (1971), *Structuralism*, London, Routledge & Kegan Paul.
Piaget, J. (1974), *Understanding Causality*, New York, W. W Norton.
Piaget, J. and Inhelder, B. (1967), *The Child's Conception of Space*, New York, W.W. Norton.
Pilbeam, D. (1980), 'Major trends in human evolution' in L-K. Königsson (ed.), *Current Argument on Early Man*, Oxford, Pergamon Press for Royal Swedish Academy of Sciences, pp. 261–85.
Pilbeam, D. R. (1986), 'The origin of *Homo sapiens*; the fossil evidence', in *MTP*, pp. 331–8.
Pilgrim, G. E. (1915), 'New Siwalik Primates and their Bearing on the Question of the Evolution of Man and the Anthropoidea', *Records of the Geological Survey of India*, vol. XIV, pp. 1–24.
Plooij, F. X. (1978), 'Tool use during chimpanzee bushpig hunt', *Carnivore* vol. 1, no. 2, pp. 103–6.
Plotkin, H. C. (1979), 'Brain-behaviour studies and evolutionary biology', in D. A. Oakley and H. C. Plotkin (eds), *Brain, Behaviour and Evolution*, London, Methuen, pp. 52–77.
Plotkin, H. C. (1982), 'Evolutionary epistemology and evolutionary theory', *LDC*, pp. 3–16.
Plotkin, H. C. and Odling Smee, F. J. (1982), 'Learning in the context of a hierarchy of knowledge gaining processes', *LDC*, pp. 443–71.
Plotkin, H. C. (ed.), (1982), see p. 330.
Popper, K. (1977), *The Poverty of Historicism*, New York, Harper & Row.
Post, L. van der (1958), *The Lost World of the Kalahari*, London, Hogarth Press.
Potts, R. (1984), 'Home bases and early hominids', *Sci.*, vol. 72, pp. 338–47.
Potts, R. and Shipman, P. (1981), 'Cutmarks made by stone tools on bones from Olduvai Gorge, Tanzania', *Nat.*, vol. 291, pp. 577–80.
Poulianos, A. N. (1982), *The Cave of the Petralonian Archanthropinae*, Athens, Library of the Anthropological Association of Greece.
Poulianos, A. N. (1984), 'Once more on the age and stratigraphy of the Petralonian Man', *JHE*, vol. 13, pp. 465–7.
Prasad, K. N. (1982), 'Was *Ramapithecus* a tool-user?', *JHE*, vol. 11, no. 1, pp. 101–4.
Prichard, J. C. (1813), *Researches into the Physical History of Man*, London, J. & A. Arch, Bristol, B. H. Barry.
Quennell, M. and Quennell, C. H. B. (1926), *Everyday Life in the Old Stone Age*, London, Batsford.
Reader, J. (1981), *Missing Links, The Hunt for the Earliest Man*, London, Collins.
Reed, E. (1975), *Woman's Evolution: From Matriarchal Clan to Patriarchal Family*, New York, Pathfinder Press.
Revesz, G. (1956), *The Origins and Prehistory of Language*, London and New York, Longmans Green.
Reynolds, V. (1976), *The Biology of Human Action*, San Francisco, W. H. Freeman.
Richards, G. D. (1984), 'Getting the intelligence controversy knotted', *Bulletin of the British Psychological Society*, vol. 37, pp. 77–9.

Ridley, M. (1983), 'Can classification do without evolution?', *N.Sc.*, 1st December, pp. 647–51.
Ritchie, C. I. A. (1979), *Rock Art of Africa*, South Brunswick and New York, A. S. Barnes; London, Thomas Yoseloff; Philadelphia, Art Alliance Press.
Roe, D. A. (1981), *The Lower and Middle Palaeolithic Periods in Britain*, London, Routledge & Kegan Paul.
Rose, M. D. (1984), 'A hominine hip bone, KNM-ER 3228, from East Lake Turkana, Kenya', *AJPA*, vol. 63, pp. 371–8.
Rose, S., Kamin, L. J., and Lewontin, R. C. (1984), *Not In Our Genes*, Harmondsworth, Penguin Books.
Rumbaugh, D. M. (ed.) (1977), *Language Learning by a Chimpanzee*, New York, Academic Press.
Rychlak, J. F. (1977), *The Psychology of Rigorous Humanism*, New York, Wiley.
Sacher, C. A. (1982), 'The role of brain maturation in the evolution of the primates', *PBE*, pp. 97–112.
Sagan, C. (1977), *The Dragons of Eden*, New York, Random.
Sahlins, M. (1972), *Stone Age Economics*, Chicago, Aldine.
Sandars, N. K. (trans.), (1971), *Poems of Heaven and Hell from Ancient Mesopotamia*, Harmondsworth, Penguin Books.
Sarich, V. M. (1983), 'Retrospective in hominoid macromolecular systematics', *NI*, pp. 137–50.
Sarich, V. M. and Wilson, A. C. (1967), 'Immunological time scale for hominoid evolution', *Sci.*, vol. 158, pp. 1200–3.
Schaller, G. B. (1963), *The Mountain Gorilla*, Chicago, Chicago University Press.
Schaller, G. B. (1964), *The Year of the Gorilla*, London, Collins.
Schmitt, T. J. and Nairn, A. E. M. (1984), 'Interpretations of the magnetostratigraphy of the Hadar hominid site, Ethiopia', *Nat.*, vol. 309, pp. 704–6.
Schwalbe, G. (1904), *Die Vorgeschichte des Menschen*, Braunschweig, Fried. Viewig & Sohn.
Schwartz, J. H. (1984), 'The evolutionary relationships of man and orang-utans', *Nat.*, vol. 308, pp. 501–5.
Seidenberg, M. S. and Petitto, L. A. (1979), 'Signing behavior in apes: a critical review', *Cognition*, vol. 7, pp. 177–215.
Senut, B. (1980), 'New data on the humerus and its joints in Plio-Pleistocene hominids', *Collegium Antropologicum*, vol. 4, no. 1, pp. 87–94.
Senut, B. (1981), 'Humeral outlines in some hominoid primates and in Plio-pleistocene hominids', *AJPA*, vol. 56, pp. 275–83.
Shackleton, N. J. and Opdyke, N. D. (1977), 'Oxygen isotope and palaeomagnetic evidence for early Northern Hemisphere glaciation', *Nat.*, vol. 270, pp. 216–19.
Shackley, M. (1980), *Neanderthal Man*, London, Duckworth.
Shapiro, H. L. (1976), *Peking Man*, London, Allen & Unwin/Book Club Associates.
Shipman, P. and Rose, J. (1983), 'Evidence of butchery and hominid activities at Torralba and Ambrona; an evaluation using microscopic techniques', *Journal of Archaeological Science*, vol. 10, pp. 465–74.
Sieveking, A. (1979), *The Cave Artists*, London, Thames & Hudson.
Sieveking, G. de, Langworth, I. H., Hughes, M. J., Clark, A. J. and Millet, A. (1973), 'A new survey of Grime's Graves – first report', *PPS*, vol. 39, pp. 182–218.

Silberbauer, G. B. (1981), *Hunter and Habitat in the Central Kalahari Desert*, Cambridge UK, Cambridge University Press.
Singer, B. (1981), 'History of the study of animal behaviour', in D. McFarland (ed.), *The Oxford Companion to Animal Behaviour*, London, Oxford University Press, pp. 255–72.
Smith, F. H. (1982), 'Upper Pleistocene Hominid Evolution in South-Central Europe', *CA*, vol. 23, no. 6, pp. 667–703.
Smith, J. M. (1978), 'Concepts of sociobiology', in *The Roots of Sociobiology*, London Past and Present Society.
Smith, J. M. (1982), *Evolution and the Theory of Games*, Cambridge UK, Cambridge University Press.
Smith, J. M. (1983), 'Game theory and the evolution of cooperation', *EMM*, pp. 445–56.
Smith, J. M. (1984), 'Game theory and the evolution of behaviour', *BBS*, vol. 7, pp. 95–125.
Smith, R. J. and Pilbeam, D. (1980), 'Evolution of the orang utan', *Nat.*, vol. 284, pp. 447–8.
Solecki, R. S. (1971 [1955]), *Shanidar*, London, Allen Lane.
Sollas, W. J. (1911, rep. 1915), *Ancient Hunters and their Modern Representatives*, London, Macmillan.
Spencer, F. (ed.), (1982), *A History of American Physical Anthropology 1930–1980*, New York, Academic Press.
Spencer, H. (1855), *The Principles of Psychology*, London, Williams & Norgate.
Stam, J. H. (1976), *Inquiries Into the Origin of Language: The Fate of a Question*, New York, Harper & Row.
Steele Russell, I. (1979), 'Brain size and intelligence: a comparative perspective', in D. A. Oakley and H. C. Plotkin (eds), *Brain, Behaviour and Evolution*, London, Methuen, pp. 126–53.
Stephan, H. (1969), 'Quantitative investigations on visual structure in primate brains', *Proceedings of the 2nd International Congress on Primatology*, vol. 3, pp. 34–42.
Stephan, H., Bauchot, R. and Andy, O. J. (1970), 'Data on size of the brain and of various brain parts in insectivores and primates', in C. R. Noback and W. Montagna (eds), *The Primate Brain*, New York, Appleton-Century-Crofts, pp. 289–97.
Stern, J. T. Jr., and Susman, R. L. (1983), 'The locomotor anatomy of *Australopithecus afarensis*', *AJPA*, vol. 60, pp. 279–317.
Stevens, A. (1975), 'Animals in Palaeolithic Cave Art: Leroi-Gourhan's hypothesis', *Antiquity*, vol. 49, pp. 54–7.
Stringer, C. B. (1974), 'Population relationships of Later Pleistocene hominids', *Journal of Archaeological Sciences*, vol. 1, pp. 317–42.
Stringer, C. B. (1982), 'Towards a solution of the Neanderthal problem', *JHE*, vol. 11, pp. 431–8.
Stringer, C. B. (1983), 'Some further notes on the morphology and dating of the Petralona hominid', *JHE*, vol. 12, pp. 731–42.
Stringer, C. B. (1984a), 'Fate of the Neanderthal', *Natural History*, vol. 93, no. 12, pp. 6–12.
Stringer, C. B. (1984b), 'Human evolution and biological adaptation in the Pleistocene', in R. Foley (ed.), *Hominid Evolution and Community Ecology*, London, Academic Press.
Stringer, C. B. (1986), 'The credibility of *Homo habilis*', *MTP*, pp. 266–94.
Stringer, C. B., Howell, F. Clark, and Melentis, J. K. (1979), 'The significance

of the fossil hominid skull from Petralona'; *Journal of Archaeological Science*, vol. 6, pp. 235–53.

Sturtevant, A. H. (1966), *A History of Genetics*, New York, Harper & Row.

Suarez, S. D. and Gallup, G. G. (1981), 'Self-recognition in chimpanzees and orangutans but not gorillas', *JHE*, vol. 110, pp. 175–88.

Sulloway, F. (1979), *Freud, Biologist of the Mind: Beyond the Psychoanalytic Legend*, New York, Basic Books and London, Burnett Books.

Tanner, N. M. (1981), *On Becoming Human*, Cambridge, UK, Cambridge University Press.

Tax, S. (ed.) (1960), *The Evolution of Man: Mind, Culture and Society*; Vol. 2. *Evolution After Darwin*, Chicago, Chicago University Press.

Taylor, R. E. and 7 others, (1983), 'Middle Holocene age of the Sunnyvale human skeleton', *Sci.*, vol. 220, pp. 1271–3.

Terrace, H. S. (1980), *Nim*, London, Eyre & Methuen.

Thibault, J. W. and Kelley, H. H. (1959), *The Social Psychology of Groups*, New York, Wiley.

Thom, A. (1971), *Megalithic Lunar Observatories*, Oxford, Clarendon Press.

Thomas, E. Marshall (1959), *The Harmless People*, London, Secker & Warburg.

Thomson, J. A. (1908), *Heredity*, London, John Murray.

Thorndike, E. L. (1898), 'Animal Intelligence: an Experimental Study of the Associative Processes in Animals', *Psychological Review*, Monograph Supplement No. 8, pp. 68–72.

Thorne, A. G. and Wolpoff, M. H. (1981), 'Regional continuity in Australasian Pleistocene hominid evolution', AJPA, vol. 55, pp. 337–49.

Thorpe, W. H. (1974), *Animal Nature and Human Nature*, London, Methuen.

Tobias, P. V. (1978), 'Position et rôle des australopithécines dans la phylogenie humaine, avec étude particulière de *Homo habilis* et des théories controversées avancées à propos des premiers hominides fossiles de Hadar et de Laetolil', in *Origines humaines et les époques de l'intelligence. Colloque international (Juin, 1977) de la Fondation Singer-Polignac*, Paris, Masson, pp. 38–77.

Tobias, P. V. (1979), 'Evolution of Human Brain, Intellect and Spirit', *1st Abbie Memorial Lecture*, Adelaide, University of Adelaide.

Tobias, P. V. (1980a), ' "*Australopithecus afarensis*" and *A. africanus*.: Critique and an alternative hypothesis', *Palaeontologica africana*, vol. 23, pp. 1–17.

Tobias, P. V. (1980b), 'From Linné to Leakey: Six Signposts in Human Evolution', in L-K. Königsson (ed.), *Current Argument on Early Man*, Oxford, Pergamon Press for Royal Swedish Academy of Sciences, pp. 1–12.

Tobias, P. V. (1981), 'The Nasopharynx: Review of structure and development with notes on speech, pharyngeal hypophysis, chordoma and the dens', *Journal of the Dental Association of South Africa*, vol. 36, pp. 765–78.

Tongue, M. H. (1909), *Bushman Paintings*, Oxford, Clarendon Press.

Topinard, P. (1877), *Anthropology*, London, Chapman & Hall.

Toulmin, S. E. and Goodfield, G. J. (1965), *The Discovery of Time*, London, Nuffield Foundation (reprinted 1967, Harmondsworth, Penguin Books).

Trinkaus, E. (1984), 'Does KNM-ER 1481A establish *Homo erectus* at 2.0 myr BP?', *AJPA*, vol. 64, pp. 137–9.

Trinkaus, E. and Howells, W. W. (1979), 'The Neanderthals', *Sci. Am.*, vol. 241, no. 6, pp. 94–105.

Trinkaus, E. and Zimmerman, M. R. (1982), 'Trauma among the Shanidar Neanderthals', *AJPA*, vol. 57, pp. 61–76.

Trivers, R. L. (1971), 'The evolution of reciprocal altruism', *Quarterly Review of Biology*, vol. 46, pp. 35–57.

Trotter, W. (1916), *Instincts of the Herd in Peace and War*, London, T. Fisher Unwin.

Turke, P. W. (1984), 'Effects of ovulatory concealment and synchrony on protohominid mating systems and parental roles', *Ethology and Sociobiology*, vol. 5, pp. 33–44.

Turnbull, C. M. (1968), *The Forest People*, New York, Simon & Schuster.

Tuttle, R. and Beck, B. B. (1972), 'Knuckle walking hand postures in the Orangutan (*Pongo pygmaeus*),' *Nat.*, vol. 236, pp. 33–41.

Tyson, E. (1699), *Orang-Outang, sive Homo Sylvestris: or the Anatomy of a Pygmie compared with that of a monkey, an ape, and a man*, London, T. Bennett and D. Brown.

Ucko, P. J. (ed.) (1977), *Form in Indigenous Art*, Canberra, London & New Jersey, Australian Institute of Aborigine Studies.

Vallois, H. V. (1949), 'The Fontéchevade fossil man', *AJPA*, vol. 7, pp. 339–62.

Vauclair, J. (1982), 'Sensorimotor intelligence in human and non-human primates', *JHE*, vol. 11, pp. 757–64.

Vauclair, J. (1984), 'Phylogenetic approach to object manipulation in human and ape infants', *Human Development*, vol. 27, pp. 321–8.

Vauclair, J. and Bard, K. A. (1983), 'Development of manipulations with objects in ape and human infants', *JHE*, vol. 12, pp. 631–45.

Vermeersch, P. M., Gijselings, G. and Paulissen, E. (1984), 'Discovery of the Nazlet Khater Man, Upper Egypt', *JHE*, vol. 13, pp. 281–6.

Vermeersch, P. M., Paulissen, E. Gijselings, G., Otte, M., Thoma, A., vanPeer, P. and Lauwers, R. (1984), '33,000-yr old chert mining site and related *Homo* in the Egyptian Nile Valley', *Nat.*, vol. 309, pp. 343–4.

Vilensky, J. A., van Hoesen, G. W. and Damasio, A. R. (1982), 'The limbic system and human evolution', *JHE*, vol. 11, pp. 447–60.

Von Franz, M–L (1972), *Patterns of Creativity Mirrored in Creation Myths*, Zurich, Spring Publications.

Von Koenigswald, G. H. R. (1956), *Meeting Prehistoric Man*, London, The Scientific Book Club.

Voorzanger, B. (1984), 'Altruism in sociobiology', *JHE*, vol. 13, pp. 33–9.

Vrba, E. S. (1983), Peer Commentary letter in B. Blumenberg (1983), p. 609.

Walker, A., Falk, D., Smith, R. and Pickford, M. (1983), 'The skull of *Proconsul africanus*: reconstruction and cranial capacity', *Nat.*, vol. 305, pp. 525–7.

Wallace, J. (1978), 'Evolutionary trends in the early hominid dentition', in C. Jolly (ed.), *Early Hominids in Africa*, London, Duckworth, pp. 285–310.

Walter, R. C. and Aronson, J. L. (1982), 'Revisions of K/Ar ages for the Hadar hominid site, Ethiopia', *Nat.*, vol. 296, pp. 123–7.

Warden, C. J. (1932), *The Evolution of Human Behavior*, New York, Macmillan.

Washburn, S. L. and DeVore, I. (1962), 'Social behavior of baboons and early man', in S. L. Washburn (ed.), *Social Life of Early Man*, London, Methuen.

Washburn, S. L. and Howell, F. C. (1959), 'Human evolution and culture', in S. Tax (ed.) (1960), pp. 33–56.

Weidenreich, F. (1943), 'The "Neanderthal Man" and the ancestors of "Homo sapiens" ', *American Anthropologist*, vol. 42, pp. 375–83.

Weiner, J. S. (1955), *The Piltdown Forgery*, London, Oxford University Press.

Weiner, J. S., Oakley, K. P. and LeGros Clark, W. E. (1953), 'The solution of the Piltdown problem', *Bulletin of the British Museum of Natural History, London, (Geology)*, no. 2, pp. 141–6.

Weismann, A. (1885), *Die Kontinuität des Keimplasmas, als Grundlage einer Theorie der Vererbung*, Jena, Fischer.
Weismann, A. (1893), *The Germ-Plasm: A Theory of Heredity*, London, W. Scot.
Wendt, H. (1972), *From Ape to Adam, The First Million Years of Man*, London, Thames & Hudson.
Wenke, R. J. (1980), *Patterns in History: Mankind's First Three Million Years*, London, Oxford University Press.
Wheeler, P. E. (1984), 'The evolution of bipedality and loss of functional body hair in hominids', *JHE*, vol. 13, pp. 91–8.
Willetts, W. (1958), *Chinese Art*, Harmondsworth, Penguin Books.
Williams, G. C. (1966, rep. 1982), 'Natural selection, adaptation, and progress', *LDC*, pp. 39–59.
Wilson, E. O. (1975), *Sociobiology: The New Synthesis*, Cambridge Mass., Harvard University Press.
Wilson, E.O. (1978), *On Human Nature*, Cambridge Mass., Harvard University Press.
Wind, J. (1976a), 'Phylogeny of the Human Vocal Tract', *OEL*, pp. 612–30.
Wind, J. (1976b), 'Human drowning – phylogenetic origin', *JHE*, vol. 5, no. 4, pp. 349–63.
Wind, J. (1978), 'Fossil evidence for primate vocalizations?' in D. J. Chivers and K. A. Joysey (eds), *Recent Advances in Primatology*, vol. 3, pp. 87–91.
Winograd, T. (1980), 'What does it mean to understand language?', *Cognitive Science*, vol. 4, pp. 209–41.
Winslow, J. H. and Meyer, A. (1983), 'The Perpetrator of Piltdown', *Science 83*, Sept., pp. 32–43.
Wintle, A. G. and Jacobs, J. A. (1982), 'A critical review of the dating evidence for Petralona Cave', *Journal of Archaeological Science*, vol. 9, pp. 39–47.
Wispé, L. G. and Thompson, J. N. Jr (eds) (1976), 'The war between the words. Biological versus social evolution and some related issues', *Am. Psy.* May, pp. 341–84.
Wittgenstein, L. (1967), *Philosophical Investigations*, Oxford, Basil Blackwell.
Wolpoff, M. H. (1971), 'Is Vertesszöllös an occipital of *Homo erectus*?', *Nat.*, vol. 232, pp. 867–8.
Wolpoff, M. H. (1982), '*Ramapithecus* and hominid origins', *CA*, vol. 23, no. 5, pp. 501–22.
Wood, B. A. (1981), 'Tooth size and shape and their relevance to studies of hominid evolution', *Philosophical Transactions of the Royal Society of London*, vol. B. 292, pp. 65–76.
Wood, B. A., Martin, L. B. and Andrews, P. (1986), *Major Topics in Primate and Human Evolution*, Cambridge, UK, Cambridge University Press.
Wood, B. A. and Stack, C. G. (1980), 'Does allometry explain the differences between "Gracile" and "Robust" Australopithecines?', *AJPA*, vol. 52, pp. 55–62.
Wood Jones, F. (1929), *Man's Place among the Mammals*, London, E. Arnold.
Wright, R. V. S. (1972), 'Imitative learning of a flaked-tool technology – the case of an orang-utan', *Mankind*, vol. 8, pp. 296–306.
*Wu, R. (1982), 'Palaeoanthropology in China, 1949–1979', *CA*, vol. 23, no. 5, pp. 473–7.
Wu, R. and Xingren, D. (1983), 'Des fossiles d'*Homo erectus* découverts en Chine', *L'Anth.*, vol. 87, no. 2, pp. 177–83.
Wu, R. and Oxnard, C. E. (1983), 'Ramapithecines from China: evidence from tooth dimensions', *Nat.*, vol. 306, pp. 258–60.

Wymer, J. (1982), *The Palaeolithic Age*, London, Croom Helm.
Wynn, T. (1979), 'The intelligence of Late Acheulean hominids', *Man*, vol. 14 (N.S.), pp. 371–91.
Wynn, T. (1981), 'The intelligence of Oldowan hominids', *JHE*, vol. 10, pp. 529–41.
Wynn, T. (1982), 'Piaget, Stone Tools and the Evolution of Human Intelligence', unpublished paper delivered to the *International Primatological Society Symposium*.
Yates, F. A. (1964), *Giordano Bruno and the Hermetic Tradition*, London, Routledge & Kegan Paul.
Yerkes, R. M. (1916), *The Mental Life of Monkeys and Apes*, Behaviour Monographs, vol. 3, no. 1, New York, Holt.
Zihlman, A. L. (1979), 'Pygmy chimpanzee morphology and the interpretation of early hominids', *South African Journal of Science*, vol. 75, pp. 165–8.
Zihlman, A. L. and Lowenstein, J. M. (1983), '*Ramapithecus* and *Pan paniscus*, significance for human origins', *NI*, pp. 677–94.
Zihlman, A. L., Cronin, J. E., Cramer, D. L. and Sarich, V. M. (1978), 'Pygmy chimpanzees as a possible prototype for the common ancestor of humans, chimpanzees and gorillas', *Nat.*, vol. 275, pp. 744–6.
Zimmer, H. (1972), *Myths and Symbols in Indian Art and Civilization*, Princeton, Princeton University Press and Bollingen Foundation.
Zuckerman, S. (1932), *The Social Life of Monkeys and Apes*, London, Kegan Paul.

* Some sources will reference Wu Rukang under Rukang. This, however, is incorrect as Chinese naming reverses the western order of given and family names.

Name index

Subject Index

For Product Safety Concerns and Information please contact our EU representative GPSR@taylorandfrancis.com Taylor & Francis Verlag GmbH, Kaufingerstraße 24, 80331 München, Germany

Batch number: 10725800

Printed by Printforce, the Netherlands